Optical Activity and Chiral Discrimination

NATO ADVANCED STUDY INSTITUTES SERIES

Proceedings of the Advanced Study Institute Programme, which aims at the dissemination of advanced knowledge and the formation of contacts among scientists from different countries

The series is published by an international board of publishers in conjunction with NATO Scientific Affairs Division

A Life Sciences B Physics	Plenum Publishing Corporation London and New York
C Mathematical and Physical Sciences	D. Reidel Publishing Company Dordrecht, Boston and London
D Behavioral and Social Sciences	Sijthoff International Publishing Company Leiden
E Applied Sciences	Noordhoff International Publishing Leiden

Series C – Mathematical and Physical Sciences

Volume 48 - Optical Activity and Chiral Discrimination

Optical Activity and Chiral Discrimination

Proceedings of the NATO Advanced Study Institute held at the University of Sussex, Falmer, England, September 10-22, 1978

edited by

STEPHEN F. MASON
Department of Chemistry, University of London King's College, London, England

D. Reidel Publishing Company

Dordrecht : Holland / Boston : U.S.A. / London : England

Published in cooperation with NATO Scientific Affairs Division

Library of Congress Cataloging in Publication Data

CIP

Nato Advanced Study Institute, University of Sussex, 1978.
Optical activity and chiral discrimination.

(NATO advanced study institutes series : Series C, Mathematical and physical sciences ; v. 48)
Bibliography: p.
Includes index.
1. Optical rotatory dispersion–Congresses. 2. Circular dichroism–Congresses. 3. Chirality–Congresses. I. Mason, Stephen Finney, 1923–
II. Title. III. Series.
QD473.N37 1978 541'.7 79-10532

ISBN 978-90-481-8355-5

Published by D. Reidel Publishing Company
P.O. Box 17, Dordrecht, Holland

Sold and distributed in the U.S.A., Canada, and Mexico
by D. Reidel Publishing Company, Inc.
Lincoln Building, 160 Old Derby Street, Hingham, Mass. 02043, U.S.A.

CONTENTS

PREFACE

For Louis Pasteur, the two distinctive properties of dissymmetric systems, optical activity and chiral discrimination, provided prime evidence for a Divine origin to the universe. Handedness appeared to be built into the macrocosm of the galaxies, each with a non-superposable mirror image by virtue of its rotation, as well as the microcosm of each molecule of most natural products. The best that the chemist in the laboratory could accomplish appeared to be the synthesis of the *détordu* internally-compensated *meso*-form and, as Pasteur ultimately came to admit, the externally-compensated *racemic*-form. In the latter case the chemist generated not merely one but two chiral structures, although parity, and secondary-symmetry generally, seemed to be conserved in the enantiomer-antipode pair.

The cosmic element in the Pasteur tradition received an augmentation in secular form from demonstrations of the non-conservation of parity in the weak interactions, and from the discovery of net circularity in the extra-terrestrial photons, such as those from the less-distant planets, particularly the photons from the Jupiter red-spot. The development of the photoacoustic circular analysers a decade ago was received in fact with as much enthusiasm by the astronomers as by the chemists. It would be just to add, however, that the majority of these circular analysers are now to be found, not in the observatories, but in the physical and chemistry laboratories devoted to the molecular aspects of the Pasteur tradition.

These aspects were enlarged near the turn of the nineteenth century by the contributions of Aimé Cotton to chiroptical spectroscopy, and of Emil Fischer to dissymmetric induction and asymmetric synthesis, requiring the systematisation provided by

Stephen F. Mason (ed.), Optical Activity and Chiral Discrimination. ix–xi.

his adventitiously-correct stereochemical convention. From that period the devotees of the Pasteur tradition, to which historical rectitude compels the addition of the names of Biot, Le Bel and van't Hoff, and Werner, as well as Cotton and Fischer, gathered from time to time in order to assess and clarify the state of the art, and to disseminate the current understanding. General discussions on Optical Rotatory Power, it may be recalled, were held in the rooms of Burlington House, London, in 1914 and 1930 by the Faraday Society and in 1966 by the Royal Society, together with similar meetings at comparable intervals held elsewhere outside the U.K.

Through the sponsorship of the Scientific Affairs Division of NATO it has been possible, in conformity with the pace of advance in the field over the past two decades, to organise a series of Advanced Study Institutes on the distinctive properties of chiral molecular systems. The first of the NATO ASI meetings in the theoretically-refurbished and experimentally-enlarged Pasteur tradition was held at Bonn in 1965, organised by Professor G. Snatzke, and the second at Tirrenia (Pisa) in 1971, under the direction of Professor F. Ciardelli and Professor P. Salvadori. The majority of the main lectures delivered at the third meeting, held at the University of Sussex, Brighton, in September, 1978, are published in the present volume in order to make the proceedings available to a wider audience than the necessarily-limited group of participants.

The 1978 Sussex meeting brought together some 150 participants from all branches of the chemical sciences and contingent disciplines, - from biochemistry, organic, inorganic, physical and theoretical chemistry, and from biophysics and chemical physics. In this interdisciplinary gathering the augmented Pasteur tradition remained coherent, centred on novel forms of optical activity and new subtleties in the chiral discriminations. Particular attention was directed at the 1978 ASI to developments since the previous meetings, notably, the theory and applications of chemical and physical differentiations between enantiomeric and diastereomeric substances, scattering and luminescence circular intensity differential studies, and circular dichroism spectroscopy over the range from the vacuum ultraviolet to the infrared wavelength region. The developments included recent advances in instrumentation, as well as basic theory and specific applications.

Nor were the quasi-philosophical elements of the tradition ignored. A discussion of the significance of parity-conservation and of time-reversal symmetry for optical activity, whether intrinsic or field-induced, was followed at a later hour in a more euphoric mood by speculations on the implication of the nuclear motions, and on the non-conservation of parity in the nuclear weak interactions, for chiral and achiral systems alike.

In essence, it appears that no racemate is truly equimolecular in its two enantiomers, because of the chiral discrimination of the weak interactions. The electron becomes persistently left-handed with antiparallel angular and linear momenta. In consequence there is always an enantiomeric excess of one of the isomers in a racemate, however small. Given the aeons of time generally allowed for biochemical evolution, the present chorus of natural products clapping with only one hand is entirely explicable. The true 'racemates', in the equimolecular-isomer sense, are the non-linear achiral systems with more than three atoms. All such molecules are achiral only at the fleeting instant occupied by the equilibrium nuclear configuration, and they are 'chiral' most of the time, albeit as 'racemic' mixtures of the two 'enantiomers' represented by the nuclear configuration at the classical turning-points of the pseudoscalar vibrational mode or modes. Even the vibrationally-deficient species, which lack a pseudoscalar normal mode, possess combination modes with pseudoscalar character. None are restricted to the flatland of the bent triatomics, nor to the more limited realm of the diatomics which, with only two ends that are invariably close together, conform in fair measure to the united-atom description.

We are much indebted to the North-Atlantic Treaty Organisation for sponsoring the 1978 Advanced Study Institute on Optical Activity and Chiral Discrimination. In addition we thank Sussex University for the provision of excellent hospitality and conference facilities. The Sussex campus is semi-rural but wholly-civilised and, lying at a sufficient remove from the great wen of London, and even from the mini-metropolis of Brighton, external distractions were minimised.

Sincere thanks are due to all of our lecturers, organisers, and other participants who made the 1978 ASI a success. The contribution of Dr. Tony McCaffery of Sussex University, who attended to all of the local arrangements, is particularly appreciated. It is sad to record the untimely death, in November, 1977, of Professor William Klyne of Westfield College, London. A veteran of the 1965 and 1971 ORD and CD meetings, Bill Klyne made a valuable contribution to the early organisation of the 1978 ASI. His wide-ranging and substantial investigations in chiroptical spectroscopy and organic stereochemistry will remain a lasting monument to him.

King's College London STEPHEN MASON
October 1978

PART I OPTICAL ACTIVITY

GENERAL MODELS FOR OPTICAL ACTIVITY

Stephen F. Mason

Chemistry Department,
King's College,
London WC2R 2LS,
U.K.

1. INTRODUCTION

Physical interpretations of optical activity ultimately derive from the general helix model introduced by Fresnel (1824) to account for optical rotation in terms of the then newly-developed transverse vibration theory of light. The discovery of the left-circular and right-circular polarization modes of radiation by Fresnel, through the introduction of the quarter-wave plate, the Fresnel rhomb, and the triprism composed of (+)- and (-)-quartz elements, enabled him to ascribe optical rotation to a circular birefringence of the active medium. Plane-polarized radiation, regarded as the resultant of superposed left and right circularly polarized components with equal amplitudes, undergoes a rotation if one component is retarded relative to the other on propagation through a transparent medium. The angle of rotation in radians, ϕ, at the wavelength, λ, for a unit path-length is given by the Fresnel equation,

$$\phi = (n_L - n_R)\pi/\lambda \qquad (1)$$

where n_L and n_R are the refractive index of the medium for left and right circularly polarized light, respectively [1].

Fresnel supposed that the molecules of an active medium have a left- or right-handed helical morphology, like the spatial forms of left or right circularly polarized transverse waves, where the particles of the luminiferous ether were assumed to rotate respectively anticlockwise and clockwise around the direction of propagation, viewed from the source. After the luminiferous ether had been invested with electro-

Stephen F. Mason (ed.), Optical Activity and Chiral Discrimination. 1–24.

magnetic properties, the helix model was developed by Drude [2] who related optical activity to the helical motion of an electron in a chiral molecule interacting with an oscillatory electromagnetic radiation field. At the time Cotton had recently discovered circular dichroism in aqueous solutions of chiral transition metal complexes [3], and Fischer had discussed the problem of assigning an absolute stereochemical configuration to chiral organic molecules [4]. For Drude, a right-handed helical charge displacement in a dissymmetric structure results in a positive optical activity whereas a left-handed helical path gives a negative rotatory power.

Under resonance conditions, where the frequency of the radiation field equals the natural oscillation frequency of the elastically-bound electron, radiative energy is absorbed, and under non-resonance conditions, refraction of the radiation ensues. Optical rotatory power takes the form, under non-resonance conditions, of circular birefringence and, under resonance conditions, of circular dichroism, the two effects being connected as the real and the imaginary parts, respectively, of the complex circular birefringence. Specifically a positive circular dichroism ($\varepsilon_L > \varepsilon_R$) at an absorption frequency implies a positive circular birefringence ($n_L > n_R$), and a consequent dextrorotation, at lower frequencies, following the rule of Natanson [5]. In principle at least, the absolute stereochemical configuration of a chiral molecule appears to be derivable in Drude's model from the sign of the optical rotatory power.

Subsequently Kuhn [6] and others criticised the Drude model on the ground that a single particle has no appreciable extension, whereas the finite extension of a molecule in relation to the much-larger wavelength of light was a factor requiring explicit consideration in the treatment of optical activity, as Boltzmann [7] had argued earlier. In the model of Kuhn [6], which is essentially the classical forerunner of the quantum-mechanical exciton models, optical activity is generated by the Coulombic coupling of two or more anisotropic charged oscillators with a finite separation in the molecule. The overall charge displacement is helical if the principal directions of the anisotropies are not coplanar, and a rotatory power is produced proportional to the intramolecular distance between the oscillators and to their oscillator strengths.

Again the absolute configuration of a chiral molecule is derivable in principle from the sign of the optical activity, but the application of Kuhn's model was circumscribed by the limited development of molecular electronic spectroscopy in the 1930s. In particular, knowledge of the polarization directions of electronic transitions with an electric dipole character in polyatomic molecules was limited, and magnetic dipole transitions remained largely uncharacterised.

2. ROTATIONAL AND DIPOLE STRENGTHS.

In the initial quantum-mechanical treatment of optical activity, Rosenfeld [8] defined a property, the rotational strength, R_{om}, of an electronic transition connecting the stationary states ψ_o and ψ_m of a chiral molecule. The rotational strength, which is observable as a CD absorption band at the transition frequency ν_{om}, is given by the scalar product of the electric dipole, $\vec{\mu}_{om}$, and magnetic dipole, $\vec{m}_{mo}$, transition moment,

$$R_{om} = \mathrm{Im}\{\langle\psi_o|\hat{\mu}|\psi_m\rangle\cdot\langle\psi_m|\hat{m}|\psi_o\rangle\} \qquad (2)$$

where Im signifies that the imaginary part is to be taken, since the magnetic dipole operator $\hat{m}$ is a pure imaginary.

The rotational strength has the symmetry properties of a pseudoscalar quantity with a sign dependent upon the parallel or antiparallel orientation of the polar and the axial vector components, the electric moment and the magnetic moment, respectively. Chiral molecules with an observable rotational strength are accordingly restricted to the point groups in which pseudoscalar quantities are totally symmetric. These are the pure rotation groups, I, O, T, Dp, and Cp, of any order, p.

For a given chiral molecule, the rotational strengths sum to zero over all transitions, and the optical rotation, which is dependent upon frequency-weighted contributions from each transition, vanishes in the limit of high or low radiation frequency [9]. The sum rule for the rotational strengths has a classical counterpart in the model of Kuhn [6].

Experimentally the rotational strength of an electronic transition is measured by the area of the corresponding CD absorption band through the expression [10],

$$R_{om} = \frac{3\hbar c2303}{16\pi^2 N} \int \frac{\Delta\varepsilon}{\beta\bar{\nu}}\, d\bar{\nu} \qquad (3)$$

where $\hbar$ is the reduced Planck constant ($h/2\pi$), c is the velocity of light, and N is the Avogadro number. In equation (3) $\Delta\varepsilon$ is the differential molar extinction coefficient ($\varepsilon_L - \varepsilon_R$) of the chiral solute for left and right circularly polarized light in litre mole^{-1} cm^{-1}, and β is the Lorentz field correction $[(n^2 + 2)/3]$ for the refractivity of the solvent n at the radiation wave-number $\bar{\nu}$(cm^{-1}). Substitution of the values of the universal constants into equation (3) gives the numerical form,

$$R_{om} = 0.248 \int (\Delta\varepsilon/\bar{\nu}\beta)\, d\bar{\nu} \quad (4)$$

with the rotational strength in units of the Debye-Bohr magneton. The Debye and the Bohr magneton have the respective values of 10^{-18} esu cm and 9.273×10^{-21} erg gauss^{-1} in the c.g.s. unit system, or 3.334×10^{-30} cm and 9.273×10^{-24} J T^{-1} in S.I. units. The atomic unit of rotational strength ($\hbar^3/m^2e^2c$) is equivalent to 5.083 Debye-Bohr magneton.

The area of the corresponding isotropic absorption band measures the square of the electric dipole moment of the transition, the dipole strength, D_{om}. The dipole strength in units of the square Debye (10^{-36} c.g.s. unit) is related to the isotropic band area by the expression [10],

$$D_{om} = 9.18 \times 10^{-3} \int (n\varepsilon/\beta^2\bar{\nu})\, d\bar{\nu} \quad (5)$$

In practice the solvent refractivity corrections, β in equations (3) and (4) and (β^2/n) in equation (5), are frequently set to unity, and the wavenumber of the band maximum is taken to represent an average value of $\bar{\nu}$ over the band.

3. QUANTUM MECHANICAL MODELS FOR OPTICAL ACTIVITY.

Following the initial general treatment of optical activity [8], more specific and detailed quantum mechanical models were developed by Kirkwood [11] and by Condon, Altar and Eyring [12]. These models and their subsequent developments [13-19] have in common the division of a chiral molecule into two or more sterically-constrained but otherwise independent groups which do not exchange electrons and interact solely through the Coulomb potential between their respective charge distributions. The division of a chiral molecule into a symmetric chromophore, or light-absorbing group, and a dissymmetric molecular environment of substituent groups, or ligands in the case of coordination compounds, is the foundation common to both of the principal independent-systems models.

An independent-systems treatment of any one-electron quantity carried to the first order of perturbation theory, specifically, a transition moment contributing to the rotational strength of a chiral molecule, rests upon one of two mutually-exclusive limiting assumptions. Either the dissymmetric substituents perturb the chromophore, or the chromophore perturbs the substituents. Either a particular charge distribution of the substituents, generally taken to be that of the electronic ground state, perturbs all of the electronic states of the chromophore, as in the one-electron static-field theory [12,18], or alternatively a particular charge distribution of the chromophore, that of the electronic transition of interest, perturbs all of the electronic states of the substituents. The latter course is adopted in the dynamic-coupling or substituent-polarization model [11,19], which is complementary, within the first-order independent-systems scheme, to the one-electron static coupling theory. The two mechanisms are distinct and additive only in the first-order, terms corresponding to mixed static and dynamic coupling appearing in the second and higher orders of perturbation theory.

In the one-electron theory it is postulated that the electric-dipole and the magnetic-dipole transitions of the primary group of interest, the chromophore, are mixed by the static Coulombic field due to the other groups in the molecule, the substituents. The one-electron model is described as a $\mu_1 - m_1$ mechanism, dependent upon a static potential V_2 from a second group or groups [18].

The dynamic coupling theory covers two distinct first-order cases, the $m_1 - \mu_2$ and the $\mu_1 - \mu_2$ mechanism, both involving an inter-group Coulomb potential V_{12} between transitional charge distributions [18,19]. In the $m_1 - \mu_2$ mechanism, a magnetic-dipole transition of the chromophore is dynamically coupled, through the potential of the phase-locked concomitant electric multipole of that transition, with a transient electric dipole induced in the substituent groups.

The second dynamic-coupling mechanism, $\mu_1 - \mu_2$, refers to an electric dipole transition of the chromophore. The Coulombic coupling of the necessarily non-coplanar chromophore and substituent dipoles results in an overall helical charge displacement with an electric and a magnetic transition moment which are functions of the stereochemically-determined separation and mutual orientation of the individual component group moments. The $\mu_1 - \mu_2$ mechanism covers a range of sub-cases, dependent upon the frequency-interval between the lower-energy transitions of the chromophore and the substituent. The zero-order interval vanishes for the dissymmetric dimer or the chiral homo-polymer, where the individual units are equally chromophoric. The degenerate or near-degenerate $\mu_1 - \mu_2$ mechanism has been termed the exciton model, since the coupling may be regarded as the outcome of the dynamic exchange of the excitation energy between two or more chromophoric units in the dimer or polymer.

In principle the independent-systems models are limited by their common neglect of electron exchange between the chromophore and the substituents, and, strictly speaking, all chiral molecules fall into the class of inherently dissymmetric chromophores. From the latter viewpoint, each chiral molecule becomes an individual problem, whereas the independent-systems mechanisms provide a general treatment of a given electronic transition in a series of related chiral molecules, and afford firm stereochemical and spectroscopic observable expectations.

4. THE ONE-ELECTRON MECHANISM.

In the development of the $m_1 - \mu_1$ mechanism it is generally assumed that primarily a particular magnetic dipole $M_o \rightarrow M_m$ and electric dipole transition $M_o \rightarrow M_e$ of the chromophore are mixed by the field of the substituent groups, which remain in their ground state, L_o. With the neglect of electron exchange between the chromophore and the substituents, the electronic state functions of the chiral molecule are expressed as simple products. Corrected to the first order, the molecular excited states become,

$$|M_mL_o\rangle = |M_mL_o) + (E_m - E_e)^{-1}(M_eL_o|V|M_mL_o)|M_eL_o) \qquad (6)$$

and

$$|M_eL_o\rangle = |M_eL_o) + (E_e - E_m)^{-1}(M_mL_o|V|M_eL_o)|M_mL_o) \qquad (7)$$

where E_m, E_e are the energies of the chromophoric excited states, and V is the Coulombic potential between the charges e_i of the chromophore M and those e_j of the substituents L, separated by the distance, r_{ij},

$$V = \sum_{i(M),j(L)} \frac{e_i e_j}{r_{ij}} \qquad (8)$$

The electric moment of the chromophore transition, $M_o \rightarrow M_e$, is taken to have its zero-order value, μ_{oe}, in the dissymmetrically substituted derivative, where the magnetic moment becomes non-vanishing on account of the first-order mixing,

$$m_{eo} = m_{mo}(M_mL_o|V|M_eL_o)(E_e - E_m)^{-1} \qquad (9)$$

Equally, the magnetic dipole transition, $M_o \rightarrow M_m$, while retaining its zero-order magnetic moment, m_{mo}, acquires a first-order electric moment,

$$\mu_{om} = \mu_{oe}(M_eL_o|V|M_mL_o)(E_m - E_e)^{-1} \qquad (10)$$

If the zero-order moments, μ_{oe} and m_{mo}, have a common α-polarization (α = x, y, or z), the two transitions acquire oppositely-signed rotational strengths, equal in magnitude,

$$R_{om} = -R_{oe} = i\mu_{oe}m_{mo}(M_eL_o|V|M_mL_o)(E_e - E_m)^{-1} \qquad (11)$$

at their individual positions E_{om} and E_{oe}, respectively, on the frequency scale.

The excited states in the perturbation matrix element $(M_eL_o|V|M_mL_o)$ of equation (11) refer only to the chromophore of the chiral molecule, and they may be classified under the symmetry representations of the, generally achiral, point group of the symmetric unsubstituted parent chromophore. By assumption, M_e and M_m are excited states to which, respectively, an electric and a magnetic dipole transition are allowed with collinear α-polarization from the ground state, M_o. The direct product of the symmetries of the states M_e and M_m, which transform as collinear components of a polar and an axial vector, respectively, is spanned by the pseudoscalar representation of the chromophore point group. If the matrix element $(M_eL_o|V|M_mL_o)$ is non-vanishing, the effective state-mixing component of the potential field deriving from the ground-state charge distribution of the substituents in turn belongs to the same pseudoscalar representation [18].

The pseudoscalar property of the substituent field affords sector rules, based on the $m_1 - \mu_1$ mechanism, which connect the position of a substituent in the molecular environment of the chromophore in a chiral molecule with the sign of the Cotton effect induced by that group in a given chromophore transition [18]. In the prototypical case of the C_{2v} carbonyl chromophore in chiral ketones, the sign of the Cotton effect of each transition is expected to exhibit at least a quadrant rule dependance upon the location of the dissymmetric substituent, or substituents, since XY is the lowest-order base function for the A_2 representation in C_{2v}. The additional nodal plane of the octant rule (-XYZ in Figure 1) for the $n \rightarrow \pi^*$ transition of chiral carbonyl compounds near 300 nm, while consistent with the pseudoscalar criterion of the one-electron mechanism, is not necessarily symmetry-determined. In the original proposal [20] the nodal planes of the octant rule were physically-based upon the nodal surfaces of the oxygen $2p_y$ non-bonding and the carbonyl π_x^* antibonding orbital, and subsequent experimental studies have established empirically general octant-rule behaviour in the $n \rightarrow \pi^*$ Cotton effects of chiral ketones [21, 22].

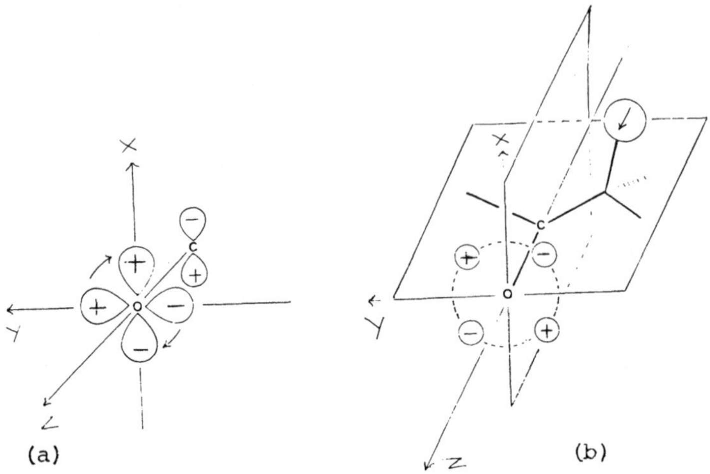

Figure 1. The 300 nm $n \rightarrow \pi^*$ transition of the carbonyl chromophore; (a) the rotatory charge displacement, $p_y \rightarrow p_x$, at the oxygen atom to an observer on the +Z axis, and, (b) the electric quadrupole transition moment, θ_{xy}, with the Coulombically-correlated charge displacement along the +Z direction in a substituent located in the upper right rear octant.

All three nodal planes of the octant rule for chiral olefins with a D_{2h} chromophore are symmetry-determined, however, and a XYZ sector rule [23] is irreducibly the simplest feasible by the $m_1 - \mu_1$ mechanism for the Cotton effects of any given electronic transition in dissymmetric ethylene derivatives. The set of basis functions for the pseudoscalar representation of a group provide the possible sector rules for the chromophores belonging to that group. These sets have been listed for the common point groups [18].

The one-electron $m_1 - \mu_1$ mechanism for optical activity implies that the coupled electric and magnetic dipole transitions of the chromophore in a chiral molecule necessarily have rotational strengths of opposite sign and equal magnitude (equation 11). The extension of CD spectroscopy to the vacuum ultraviolet region permits the investigation of this expectation.

The carbonyl $n \rightarrow \pi^*$ transition of chiral ketones near 300 nm is magnetic dipole allowed, involving the rotatory charge displacement, $2p_y \rightarrow 2p_x$, at the oxygen atom, and it is well separated on the frequency scale from the electric

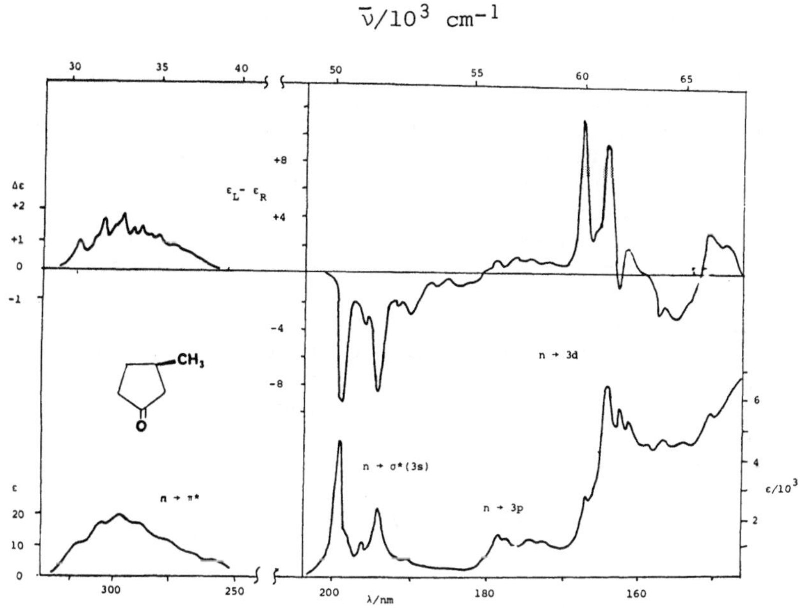

Figure 2. The vapour-phase absorption (lower curve) and CD spectrum (upper curve) of (S)-(+)-3-methylcyclopentanone over the quartz and vacuum ultraviolet region.

dipole transitions of the carbonyl chromophore, which lie in the VUV region. One-electron treatments of the carbonyl $n \rightarrow \pi^*$ optical activity postulate the mixing in of a z-polarized electric dipole component from the $2p_y \rightarrow 3d_{yz}$ Rydberg transition [24,25] or the carbonyl $\pi \rightarrow \pi^*$ transition [26]. The $2p \rightarrow 3d$ Rydberg transition of simple carbonyl compounds lies near to 165 nm, but the $\pi \rightarrow \pi^*$ transition remains uncharacterised, probably lying at a substantially higher energy [27]. The CD spectrum of (S)-(+)-3-methylcyclopentanone over the quartz and VUV region shows (Figure 2) that the $n \rightarrow \pi^*$ transition near 300 nm and the Rydberg $2p \rightarrow 3d$ transition in the 165 nm region have rotational strengths of the same sign, both being positive, so that the mixing of these two transitions by the $m_1 - \mu_1$ mechanism may be eliminated as the primary origin of the $n \rightarrow \pi^*$ Cotton effect [28].

5. THE DYNAMIC COUPLING $m_1 - \mu_2$ MECHANISM.

The one-electron static-coupling mechanism for optical activity contains the assumption that, while the chromophore is soft and perturbable, the substituent is hard and remains in its ground state unmodified throughout the light-absorption process. All substituent groups are refractive and polarizable, becoming chromophoric at least in the VUV region, and the optical perturbality of the substituent is explicitly taken into account in the dynamic coupling $m_1 - \mu_2$ and $\mu_1 - \mu_2$ mechanisms.

In the first-order $m_1 - \mu_2$ treatment [19] the molecular state functions are made up of simple product configurations containing either the ground state M_o or the magnetic dipole state M_m of the chromophore and one of the complete set L_ℓ of substituent functions. Corrected to first-order, the molecular ground-state function is given by,

$$\langle M_o L_o| = (M_o L_o| + \sum_{\ell \neq o} (-E_m - E_\ell)^{-1} (M_o L_o|V|M_m L_\ell)(M L_\ell| \tag{12}$$

and the corrected excited state becomes,

$$|M_m L_o\rangle = M_m L_o) + \sum_{\ell \neq o} (E_m - E_\ell)^{-} (M_o L_\ell|V|M_m L_o)\ M_o L_\ell) \tag{13}$$

where the state energies E_m and E_ℓ are measured relative to that of the ground state E_o as zero.

The magnetic moment of the chromophore transition, $M_o \rightarrow M_m$, is taken again, as in the static-field mechanism, to have its zero-order magnitude and particular polarization, m^α_{mo}, with α = x,y, or z, in the chiral derivative. The collinear first-order electric dipole moment is now located, in the dynamic-coupling mechanism, wholly in the substituent, having the form,

$$\mu^\alpha_{om} = \sum_{\ell \neq o} [(E_m - E_\ell)^{-1} (M_o L_\ell|V|M_m L_o)\mu^\alpha_{o\ell}$$

$$- (E_m + E_\ell)^{-1} (M_o L_o|V|M_m L_\ell)\mu^\alpha_{\ell o}] \tag{14}$$

The perturbation matrix-elements of equation (14) with the rearranged form $(M_o M_m|V|L_o L_\ell)$ represent the Coulomb potential (equation 8) between two transitional charge distributions $(M_o M_m)$ centred on the chromophore origin and $(L_o L_\ell)$ located on the substituent at a distance R from that origin. An approximation to the potential is afforded by a multipole expansion of each of these charge distributions, centred upon their respective particular coordinate-origins, and the truncation of each series after the leading term. The term retained for the charge distribution of the substituent transition is the electric dipole, $\mu_{o\ell}$. As the chromophore transition is magnetic dipole allowed, with a one-centre $\ell \rightarrow \ell$ atomic orbital character, the leading electric moment is an even 2^n-pole, with $n = 2,4,\cdots 2\ell$, each having $(2n + 1)$ components.

For the paradigmatic carbonyl $n \rightarrow \pi^*$ transition, with $p_y \rightarrow p_x$ character at the oxygen atom, the leading electric moment is the xy-component of a quadrupole, θ_{xy}, with a charge distribution which is phase-locked to the charge-rotation giving the concomitant z-polarized magnetic dipole of the transition, m_z (Figure 1). In this case the perturbation matrix-element (equation 14) becomes,

$$(M_o M_m|V|L_o L_\ell) = \sum_{\beta=x,y,z} \mu_{o\ell}^{\beta}\, \theta_{om}^{xy}\, G_{\beta,xy} \tag{15}$$

where the geometric tensors, $G_{\beta,xy}$ represent the radial (R) and angular $[XYZ/R^3]$ factors governing the Coulomb potential between the quadrupole, θ_{xy}, centred on the chromophore origin, and the β-component of the electric dipole, $\mu_{o\ell}$, located in the substituent at the position X,Y,Z in the chromophore frame. The geometric tensors afford the dynamic-coupling sector rules [19].

A general symmetry condition for a non-vanishing matrix element $(M_o M_m|V|L_o L_\ell)$ consists of the criterion [29] that the leading multipole component derived from the respective expansions of the charge distributions $(M_o M_m)$ and $(L_o L_\ell)$ transform under the same irreducible representation, or the same row of a degenerate representation, in the point group to which the chiral molecule belongs (not that of the chromophore alone, which generally has higher symmetry). While formulated in a more general connection [29], this symmetry condition provides the dynamic coupling mechanism with an analogue of the pseudoscalar criterion for the corresponding perturbation matrix elements of the one-electron static-field mechanism [18], although the condition is more detailed and specific in its applications.

A number of chiral ketones possess C_2 symmetry. In the C_2 group, the z-component of a dipole, electric or magnetic, transforms with the xy-component of an electric quadrupole under the A representation, whereas the x- and the y-components of a dipole are spanned by the B representation. The dynamic-coupling matrix element of equation (15) accordingly is restricted to a single term for the $n \rightarrow \pi^*$ transition of ketones with C_2 symmetry, namely, the term expressing the potential between the chromophore quadrupole moment, θ_{xy}, and the substituent dipole component, μ_z, for which the geometric tensor has the form,

$$G_{z,xy} = -15[XYZ/R^7] \tag{16}$$

Equation (16) provides a dynamic-coupling sector rule for the $n \rightarrow \pi^*$ transition of chiral carbonyl compounds with the true octant form and the correct sign [19]. Viewed from the +z direction, a clockwise charge displacement $p_y \rightarrow p_x$ (Figure 1a) is associated with a quadrupolar transition charge distribution in the chromophore which gives rise to a low-energy Coulombic correlation in the form of a charge displacement within the substituent in the +z direction if the substituent is located in one of the negative octants, e.g. near upper-right (Figure 1b). Overall the Coulombically-correlated charge displacements in the chromophore and substituent are helically left-handed, giving a negative Cotton effect (Figure 1).

The development of the $m_1 - \mu_2$ mechanism for the $n \rightarrow \pi^*$ optical activity of chiral ketones proceeds by noting that the substituent dipole making up the first-order electric moment of the $M_o \rightarrow M_m$ transition (equation 14) is restricted in all cases to the z-component by the condition of collinearity with the zero-order magnetic moment of the chromophore transition, m^z_{mo}. In consequence the explicit summation of equation (14), following the substitution of equations (15) and (16), gives the expression [19],

$$R_{om} = -15\, i\, m^{z}_{mo}\, \theta^{xy}_{om}\, \bar{\alpha}(L)\, [XYZ/R^7]_{(L)} \tag{17}$$

for the rotational strength of the $n \rightarrow \pi^*$ transition of all ketones with C_2 symmetry and C_1 ketones containing substituents with an effective isotropic polarizability, $\bar{\alpha}(L)$, at $\bar{\nu}_{n\pi^*}$

In the general case of ketones devoid of symmetry elements, with a substituent group or groups possessing an anisotropic polarizability tensor, terms additional to equation (17) contribute to the $n \rightarrow \pi^*$ rotational strength. The additional terms are dependent upon the anisotropy of the polarizability $(\alpha_{\parallel} - \alpha_{\perp})(L)$, on the assumption of cylindrical symmetry around the bond linking the substituent to the molecular frame of the parent symmetrical ketone, and upon the geometric tensors, $G_{x,xy}$ and $G_{y,xy}$, of equation (15) governing the potential between the chromophore quadrupole, θ_{xy}, and the x- and the y-component, respectively, of the substituent transition dipole, $\mu_{o\ell}$.

These two additional geometric tensors afford sector rules (Figure 3) determining the contribution of substituents, as a function of their anisotropy $(\alpha_{\parallel} - \alpha_{\perp})(L)$, and their position and orientation in the chromophore coordinate frame, to the $n \rightarrow \pi^*$ rotational strength, in addition to the isotropic contribution, dependent upon the octant rule (equation 17) [19]. The anisotropic contribution may dominate the $n \rightarrow \pi^*$ rotational strength of a chiral ketone, giving rise to a dissignate or anti-octant effect if the sign is opposed to that of a weaker isotropic contribution (Figure 3a), as in the case [30] of a number of β-axially substituted adamantanones (I) (Table). Where isotropic and the anisotropic contributions have the same sign (Figure 3b), as in the case [30] of most β-equatorially substituted adamantanones (II), the consignate or general octant behaviour is enhanced (Table).

Apparent dissignate behaviour, according to the original octant-rule prescription [20], emerges from equation (17) for chiral ketones containing a substituent with a smaller mean or isotropic polarizability than that of the atom or group which has been replaced in the parent symmetric ketone. The dissignate behaviour of the fluoro-substituent in the $n \rightarrow \pi^*$ Cotton effect of fluoro-ketones has long been known [20] and, more recently, it has been established [31] that, in isotopically-chiral ketones, the heavier isotope is relatively dissignate, e.g. deuterium relative to hydrogen, and the CD_3 and $^{13}CH_3$ group relative to the $^{12}CH_3$ group. These observations are in conformity with equation (17), the atomic polarizabilities of hydrogen, deuterium, and the fluoro-substituent being 0.41, 0.40, and 0.32Å^3, respectively [32,33], while the molecular polarizability of CH_4 and CD_4 is 2.640 and 2.605Å^3, respectively [34].

The dissignate effect of a substituent with a smaller mean polarizability than that of the atom or group replaced in the parent symmetric ketone remains in the corresponding disubstituted chiral ketone with C_2 symmetry, since the octant rule itself has twofold rotational symmetry (equations 16 and 17). However, the dissignate effects due to the anisotropic polarizability contribution, dependent upon $(\alpha_{\parallel} - \alpha_{\perp})(L)$, necessarily vanish in the corresponding disubstituted ketone with C_2 symmetry, since each of the anisotropic conical sector rules lack the $C_2(Z)$ element of symmetry (Figure 3), as follows in addition from the general symmetry condition imposed on a non-vanishing perturbation matrix element in the dynamic coupling mechanism [29] (equation 15).

Hence, a firm and distinctive expectation of the dynamic coupling $m_1 - \mu_2$ mechanism for the $n \rightarrow \pi^*$ carbonyl Cotton effect derives from the conclusion that a *dissignate* β-axially

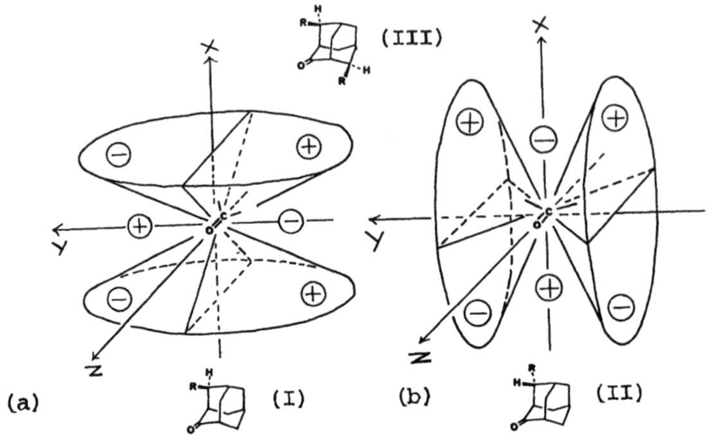

Figure 3. Sector rules for the contribution of the polarizability anisotropy ($\alpha_{\parallel} - \alpha_{\perp}$) of a substituent to the carbonyl $n \rightarrow \pi^*$ rotational strength by the dynamic coupling $m_1 - \mu_2$ mechanism for (a) a 2-axially substituted adamantan-4-one (I) and (b) the corresponding 2-equatorial derivative (II).

Table. The $n \rightarrow \pi^*$ Cotton effect of the 2-axially substituted adamantan-4-one derivatives (I), and of the corresponding 2-equatorial analogues (II), together with the atomic or group mean polarizability of the substituent $\Delta\bar{\alpha}$(R) taken relative to that of hydrogen ($\bar{\alpha}$(H) = 0.41Å^3) as zero, and the polarizability anisotropy of the R-C bond ($\alpha_{\parallel} - \alpha_{\perp}$) in Å^3 from [32,33].

R	$\Delta\bar{\alpha}$(R)	($\alpha_{\parallel} - \alpha_{\perp}$)	β-axial(I)		β-equatorial(II)		Ref
			λ/nm	Δε	λ/nm	Δε	
F	-0.09		278	+0.033	294	-0.19	a
D	-0.01		294	-0.017	295	-0.09	b
CH_3	+1.83	+0.72	303	-0.093	300	+0.78	a
Cl	+1.91	+1.0	303	-0.41	306	+4.15	a
Br	+3.05	+1.5	303	-0.45	305	+7.88	a
I	+5.07	+2.1	304	-0.76	306	+14.62	a

(a) G. Snatzke and G. Eckhardt, Tetrahedron, 24, 4543 (1968); G. Snatzke, B. Ehrig, and H. Klein, Tetrahedron, 25, 5601 (1969).

(b) H. Numan and H. Wynberg, J. Org. Chem., 43, 2232 (1978).

substituted adamantanone (I) necessarily has a *consignate* disubstituted analogue (III) in which the two β-axial groups are related by the $C_2(z)$ axis, each with the same stereochemical configuration at the substituted carbon atom as that of the monosubstituted derivative (I). The corresponding expectation from the one-electron static-field $m_1 - \mu_1$ mechanism [18] or from general MO treatments of the CNDO type [35] is that a given $n \rightarrow \pi^*$ octant rule effect, dissignate or consignate, in a monosubstituted ketone, including (I), appears with the same sign and approximately double the magnitude in the corresponding disubstituted derivative with C_2 symmetry, such as (III), with a common absolute stereochemistry at the substituted carbon atoms. As yet no experimental investigation of these symmetry-based expectations have been reported, and insufficient CD data are available, even from the extensively-studied chiral ketone field, to distinguish between the alternatives.

6. THE NON-DEGENERATE $\mu_1 - \mu_2$ MECHANISM.

In the $\mu_1 - \mu_2$ mechanism for optical activity, as in the classical coupled oscillator model [6], an overall helical charge displacement derives from two or more non-coplanar linear charge displacements with a finite separation. At a distant point the linear charge displacement of an electric dipole transition moment of a chromophore, $\vec{\mu}_t$, generates a transient magnetic dipole, $\vec{m}_t$, proportional to the position vector, $\vec{R}$, of the charge from the point, and to the current density, the product of the charge and its velocity. Classically and quantum-mechanically the charge velocity of an electric dipole transition is related to the transition frequency, ν_t, and the magnetic moment produced at an external point is given by the expression,

$$\vec{m}_t = i\pi(\nu_t/c)(\vec{R} \times \vec{\mu}_t) \qquad (18)$$

The electric moment of an isolated chromophore gives no overall magnetic moment since the contributions at two external points related by reflection through any plane containing the electric moment sum to zero. In a two-chromophore system the second group provides a unique external origin for the magnetic moment produced by the electric moment of the first group and *vice versa*. For a dimer with non-coplanar chromophores the magnetic moment generated by the one group has a component collinear with the electric moment of the other, producing a rotational strength proportional to the triple scalar product, $[\vec{R}_{21} \cdot \vec{\mu}_2 \times \vec{\mu}_1]$, where $\vec{R}_{21} = \vec{R}_2 - \vec{R}_1 = -\vec{R}_{12}$.

The non-degenerate $\mu_1 - \mu_2$ mechanism provides sector rules for the allowed electric dipole transitions of a symmetric chromophore in a dissymmetric molecular environment [19,36]. In the model chiral molecule considered, the electric dipole transition $M_o \rightarrow M_n$ has a zero-order moment μ_{on}^z located in the chromophore, defining the Z-coordinate of the chromophore frame. In addition the transition has a first-order electric moment located in the substituent, L, given by the analogue of equation (14). It is assumed that the substituent group has cylindrical symmetry, and the chromophore frame is rotated around the pre-established Z-axis so that the principal direction (z') of the substituent is perpendicular to the Y-axis, lying in a plane parallel to the chromophore XZ plane [18,19,36]. The choice of chromophore axes sets to zero the contribution of the Y-component of the first-order substituent dipole to the optical activity (α = y in equation 14), and reduces the number of terms in the expression for the rotational strength of the perturbed $M_o \rightarrow M_n$ transition to,

$$R_{on} = (\pi/2)\bar{\nu}_{on}|\mu_{on}^{z}|^2[Y(3Z^2-R^2)\sin 2\theta - 6XYZ\sin^2\theta]R^{-5}(\alpha_{\parallel} - \alpha_{\perp})(L) \quad (19)$$

where the polarizability anisotropy of the substituent $(\alpha_{\parallel} - \alpha_{\perp})(L)$ refers to the chromophore transition wavenumber $\bar{\nu}_{on}$, and θ is the angle between the Z-axis of the chromophore and the principal direction (z') of the cylindrically-symmetric substituent. In equation (19) X,Y,Z refer to the Cartesian coordinates of the substituent, at a distance R from the origin, in the right-handed chromophore frame.

Two superimposed sector rules are generally required by equation (19) for the $\mu_1 - \mu_2$ Cotton effects of a symmetric chromophore in a chiral molecular environment [18,19,36]. A class of exceptional cases is that of the chiral molecules with non-trivial rotational symmetry about the direction of the chromophore transition moment, such as the $\pi \rightarrow \pi^*$ transition, with Z-polarization in chiral olefins with $C_2(Z)$ or D_2 symmetry. In these cases the general symmetry restriction [29] on a non-vanishing perturbation matrix element $(M_oM_n|V|L_oL_\ell)$ limits the surviving term in the multipole expansion of the substituent charge distribution (L_oL_ℓ) to the dipole component, $\mu_{o\ell}^z$. Thereby the octant rule term, dependent upon the substituent coordinate function (XYZ) in equation (19), is eliminated. Thus equation (19) is reduced to the conical regional rule, dependent upon the function $[Y(3Z^2-R^2)]$ of the coordinates of the substituents, for the $\pi \rightarrow \pi^*$ transition of chiral olefins with $C_2(Z)$ or D_2 symmetry, and for analogous transitions in other chiral molecules with two-fold or higher rotational symmetry. In the general case of chiral molecules devoid of symmetry elements the conical regional rule is the more important for small values of the angle θ and the octant sector rule for the larger values of θ, other factors being equal [18,19,36].

7. THE DEGENERATE $\mu_1 - \mu_2$ MECHANISM.

The degenerate $\mu_1 - \mu_2$ mechanism for optical activity applies to chiral molecules in which the substituent, L, has become a light-absorbing group equivalent to the chromophore, M, with an excited state or states, L_n, of the same energy as that of the corresponding state or states of the chromophore, M_n. The individuality of a particular monomer chromophore transition is lost in the CD spectrum of the corresponding dissymmetric dimer, where it is represented by a characteristic pair of CD bands with opposite signs and comparable band areas, the exciton CD couplet. The two CD bands arise from transitions to dimer excited states formed by the symmetric and the anti-symmetric combination of the two excited configurations $|M_nL_o)$ and $|M_oL_n)$ which are degenerate in the zero order. Where the dimer has two-fold rotational symmetry or, more generally, where the two transition dipoles, $\mu_{on}(M)$ and $\mu_{on}(L)$, are inter-related by a C_2 axis, the combinations transform respectively under the A and the B representation of the C_2 group. The forms of the dimer excited states are,

$$\psi_{\pm} = (1/\sqrt{2})\,[\,|M_nL_o) \pm |M_oL_n)\,] \tag{20}$$

with the ground state ψ_o represented by the simple product $|M_oL_o)$ on the assumption that there is no electron exchange between the two chromophores.

In a coordinate frame based upon the $C_2(Z)$ axis with the X-axis directed along the line of centres of the individual chromophore transition dipoles (Figure 4), the first-order rotational strengths of the two dimer transitions are given by the relation,

$$R_{o+} = R_{o-} = \pi\bar{\nu}_{on}|\mu_{on}|^2 R_{ML}\,\cos(z)\cos(y) \tag{21}$$

where R_{ML} is the distance between the centres of the two chromophores, and cos(z) and cos(y) are the direction cosines of the zero-order transition dipole of the chromophore on the positive X-axis in the dimer frame (Figure 4).

If the energy-interval between the two dimer transitions is zero, no optical activity is observable, as in the case of molecules with a S_4 rotation-reflection axis, such as

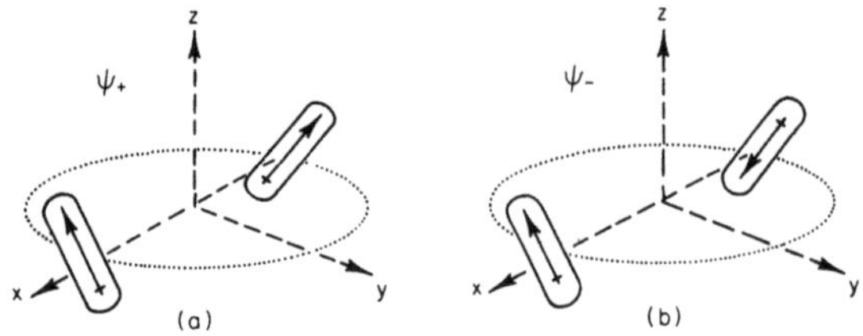

Figure 4. The symmetric (a) and antisymmetric (b) coupling modes of the monomer excitation dipole moments in a dissymmetric dimer.

perpendicular biphenyl, in which the rotational strengths of the two components of each degenerate transition mutually cancel pairwise. In dimers devoid of secondary symmetry elements the transitions to the two excited states with a common configurational parentage (equation 20) have different energies, with a wavenumber interval, $\Delta\bar{\nu} = \bar{\nu}_{o+} - \bar{\nu}_{o-}$, given by,

$$\Delta\bar{\nu} = 2|\mu_{on}|^2[\cos^2(z) - \cos^2(y) + 2\cos^2(x)]/[hcR_{ML}^3] \quad (22)$$

The absolute stereochemical configuration of a chiral dimer is obtained non-empirically from the relative signs and the relative frequency order of the two CD bands in an exciton couplet through equations (21) and (22). In general a knowledge of the polarization direction in the monomeric chromophore of the transition considered is essential for the determination of stereochemical configuration.

In favourable cases it suffices to know that the monomer transition is polarized in a given plane, the precise orientation of the moment in the plane being inconsequential for the stereochemical assignment. It is established that the $'L_a$ and $'L_b$ transitions of the aniline chromophore, near 250 and 310 nm, respectively, are polarized along the C_2 axis of the chromophore and along the in-plane direction perpendicular to that axis, respectively, but it is sufficient to know that these two transitions, being $\pi \rightarrow \pi^*$ in character, are polarized in the molecular plane in order to determine the absolute configuration of the alkaloid, calycanthine (IV), containing two aniline chromophores, from the CD spectrum [37,38]. For either the longer or the shorter in-plane direction of polarization in each of the individual aniline chromophores, the expectations from equations (21) and (22) are a positive CD band followed by a negative CD band at higher frequency for the configuration (IV), as is observed for each of the exciton CD couplets,

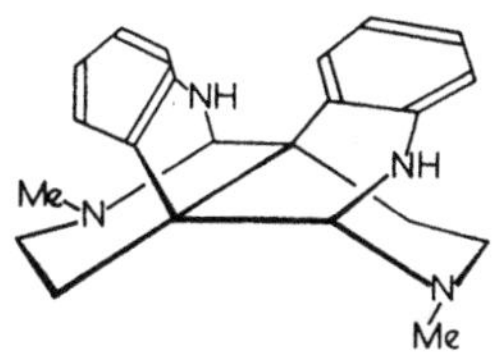

(IV) Absolute Configuration of Calycanthine

associated with the $'L_b$ and the $'L_a$ band systems, respectively, in the CD and absorption spectrum of calycanthine [37,38].

In unfavourable cases, such as (-)-1,5-diamino-9,10-dihydro-9,10-ethenoanthracene (V), which also contains two aniline chromophores, a knowledge is required for the assignment of the dimer configuration, not only of the polarization-direction of the zero-order transition, but also of its location or centre of gravity in the monomeric chromophore. A determination of the absolute stereochemical configuration by the Bijvoet X-ray method [39], and by an exciton analysis of the CD spectrum based upon a dipole-length calculation of the position of the $'L_b$ transition moment in the aniline chromophore, were found to give contrary stereochemical assignments for (V) [40]. From the apparent discrepancy it was concluded for a period that the original Bijvoet method was erroneous in sign [40], despite the general agreement between the X-ray and the CD methods for determining absolute configuration in all previously-studied cases [41]. The discrepancy stimulated the determination of the absolute configuration of calycanthine by the Bijvoet X-ray method, which gave [42] the stereochemistry (IV) in agreement with the CD assignment.

A number of methods are available for the calculation of an electric dipole transition moment, the principal methods employed for polyatomic molecules being the dipole-length and the dipole-velocity procedure. In dipole-length form the one-electron dipole moment connecting the MOs ψ_i and ψ_j is expressed by,

$$\mu_{ij}(DL) = \langle\psi_i|e\vec{r}|\psi_j\rangle \tag{23}$$

where r is the position vector of the electron from an arbitrary origin. The corresponding dipole-velocity form is given by the relation,

$$\mu_{ij}(DV) = (\beta_M/\pi\bar{\nu}_{ij})\langle\psi_i|\vec{\nabla}|\psi_j\rangle \tag{24}$$

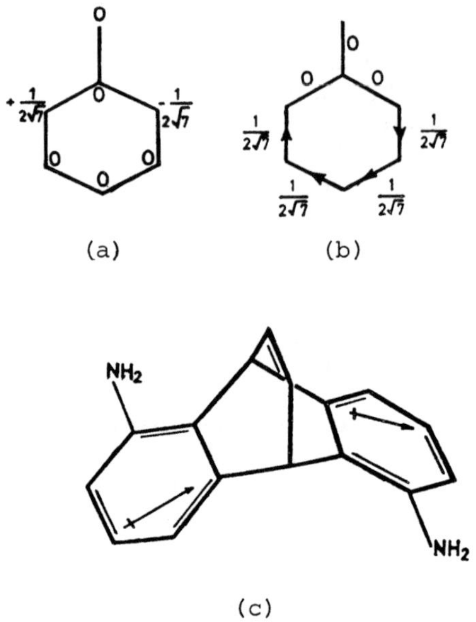

Figure 5. (a) The π-electron charge distribution of the lowest-energy transition in the benzyl anion calculated by the dipole-length method.

(b) The corresponding vectorial bond-order changes given by the dipole-velocity method, representing the transitional charge-momentum along each bond.

(c) The lower-frequency coupling mode, giving a negative rotational strength in 9S,10S-1,5-diamino-9,10-dihydro-9,10-ethenoanthracene (V), of the lowest-energy π-excitation of the two aniline chromophores calculated in the dipole-velocity representation.

where $hc\bar{\nu}_{ij}$ is the transition energy and $(\hbar/i)\vec{\nabla}$ is the momentum operator, β_M being the Bohr-magneton. If ψ_i and ψ_j are eigenfunctions, equations (23) and (24) give identical locations, magnitudes, and polarization directions for μ_{ij}, which is then an eigenvalue.

The majority of chiral molecules are relatively-large polyatomic systems for which, as yet, the MO wave-functions are inevitably inexact, and μ_{ij} has only a mean expectation-value. Early in the development of the quantum-mechanical approach to optical activity, it was indicated [43] that rotational strengths calculated by the dipole-length method are, in general, origin-dependent. Subsequently it was shown that origin-independent rotational strengths are ensured by working consistently in the dipole-velocity representation [44]. However, the sum-rule for the rotational strengths of a chiral molecule is generally violated by the dipole-velocity, but not by the dipole-length procedure [45].

In stereochemical applications the reduced validity of the sum-rule for rotational strengths is expected to be generally less of a hazard that the origin-dependency of those strengths, since the signs and the frequency-order of the bands in the CD spectrum (equations 21 and 22) are primary in configurational assignments. Weak non-degenerate $\mu_1 - \mu_2$ optical activity is generally superimposed on the stronger degenerate $\mu_1 - \mu_2$ rotational strengths, and the sum-rule of equation (21) is not expected to be quantitatively exact within a given exciton CD couplet on this ground alone.

The origin-dependency of the rotational strengths calculated by the dipole-length method leads to the incorrect stereochemical assignment in the case of the aniline-chromophore dimer (V). Aniline and other benzene derivatives with a conjugated exocyclic atom have the benzyl anion as a common 8 π-electron analogue. In the corresponding-carbanion model [46] the lowest-energy π-excitation of aniline, giving the $'L_b$ band near 310 nm, becomes an analogue of the transition of a π-electron from the non-bonding MO to the lowest unoccupied MO of the benzyl anion. The charge densities of this transition show (Figure 5a) that the electric moment obtained by the dipole-length method is directed along the line joining the two *ortho*-carbon atoms, whereas the vectorial bond-order changes in the transition indicate (Figure 5b) that the electric moment in the dipole-velocity representation is directed along the line through the midpoints of the two *meta-para* carbon-carbon bonds. That is, the location of the electric dipole transition moment is displaced from the centre of the benzene ring towards the exocyclic group by the dipole-length method, but away from that group toward the *para*-carbon atom by the dipole velocity procedure [41]. Similar displacements of the lowest-energy electric moment in the aniline chromophore are obtained by more exact methods, notably in the π-SCF calculations on calycanthine (IV) [38].

The Coulombic coupling of the lowest-energy electric moment of the two aniline chromophores, obtained by the dipole-velocity method, gives through equations (21) and (22) a negative and a positive rotational strength at a lower and a higher frequency, respectively, for the 9S,10S-configuration of 1,5-diamino-9,10-dihydro-9,10-ethenoanthracene (Figure 5c), as is observed for the (-)-isomer [40], in agreement with the

X-ray configurational assignment. The exciton CD analysis employing the aniline excitation moments calculated by the dipole-length formalism assigns to the (-)-isomer the enantiomeric 9R,10R-configuration [40], providing a paradigm case of the long-suspected hazard of employing the dipole-length procedure in the calculation of rotational strengths [41].

The degree to which the rotational strengths of a chiral system calculated by the dipole-length and by the dipole-velocity procedure are in agreement with one another provides a criterion for the quality of the wavefunctions employed [47]. In a comparison of various MO methods, ranging from the Hückel level to the *ab initio*, the ratio of the rotational strengths obtained by the two procedures for the $\pi \rightarrow \pi^*$ transition of a twisted ethylene model was found to range from 9.73 for the CNDO method with configurational interaction to 1.49 for a non-empirical INDO method with improved virtual orbitals [47].

REFERENCES

[1] A. Fresnel, Ann. Chim., 28, 147 (1825).

[2] P. Drude, Ann. d. Physik, 48, 536 (1896); The Theory of Optics, trans. C.R. Mann and R.A. Millikan, (1902), Dover reprint, New York (1959) p.400.

[3] A. Cotton, Compt. Rend. Acad. Sci. Paris, 120, 989 and 1044 (1895).

[4] E. Fischer, Chem. Ber., 24, 2683 (1891).

[5] L. Natanson, J. de Phys., 8, 321 (1909).

[6] W. Kuhn, Trans. Faraday Soc., 26, 293 (1930); Ann. Rev. Phys. Chem., 9, 417 (1958).

[7] L. Boltzmann, Pogg. Ann. Phys. Chem., Jubelband, p.134 (1874).

[8] L. Rosenfeld, Z. Phys., 52, 161 (1928).

[9] E.U. Condon, Rev. Mod. Phys., 9, 432 (1937).

[10] J.A. Schellman, Chem. Rev., 75, 323 (1975).

[11] J.G. Kirkwood, J. Chem. Phys., 5, 479 (1937).

[12] E.U. Condon, W. Altar, and H. Eyring, J. Chem. Phys., 5, 753 (1937).

[13] W.J. Kauzmann, J.E. Walter and H. Eyring, Chem. Rev., 26, 339 (1940).

[14] W. Moffitt, J. Chem. Phys., 25, 467 (1956).

[15] W. Moffitt and A. Moscowitz, J. Chem. Phys., 30, 648 (1959).

[16] A. Moscowitz, Adv. Chem. Phys., 4, 67 (1962).

[17] I. Tinoco, Jr., Adv. Chem. Phys., 4, 113 (1962).

[18] J.A. Schellman, J. Chem. Phys., 44, 55 (1966); Acc. Chem. Res., 1, 144 (1968).

[19] E.G. Höhn and O.E. Weigang, Jr., J. Chem. Phys., 48, 1127 (1968); E.C. Ong, L.C. Cusacks, and O.E. Weigang,Jr., J. Chem. Phys., 67, 3289 (1977).

[20] W. Moffitt, R.B. Woodward, A. Moscowitz, W. Klyne and C. Djerassi, J. Amer. Chem. Soc., 83, 4013 (1961).

[21] D.N. Kirk, W. Klyne, and W.P. Mose, Tetrahedron Letters, 1315 (1972).

[22] D.A. Lightner and D.E. Jackman, J.C.S. Chem. Comm., 344 (1974); D.A. Lightner and T.C. Chang, J. Amer. Chem. Soc., 96, 3015 (1974).

[23] A.I. Scott and A.D. Wrixon, Tetrahedron, 26, 3695 (1970).

[24] T. Watanabe and H. Eyring, J. Chem. Phys., 40, 3411 (1964).

[25] T.D. Bouman and A. Moscowitz, J. Chem. Phys., 48, 3115 (1968).

[26] G. Wagnière, J. Amer. Chem. Soc., 88, 3937 (1966).

[27] M.B. Robin, Higher Excited States of Polyatomic Molecules, Vol. II, Academic Press, New York, (1975).

[28] A.F. Drake and S.F. Mason, J. de Physique, 39, sup. 7, C4-212 (1978).

[29] S.F. Mason, R.D. Peacock and B. Stewart, Chem. Phys. Letters, 29, 149 (1974); Molec. Phys., 30, 1829 (1975).

[30] G. Snatzke and G. Eckhardt, Tetrahedron, 24, 4543 (1968); G. Snatzke, B. Ehrig, and H. Klein, Tetrahedron, 25, 5601 (1969).

[31] C.S. Pak and C. Djerassi, Tet. Letters, No.45, 4377 (1978); and references therein.

[32] R.J.W. Le Fèvre, Adv. Phys. Org. Chem., 3, 1 (1965).

[33] N.J. Bridge and A.D. Buckingham, Proc. Roy. Soc., 295A, 334 (1966).

[34] O. Hassel and E. Hetland, Physik Z., 40, 29 (1939).

[35] E.E. Ernstbrunner and M.R. Giddings, J. Chem. Soc. Perkin II, 989 (1978); and references therein.

[36] O.E. Weigang, Jr., and E.C. Ong, Tetrahedron, 30, 1783, (1974); O.E. Weigang, Jr., 'An Amplified Sector Rule for Electric Dipole Allowed Transitions', Biophysical Society, Washington, D.C., 1978.

[37] S.F. Mason, Proc. Chem. Soc., 362 (1962); S.F. Mason and G.W. Vane, J. Chem. Soc.(B), 370 (1966).

[38] W.S. Brickell, S.F. Mason and D.R. Roberts, J. Chem. Soc. (B), 691 (1971).

[39] J.M. Bijvoet, Proc. Acad. Sci. Amsterdam, 52, 313 (1949).

[40] J. Tanaka, C. Katayama, F. Ogura, H. Tatemitsu, and M. Nakagawa, J. Chem. Soc. Chem. Comm., 21 (1973).

[41] S.F. Mason, J. Chem. Soc. Chem. Comm., 239 (1973).

[42] A.F. Beecham, A.C. Hurley, A.M. Mathieson and J.A. Lamberton, Nature Phys. Sci., 244, 30 (1973).

[43] E. Gorin, J. Walter, and H. Eyring, J. Chem. Phys., 6, 824 (1938).

[44] W. Moffitt, J. Chem. Phys., 25, 467 (1956).

[45] R.A. Harris, J. Chem. Phys., 50, 3947 (1969).

[46] S.F. Mason, J. Chem. Soc., 1253 (1959); 219 (1960).

[47] A. Rauk, J.O. Jarvie, H. Ichimura and J.M. Barriel, J. Amer. Chem. Soc., 97, 5656 (1975).

CHIROPTICAL PROPERTIES OF ORGANIC COMPOUNDS: CHIRALITY AND SECTOR RULES

G. Snatzke

Lehrstuhl für Strukturchemie
Ruhruniversität Bochum
Postfach 10 21 48
D 4630 Bochum 1 (FRG)

ABSTRACT. A general discussion of sector, chirality, and helicity rules is given. Using qualitative MO-theory the sign of a CD-band can be predicted relatively easily for inherently chiral chromophores. Application to inherently (locally) achiral chromophores which are chirally perturbed by their surroundings is possible by applying qualitatively perturbation theory. A generalization of sector rules is proposed.

INTRODUCTION

In the chapters on the theory of CD it has been shown that the experimental magnitude "rotational strength R", i.e. the (wavenumber weighted) area under a CD-band multiplied by a factor of 0.229×10^{-38}, is theoretically given by the product $R = \mu . m . \cos(\vec{\mu},\vec{m})$, where $\vec{\mu}$ stands for the electric transition moment vector (translation of electron charge), $\vec{m}$ for the magnetic one (rotation of charge) during the transition in question. The determination of the sign of a CD-band is thus reduced to the determination of the angle between these two transition moments: if it is acute (extreme case: parallel orientation) the cosine is positive and so is the Cotton effect; if it is obtuse (extreme case: antiparallel orientation) the rotational strength is negative. As group theory shows, components of both $\vec{\mu}$ and $\vec{m}$ along the same line can occur only if the chromophoric system itself is inherently chiral (Class I of Moscowitz (1)), otherwise (Class II of Moscowitz (1)) one of the two moments is identical zero, or they are perpendicular to each other (the cosinus factor is zero!). Very often in

Stephen F. Mason (ed.), Optical Activity and Chiral Discrimination. 25–41.

Class I of inherently chiral chromophores it is possible to determine the relative directions of both transition moment vectors by application of simple qualitative MO theory, and some examples are given below. In case of inherently achiral chromophores, which are chirally perturbed by their environment (Class II of Moscowitz; if the rest of the molecule is chiral, we have "natural optical activity", as in Case I. ORD or CD can, however, also be measured of inactive compounds if other chiral molecules, e.g. the solvent molecules, are present, as long as one enantiomer predominates: "medium-induced optical activity"), the missing transition moment has to be "stolen" from another transition. Assumptions about the mechanism of such a "stealing" have to be made, but again very few basic rules from perturbation theory (2) can help in obtaining the sign of the CD without detailed calculations. Even the magnitude of the rotational strength can be estimated from these qualitative MO-considerations.

It is obvious that in general for a molecule with an achiral chromophore the chiral perturbation by a bond will be the more pronounced the nearer this chiral bond is situated to the chromophore. This led to the following pragmatic classification of chiral molecules (3): Beginning with the chromophore the molecule is divided into areas or "spheres", the chromophore itself being per definitionem the first sphere. A ring into which the chromophore may be built in forms the second sphere, and further rings or substituents, connected with the second sphere, build up the third sphere, etc.. It has been observed experimentally in most cases that that chiral sphere, which is closest to the chromophore, determines mainly the signs and even the magnitudes of the Cotton effects. Chiral groups which are farther distant from the chromophore have a more pronounced influence upon the Cotton effect only if there exists a planar zig-zag way connecting the perturbing bond (lone pair) with an orbital involved in the excitation (4). The ring size of the second sphere is of no vital importance; the separation of the second chiral sphere from the other cases of Moscowitz's Class II is justified, however, by the fact that such a chiral ring will fix the torsion angle around the bond connecting the chromophore with the nearest chirally disposed moiety within the molecule; this torsion angle is thus an equivalent characterization of the second sphere.

In case of a chiral first or a chiral second sphere the inherent chirality of the chromophore, or the chiral orientation of the nearest (noncoplanar) bond to the chromophore determines the Cotton effect, and we obtain "chirality rules" or "helicity rules". Actually one should restrict the name "helicity rule" to such a case where one looks along a C_n-axis of the first or

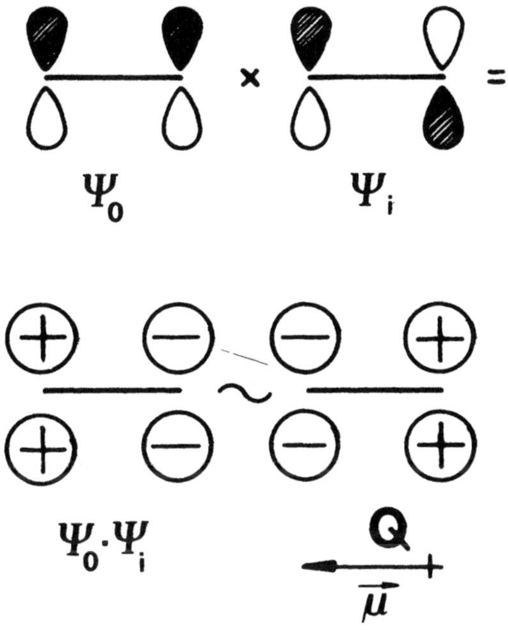

Fig. 1. Formal multiplication of orbitals for a $\pi\rightarrow\pi^*$-transition of ethylene. Q refers to the developped quadrupole, $\vec{\mu}$ is the electric transition moment.

second sphere. Although for substituents farther distant from the chromophore their orientation may also be important for the magnitude (and even for the sign) of their contributions to the Cotton effects, it is in practice mostly sufficient to take into consideration only the positions of the perturbing atoms, and this leads then to the various types of "sector rules". Throughout this chapter the term "sector rule" is, therefore, consequently applied only to cases of achiral first and second, but chiral third sphere. The space around the chromophore is divided into sectors by "nodal planes" or "nodal surfaces"; any atom lying in such a nodal surface gives no contribution to the Cotton effect, the contributions of atoms (of the same type) in adjacent sectors have opposite signs. According to Schellman (5) the simplest sector rule for a given chromophore is that in which the (local) symmetry planes of the chromophore are nodal planes. To these "symmetry-determined" planes one has, however, often to add "orbital-determined" nodal surfaces of the orbitals involved in the transition. Whereas for different transitions within one chromophore the symmetry-determined planes are always the same, the orbital determined nodal surfaces may be different.

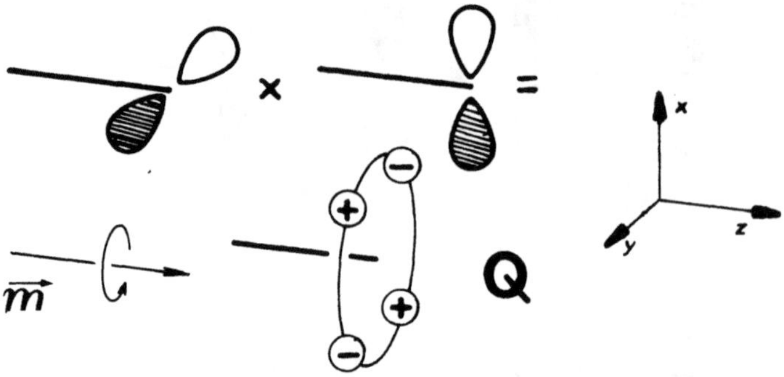

Fig. 2. Formal multiplication of orbitals for a $p_y \to p_x$-transition (e.g. $n \to \pi^*$ of carbonyl). Q refers to the developped quadrupole, $\vec{m}$ is the corresponding magnetic transition moment.

The "recipe" for finding $\vec{\mu}$ and $\vec{m}$ is quite simple (6), one has only to multiply formally together the two orbitals involved in the transition $\Psi_o \to \Psi_i$. Fig. 1 shows this for a $p_x \to p_x$-transition, as e.g. in case of the $\pi \to \pi^*$-transition of olefins or ketones. The physical meaning of such product is electron density (or probability to find the electron in a given volume element), a positive (negative) sign of $\Psi_o \cdot \Psi_i$ corresponds thus to a negative (positive) charge. For the $p_x \to p_x$-transition a quadrupole is obtained which corresponds at the same time to an overall dipole (charge translation) along the double bond. On the other hand a $p_y \to p_x$-transition (Fig. 2) leads to a quadrupole whose two individual dipoles compensate exactly, and no electric transition dipole $\vec{\mu}$ is created. But now the excitation from p_y into p_x involves a rotation of electron charge. Such a rotation of charge is equivalent to a magnetic transition moment $\vec{m}$, whose direction is obtained by the right-hand-thumb rule ($\vec{m}$ corresponds to the direction of the thumb if the other fingers follow the rotation of the charge). No such magnetic transition moment is connected with the $\pi \to \pi^*$-transition (Fig. 1). The directions of $\vec{\mu}$ and of $\vec{m}$ depend on the arbitrary choice of the signs for the two lobes of the p-orbitals and is of no physical significance; only the angle between the two moments is of importance for determining the sign of the Cotton effect. It should be kept in mind furthermore, that (whatever the choice of orbital signs may be) there exists an unequivocal correlation between the direction of $\vec{m}$ and the sign pattern of the corresponding quadrupole (Fig. 2).

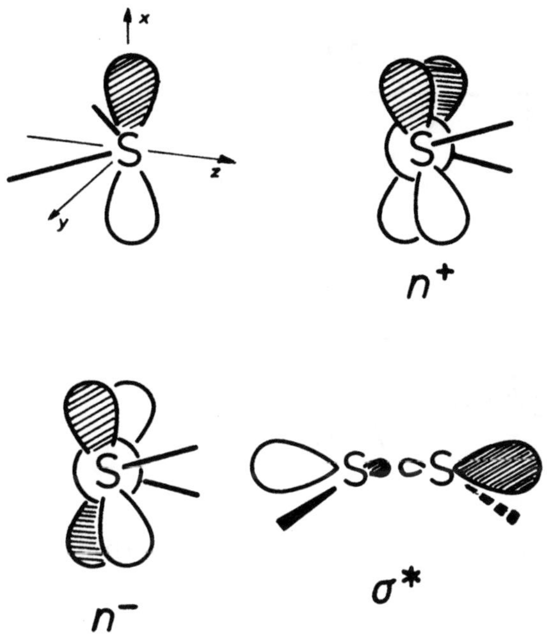

Fig. 3. Orbitals involved in the absorption of disulfides.

THE CHIRAL FIRST SPHERE

The first sphere can become chiral if either the chromophore belongs to a chiral point group (no S_n-axis, including n=1 and 2, allowed), or if two or more locally achiral chromophores are chirally arranged in close neighbourhood to each other, and if the transition considered is connected with a strong electric transition moment (Exciton interaction, Davydov splitting). As an example of the first type I choose the disulfide chromophore (6,7). The HOMO of a thioether is the nonbonding p-orbital on the sulfur perpendicular to the -S- plane. In a disulfide such two AO combine to give $n^+ = N^+.(p_1 + p_2)$ and $n^- = N^-.(p_1 - p_2)$; if we assume an acute torsion angle around the S-S bond n^- has a nodal plane perpendicular to this bond and is thus of higher energy as n^+, so it forms the HOMO. The LUMO consists of the σ^*-AO of the S-S bond. The formal multiplication of HOMO into LUMO (Fig. 4) shows the formation of two quadrupoles which are, however, not compensated as in the case of the $p_y \rightarrow p_x$- transition (Fig. 2): with the signs chosen deliberately as in Fig. 3 at each

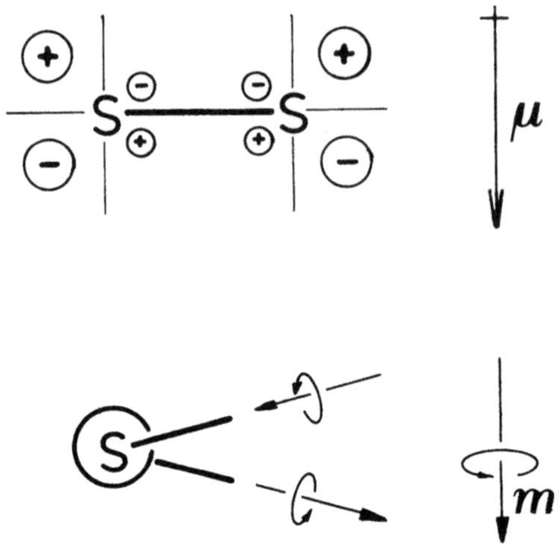

Fig. 4. Relative direction of $\vec{\mu}$ and $\vec{m}$ for a positive torsion angle of a disulfide ($n\rightarrow\sigma^*$-band).

sulfur atom the partial electric transition moment vectors point downwards, and so must their sum vector. If we follow the charge rotation around the two S-atoms and apply the right-hand-rule we find that the two individual transition moment vectors point also downwards, so the overall $\vec{m}$ is parallel to $\vec{\mu}$ for the chirality chosen (positive torsion angle). The same result was obtained by several more or less sophisticated calculations (7) and it is in agreement with the experiments. In the same manner one can explain 1) that the next CD-band at shorter wavelength must have opposite (in our case: negative) sign, and 2) that this helicity rule should be inverted if the torsion angle is obtuse (8).

The "multiplication recipe" gives in this and in several other simple cases a helicity rule which correlates the absolute conformation (the absolute conformation of a partial structure is determined if we know the sign(s) of the torsion angle(s)) of the inherently chiral chromophore with the sign(s) of the Cotton effect(s). By this we do, however, in general not yet know the absolute configuration of the whole molecule. It needs a second information (conformational analysis, X-ray data, NMR-spectra, etc., or sometimes a second CD-argument, as wavelength of the band, or magnitude of the CD) to correlate absolute conformation with absolute configuration. Very frequently this fact is overlooked: at first hand chiroptical measurements of a substance give information only about the absolute conformation of the molecules!

It is often very helpful to check the results of this multiplication procedure by group theoretical considerations. In the point group C_2 of the twisted disulfide chromophore and with the coordinate system chosen as in Fig. 3 n^- is of A-symmetry, σ^* belongs to the irreducible representation B, the excited state has thus also B-symmetry, so the transition from the ground state is associated with an electric as well a magnetic transition moment lying in the (x,y)-plane, as read off the character table. Such considerations do, however, give no information about the magnitudes of $\vec{\mu}$ and/or $\vec{m}$.

The exciton theory ("coupled oscillator theory") has been first introduced by W. Kuhn (9) in his classical papers on optical activity and was later revived especially by S.F. Mason (10). It is apparently simple and its uncritical application has thus sometimes lead to wrong results. It should furthermore be remembered that the exciton term gives only one contribution to the overall CD within an absorption band and this must not be the predominant one. In the following an example is given from biochemistry. The amide group gives two absorptions in the usually accessible wavelength range, the $n{\rightarrow}\pi^-$- and the $\pi^0{\rightarrow}\pi^-$-band. Only the latter is associated with a strong electric transition moment which is appr. parallel to the line connecting the two heteroatoms (Fig. 5). In a dioxopiperazine (Fig. 6) two such locally achiral chromophores are chirally disposed to each other, and according to the exciton theory the excited state is split into two, which

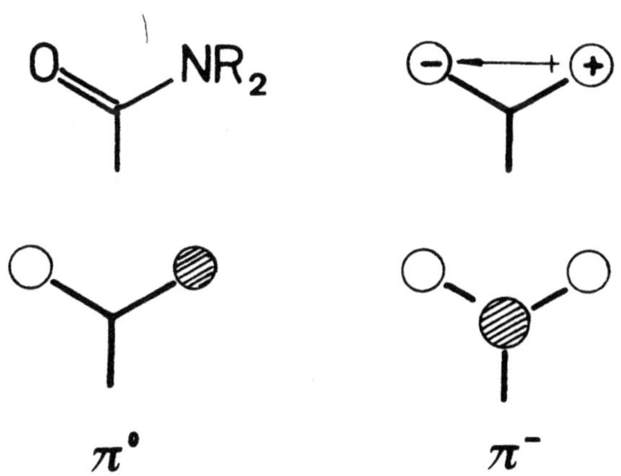

Fig. 5. π^0 and π^- of the amide chromophore and direction of the electric transition moment for the $\pi^0{\rightarrow}\pi^-$-band. (This does not imply that $\vec{\mu}$ points from N towards O !).

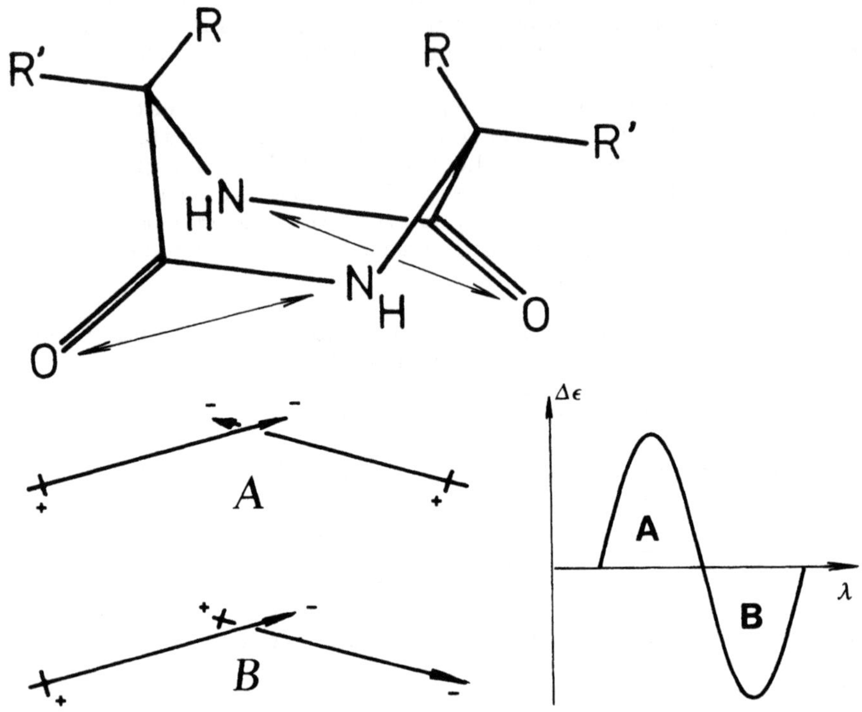

Fig. 6. Exciton coupling between the two amide chromophores ($\pi^0 \rightarrow \pi^-$-transition). For details see text.

correspond to the two possible combinations (in-phase, out-of-phase) of the two individual electric transition moments. The overall motion of the excited electron follows a righthanded screw, for transition A, a lefthanded one for B (Fig. 6), if the absolute conformation of the ring is as depicted. The CD within band A must thus be positive, within B negative (again a helicity rule). The relative band positions can also be deduced from Fig. 6: after excitation the strongest interaction of the developped charges is between poles of same signs for A, i.e. repulsive, whereas it is attractive for B; state B has thus lower energy than A. The negative CD (of B) appears then at longer, the positive one (of A) at shorter wavelengths, we obtain a "negative CD-couplet". Only the low-energy wing can be recorded in general (11), i.e. a negative Cotton effect for the chirality indicated in Fig. 6. Once again CD gives us thus only the absolute conformation of the dioxopiperazine ring, but no information about the absolute configuration at the centers of chirality!

Sometimes it is difficult to find out the helicity of the screw determined by two such vectors. One of the following hints might then be helpful. 1) It is in general quite easy to determine the direction of the sum vector; we make this an axis of a cylinder and consider the two individual vectors as tangents for a helix wound onto this same cylinder. 2) We connect point 1 of the first vector (1 → 2) with point 3 of the second one (3 → 4), thus obtaining a continuous line 2 ← 1 - 3 → 4. The second vector is now parallel shifted so that its end 3 coincides with point 1. If then the three vectors 1 → 2, 1 → 3, and 3 → 4 form a righthanded (lefthanded) coordinate system (in general not Cartesian) the helicity of the system is righthanded (lefthanded).

THE CHIRAL SECOND SPHERE

As mentioned above the essential property of a chiral ring is the fixation of torsion angles other than 0° or 180° next

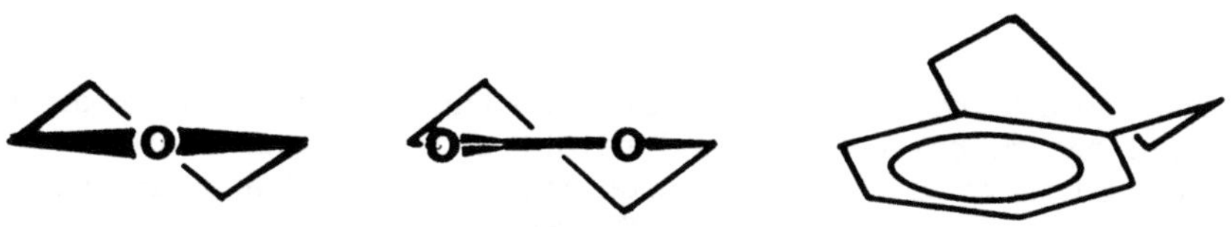

Fig. 7. Examples of molecules with chiral second sphere: cycloalkanones in twist-conformation, pentanolide in half-chair conformation, and tetralin with half-chair conformation of chiral cyclohexene ring (left to right).

to the chromophore. Fig. 7 shows 3 examples: the cycloalkanones in a twist conformation, the lactone chromophore ($n \to \pi^*$ band in both cases), and the 270 nm band of a tetralin or tetrahydroisoquinoline. In all three examples the given absolute conformation leads to a positive CD, and we can again speak of helicity (alkanones, tetralins) or chirality rules (lactones). From the theoretical point of view there is, however, no essential difference between these molecules and those in which first and second spheres are achiral: only one of the two necessary transition moment vectors is non-zero and the other one has to be stolen either from another transition. This can be done either from a second transition within the same chromophore ("intrachromophoralic mechanism"), or from another chromophore within

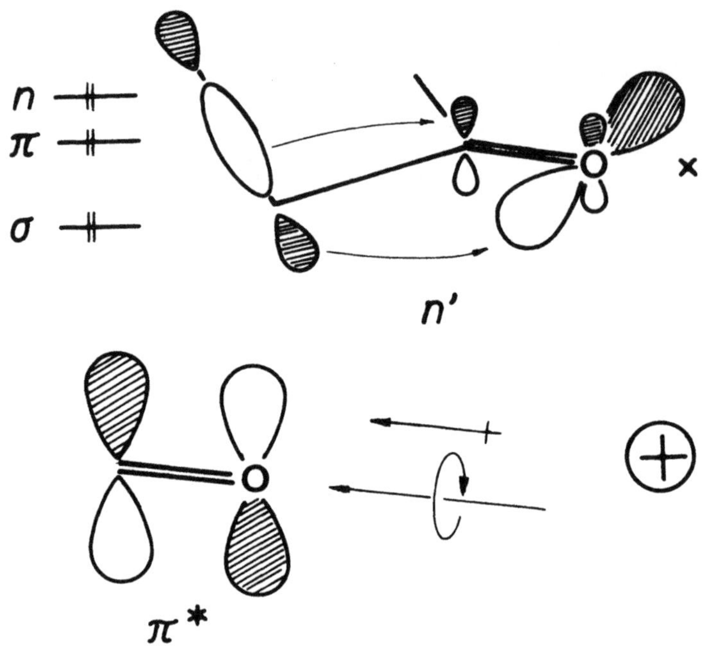

Fig. 8. Admixture of π-character into the n-orbital of a ketone by a nearby chiral σ-bond, and relative directions of the transition moments for excitation into π^*. The drawn absolute conformation gives a positive CD around 300 nm.

the same molecule ("interchromophoralic mechanism"). As the simplest example I will discuss a twisted cycloalkanone (Fig. 8) (6). The $n\to\pi^*$ transition is electrically forbidden, and that it is detectable at all in a spectrum is in general ascribed to the vibronic coupling with an allowed transition from which it can steel electric transition moment. On the other hand it is magnetically allowed, $m_z \neq 0$ (point group C_{2v}, Fig. 2). In order to

obtain rotational strength we have thus to steel an electric transition moment in the same direction, and the obvious choice is combination with the $\pi\to\pi^*$ transition, which has $\mu_z \neq 0$.

Admixture of π-character to the n-AO is strictly forbidden in point group C_{2v}, but becomes weakly allowed by lowering the

symmetry during some types of vibrations and by interaction with a chiral environment. The nearest chiral bond will have the greatest influence, and in a twisted cycloalkanone (chiral second sphere) this is the C_α - C_β - bond. The ordering of (rising) orbital energies is σ-π-n, so according to perturbation theory admixture of σ into π as well as n will rise their respective energies, i.e., in both cases we have to use the energetically unfavoured combinations. Choosing the phasing of the σ-orbital arbitrarily as in Fig. 8 determines thus the phasing of π and n in such a way that negative overlap results. Whatever phasing of the π^*-orbital we now use, the relative orientation of the resulting transition moment vectors will remain the same; in our example they are parallel, the CD must, therefore, be positive, and this is the well known experimental helicity rule (12). In just the same manner one can explain (6) why an axial bond in ß-position to the carbonyl on a chair-form of cyclohexanone gives a contribution of the "anti-octant" type (13).

THE CHIRAL THIRD SPHERE

Again I take the carbonyl $n \rightarrow \pi^*$-band as an example, which is connected with a magnetic transition moment in z-direction. The missing electric transition moment μ_z can either be taken from the same chromophore, or from another one. In the first case the $\pi \rightarrow \pi^*$-transition is the obvious candidate, and the excited state is then not any more the pure $n \rightarrow \pi^*$ state (S_1) but has admixed to it a small percentage of the $\pi\pi^*$-state. This mixing is caused by the surrounding chiral field and can be treated theoretically by perturbation theory (PT). First order PT (14) leads to the result that the sign of the contribution to the Cotton effect depends on the sign of the charge of the perturbing "pole"; experimental results, e.g. for adamantanone carboxylic acids (15) revealed, however, that the negative charge

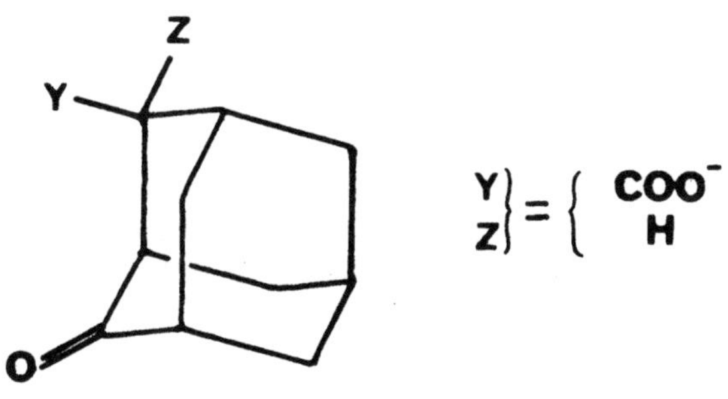

of the COO^- ion gives a contribution of the same sign as any other group. Applying second-order PT does not give such a sign-dependence, so it will be used here.

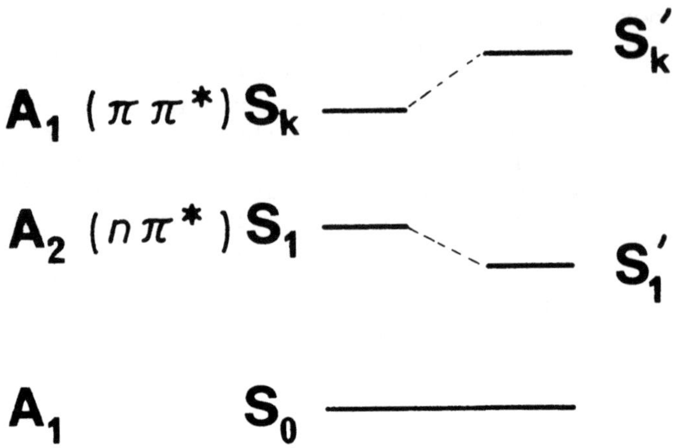

Fig. 9. Interaction between the $n\pi^*$ (S_1)- and the $\pi\pi^*$-singlet (S_k) states of a carbonyl chromophore caused by its chiral environment.

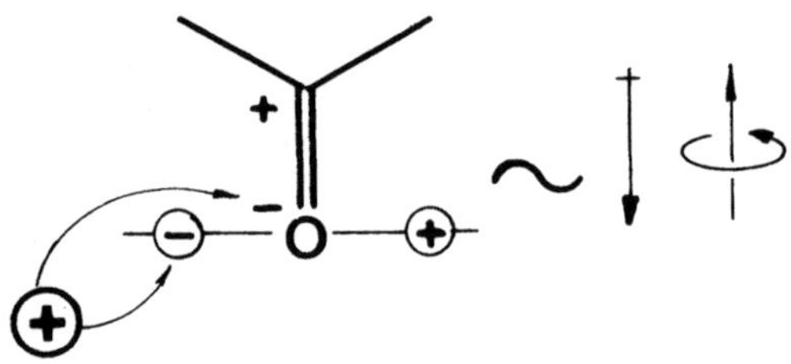

Fig. 10. Induction of the charge patterns for a transition dipole along the C=O - bond and a quadrupole around the O-atom of a carbonyl by a positive charge in the left upper front octant. The absolute conformation drawn gives a negative Cotton effect.

The by the perturbation slightly modified first excited state S_1' is the one which corresponds to the energetically favoured combination of the S_1 and the S_k state (Fig. 9) because S_1 is the excited state of lowest energy. Any perturbing charge will now interact with the electric transition dipole as well as with the quadrupole which is developped during a magnetically allowed transition, and using the energetically favoured combination means, that we have to take in both cases either only the attractive, or only the repulsive interaction. Let's assume we have a positive perturbing charge in the left upper front octant (Fig. 10) and let's furthermore take the attractive interaction, then (according to Coulomb's law) the electric transition dipole points from C towards O, and the signs of the quadrupole around the oxygen atom are such that this corresponds to a magnetic transition moment vector pointing in the opposite direction (cf. also Fig. 2). The CD should then be negative. Whenever we go with the perturbing charge into an adjacent octant either $\vec{\mu}$ or $\vec{m}$ will change its direction and we obtain indeed the octant rule with the experimentally proved signs (16). If we would have used a negative charge instead, or one of the poles of a dipole (as does the dynamic coupling model) we would have obtained the same result.

In a similar way we can deduce the sign for contributions coming from interchromophoralic interactions. The transition in question might e.g. be of the C-C or C-H $\sigma \rightarrow \sigma^*$-type, which is connected with an electric transition moment along that bond. If we place such a transition moment e.g. in the left upper back octant in the (-z)-direction (Fig. 11) and take again the energe-

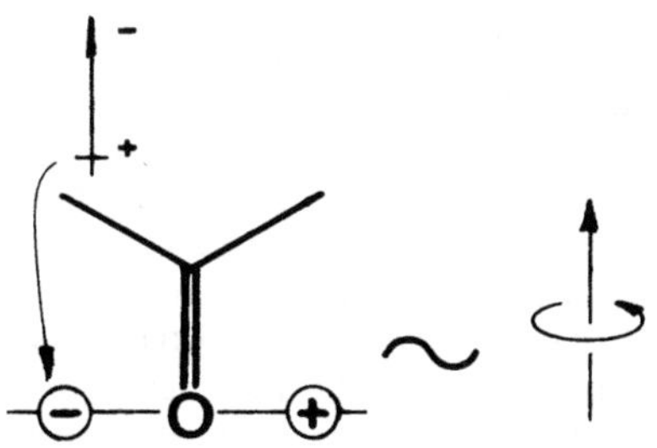

Fig. 11. Induction of the sign pattern of a quadrupole around O of a carbonyl by a transition dipole in (-z)-direction in the left upper back octant. This arrangement gives rise to a positive Cotton effect.

tically favoured interaction for deriving at the signs of the quadrupole around the oxygen (any $\sigma\sigma^*$-state has also definetely higher energy than the $S_1(n\pi^*)$-state of the carbonyl chromophore) we obtain the result given in Fig. 11: the two transition moment vectors are parallel to each other, the CD becomes positive, again in accordance with the well known octant rule (16). One can see, however, that now the direction of the perturbing C-X bond plays also some role, and if the transition dipole is placed very close to the z-axis then sign inversion may occur. Thus even refinements of the octant rule (change of sign within some areas, cf. e.g. the additional nodal surface or "enclaves" predicted by some theoreticians (17)) may be indicated by this simple qualitative MO-treatment.

In the case of the carbonyl group we have then obtained the result that different contributions to the CD (second sphere contribution, intrachromophoralic and interchromophoralic terms, perturbations by "monopoles" as well as by dipoles) all give the same sign, as long we use second-order PT. Not always we have, however, such a favourable situation and then it might become difficult or impossible to predict the sign of the CD. But even in such cases this qualitative MO-treatment is valuable because it shows the scope and limitations of potential rules for a given chromophore.

As can be seen from these discussions it is essential to know the type of transition which corresponds to a Cotton effect if one wants to apply qualitative MO-theory. Of great help is the determination of the g-value, which was defined by W.Kuhn (9) as $g = \Delta\varepsilon/\varepsilon$. The magnitude of this value, which is a function of the wavelength, can be estimated in the following way. For bands within g is constant this value is the same as $g^I = 4R/D = 4.m.\cos(\vec{\mu},\vec{m})/\mu$. Now the natural unit of $\vec{\mu}$ is the DEBYE, of $\vec{m}$ the BOHR magneton. Inserting their magnitudes leads to $g^I = 10^{-2}.4.\cos(\vec{\mu},\vec{m})$. The order of magnitude of the g-value for a transition which is electrically and magnetically allowed is then appr. 10^{-2}, and this is indeed very often found (10). Transitions within inherently achiral chromophores which are magnetically allowed but electrically forbidden will have a $\vec{\mu}$ which is only few percent of a DEBYE, the g-value should then be greater than 10^{-2}. On the other hand, transitions which are electrically allowed but magnetically forbidden will give rise to a very small g-value. Determination of this g-value helps thus in identifying electronic transitions.

GENERALIZATION OF SECTOR RULES

Applying the same qualitative MO-treatment as well as making use of the many experimental data available for different chromophores we found (18) that the sign pattern for such sector rules can be transferred from one case to another if the following conditions are met: 1) the two chromophores compared must belong to the same point group, and 2) the two orbitals from which the electron is excited must belong to the same irreducible representation, as must the two corresponding virtual MOs into which the electron is promoted. If the two systems differ in the shape and/or number of nodal surfaces (i.e., in the number of sectors) then we start to put identical signs into those sectors which are

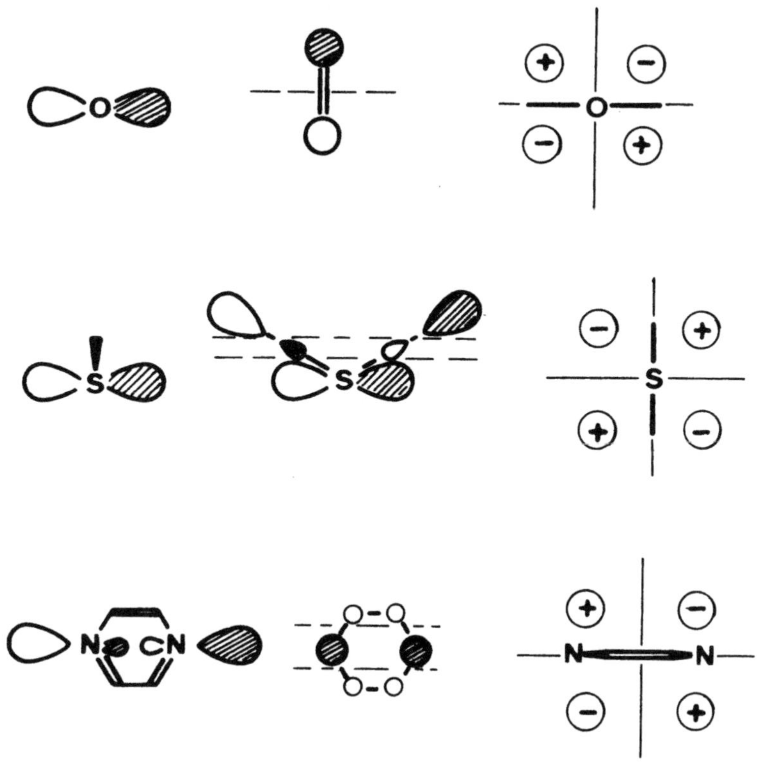

Fig. 12. Generalized Sector Rule, $A_1 \rightarrow A_2$-transitions for chromophores of point group C_{2v}: TOP: carbonyl $n \rightarrow \pi^*$-band (300 nm); MIDDLE: sulfide (oxathiolane) $n \rightarrow \sigma^*$-band (240 nm); BOTTOM: 1.4-pyrazine $n^- \rightarrow \pi^*_4$-band (227 nm). The signs given refer to back sectors, the orbital from which the electron is excited has to be oriented horizontally.

around that atom where the MOs are most similar. As often we go then into an adjacent sector we have to change the sign for the contribution to the CD of a group positioned in it. As an illustration we may compare the carbonyl $n \rightarrow \pi^*$-band, the 1.4-pyrazine $n^- \rightarrow \pi_4^*$-transition, and the rule for the $n \rightarrow \sigma^*$-transition of thio ethers (18). Although the virtual orbitals are quite different in shape the MOs from which the electron is excited are very similar, so we take the same sign pattern around the oxygen atom of the C=O, around the two nitrogens of the pyrazine, and around the S of the thio ether (Fig. 12). Proceeding to "back octants" we have to change once the sign in the first two cases, but twice for the thio ether, and we arrive then at those sector rules which have been found experimentally; Fig. 12 gives the signs for the back sectors. Newer surveys of such sector rules may be found in (19).

Acknowledgement - Financial support from Deutsche Forschungsgemeinschaft and Fonds der Chemischen Industrie is highly appreciated.

REFERENCES

1. A.Moscowitz, Tetrahedron 13, 48 (1961). - A.Moscowitz, K.Mislow, M.A.W.Glass, and C.Djerassi, J.Am.Chem.Soc. 84, 1945 (1961).
2. e.g. M.J.S.Dewar and R.C.Dougherty, The PMO Theory of Organic Chemistry, Plenum Press, New York (1975).
3. G.Snatzke, Tetrahedron 21, 413 (1965).
4. G.P.Powell and J.Hudec, J.C.S.,Chem.Comm. 806 (1971).- D.N.Kirk and W.Klyne, J.C.S.,Perkin Trans. I 1076 (1974).
5. J.A.Schellman, J.Chem.Phys. 44, 55 (1966).
6. cf. G.Snatzke, Angew.Chem., in press.
7. J.Linderberg and J.Michl, J.Am.Chem.Soc. 92, 2619 (1970). - J.Webb, R.W.Strickland, and F.S.Richardson, ibid. 95, 4775 (1973). - D.B.Boyd, Int.J.Quant.Chem.: Quantum Biology Symp. No. 1, 13 (1974).
8. cf. V.Ludescher and R.Schwyzer, Helv.chim.Acta 54, 1637 (1971).
9. cf. W.Kuhn, in K.Freudenberg (Ed.), Stereochemie, p. 317, F.Deuticke, Leipzig (1933).
10. S.F.Mason, Quart.Rev. Chem.Soc. 17, 20 (1963).
11. K.Bláha and I.Frič, Collect.Czechoslov.Chem.Commun. 35, 619 (1970).
12. C.Djerassi and W.Klyne, Proc.Natl. Acad.Sci.USA 48, 1093 (1962).
13. G.Snatzke, B.Ehrig, and H.Klein, Tetrahedron 25, 5601 (1969). - H.J.C.Jacobs and E.Havinga, ibid. 28, 135 (1972).
14. A.Moscowitz, Adv.Chem.Phys. 4, 67 (1968).
15. G.Snatzke and G.Eckhardt, Tetrahedron 24, 4543 (1968).
16. W.Moffitt, R.B.Woodward, A.Moscowitz, W.Klyne, and C.Djerassi, J.Am.Chem.Soc. 83, 4013 (1961).
17. Y.-H.Pao and D.P.Santry, J.Am.Chem.Soc. 88, 4157 (1966). - T.D.Bouman and D.A.Lightner, ibid. 98, 3145 (1976).
18. G.Snatzke and Gy.Hajós, in B.Pullman and N.Goldblum (Eds.), Excited States in Organic Chemistry and Biochemistry, p. 295, Reidel, Dordrecht (1977). - P.Welzel, I.Müther, K.Hobert, F.-J.Witteler, T.Hartwig, and G.Snatzke, Liebigs Ann. Chem., in press.

19. M.Legrand and M.J.Rougier, in H.B.Kagan (Ed.), Stereochemistry, Vol. 2, p. 33, Thieme, Stuttgart (1977). - F.Snatzke and G.Snatzke, in H.Kienitz, R.Bock, W.Fresenius, W.Huber, and G.Tölg (Eds.), Analytiker Taschenbuch 1979, Springer, Heidelberg (1979).

ADDENDUM (Correction).

In (11) the CD-bands have been incorrectly assigned; actually the Cotton effect around 215-220 nm belongs to the $n \to \pi^*$-transition, and the next band of opposite sign is then the first wing of the exciton couplet (20). The application of the exciton theory in its simplest manner, which works well e.g. for chiral dibenzoates (21) leads thus to a wrong prediction. Similar results have been obtained e.g. for the very strong couplet which is observed in many germacradienolides, as costunolide and others (22), and have also been discussed during this ASI. One obtains, however, in the aforementioned cases the right signs for these CD-couplets if one takes into account (cf. Fig. 1) that actually not the Coulomb interaction between dipoles but between two quadrupoles has to be considered, with their individual poles being appr. 0·7 - 0·9Å distant from the plane of the chromophore. This modification might also in other reported cases relieve the discrepancies between predictions and experiments, and should be tried always then when the two chromophores are very close to one another.

20. T.M.Hooker jr., P.M.Bayley, W.Radding and J.A.Schellman, Biopolymers 13, 549 (1974).
21. N.Harada, S.-M. Lai Chen and K.Nakanishi, J.Am.Chem.Soc. 97, 5345 (1975).
22. M.Suchý, L.Dolejš, V.Herout, F.Šorm, G.Snatzke and J.Himmelreich, Coll. Czechoslov.Chem.Commun. 34, 229 (1969).

CHIROPTICAL PROPERTIES OF ORGANIC COMPOUNDS: SOME APPLICATIONS TO UNSATURATED SYSTEMS

G.Snatzke

Lehrstuhl für Strukturchemie
Ruhruniversität Bochum

Postfach 10 21 48
D 4630 Bochum 1 (FRG)

ABSTRACT. Qualitative MO-theory is applied to 2π- and 4π-systems (ethylene, butadiene, conjugated enones) to explain the different rules used at present for such chromophores.

THE 2π-SYSTEM.

The simplest 2π-system is the ethylene chromophore, which, however, gave already rise to a lot of controversy in the literature during the last years (1,2). Originally only valence states have been considered, but it has been proved (2) that also Rydberg-bands appear in the CD-spectra of chiral olefins. They are strongly blueshifted or disappear at all in polar solvents or at low temperature, so that they can be now easily identified, and they will not be discussed in the following.

An ethylene becomes inherently chiral ("chirality of the first sphere") if the system is twisted, and in such a case electron charge is also rotated during the excitation from the π-into the π^*-orbital. If the angle of twist is positive these combined movements correspond to a righthanded helix, the CD should thus be positive. This result was also obtained by more detailed calculations (3), is, however, apparently not in agreement with the experimental value obtained for trans-cyclooctene (1). At least two reasons can be made responsible for this: a) the measured CD-band could correspond to another transition than the usually assumed $\pi\rightarrow\pi^*$, and b) contributions from second and/or third sphere might be stronger than this one from the chiral first sphere. This latter view is not so unrealistic, as electron diffraction measu-

Stephen F. Mason (ed.), Optical Activity and Chiral Discrimination. 43–55.

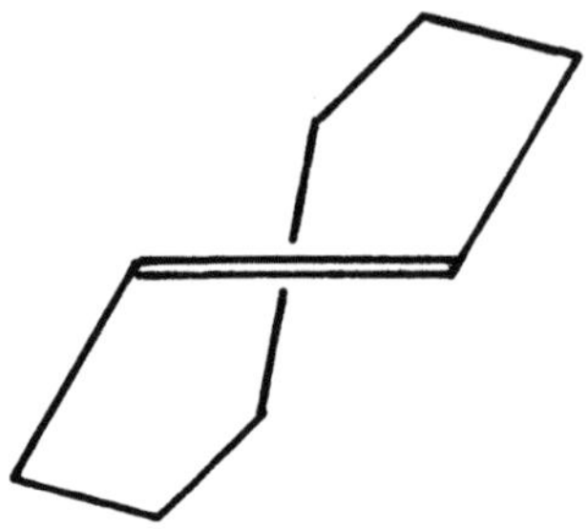

1

rements have shown (4) that considerable rehybridization takes place in the chromophore to avoid too much BAEYER- and/or PITZER-strain. By this the torsion angle between the p-orbitals forming (mainly) the π- and π^*-MOs becomes, however, appreciably smaller than the angle which describes the deviation of the C-C=C-C moiety from coplanarity.

In case of achiral first sphere the $\pi\rightarrow\pi^*$-transition remains electrically allowed but is not associated with a magnetic transition moment. To discuss the mechanism for stealing the necessary $\vec{m}$ we can again use qualitative MO-theory as in my first

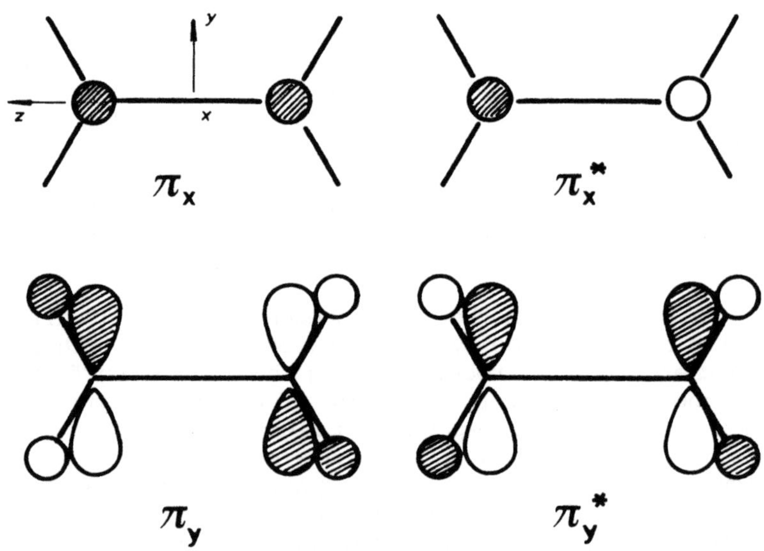

Fig. 1. Orbitals π_x, π_x^*, π_y, and π_y^* of an ethylene, projected onto the plane of the chromophore and viewed in the (-x)-direction.

lecture (5) , and the MOs of importance for us are shown in Fig.1. What we usually call "the" π- and "the" π^*-MO should then better be called the π_x- and the $\pi_x{}^*$-MO. They are the HOMO and LUMO for the ethylene chromophore. The next lower to the HOMO is an orbital which is bonding (in ethylene) along all four C-H - bonds, but antibonding along the C-C - bond, and is called π_y. Next to the HOMO is the virtual orbital $\pi_y{}^*$ (cf. also Fig. 1). Applying group theory or the formal multiplication method of the orbitals one finds that the $\pi_x \rightarrow \pi_x{}^*$-transition is z-polarized, so we need to mix in by the chiral environment orbitals (or states) which will bring the missing m_z. As is seen from Fig. 1 both the $\pi_x \rightarrow \pi_y{}^*$ - as the $\pi_y \rightarrow \pi_x{}^*$ -transition is associated with such a magnetic moment. If we use the approach of mixing in orbitals, then we have either to admix some π_y to π_x, or some $\pi_y{}^*$ to $\pi_x{}^*$, and we can try to find out from this what will be the contribution to the CD from the chiral second sphere in just the same way as was done earlier for the carbonyl chromophore (5), making use of a chiral allylic

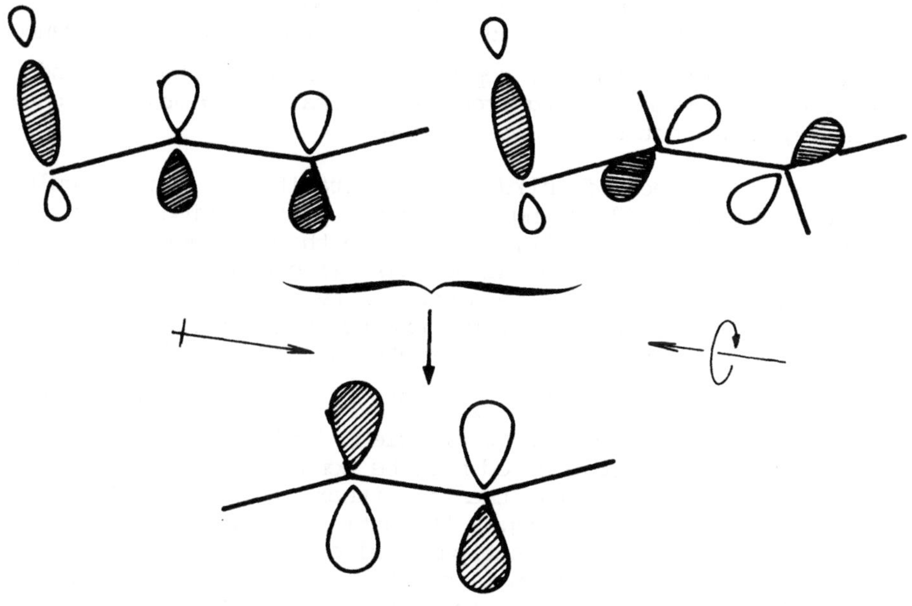

Fig. 2. Admixture of some π_y-character into π_x by a nearby chirally arranged σ-bond, and relative directions of the transition moments for the excitation into $\pi_x{}^*$. For the drawn absolute conformation the CD is negative. (Cf., however, also text!).

σ-bond as main perturber (Fig.2) in the first of the two mentioned cases. For an allylic bond extending into a negative octant the resulting two transition moment vectors are antiparallel, the CD should become negative, i.e. consignate. This result is, however, by no means unequivocal, because the phasing in Fig. 2 is based on the assumption that the energy of the "inducing" σ-orbital is lower than both the energies of π_x and of π_y. The latter may, however, not be the case, and -even worse- it may depend on the substitution of this allylic bond (and its nature, e.g. C-C vs. C-O, etc.), what is the relative energy of these two orbitals. Any exchange of their positions on the energy scale leads to sign inversion for the CD.

Drake and Mason (2) calculated that the $\pi_x\pi_y{}^*$-state is nearly degenerated with the $\pi_x\pi_x{}^*$-state, whereas the $\pi_y\pi_x{}^*$-state is of higher energy. For the latter case induction from the σ-orbital into the virtual orbitals is not very realistic, but one can discuss the interaction between σ^*, $\pi_x{}^*$, and $\pi_y{}^*$ in similar manner and will derive then at a dissignate rule for the contribution to the CD, if the σ^*-orbital energy is assumed to be higher than the energies of both the $\pi_x{}^*$- and the $\pi_y{}^*$-orbitals. Also this cannot easily be estimated and may depend on the substitution. Second sphere contributions coming from this intrachromophoralic mechanism are, thus, not easily predictable.

The effect of third-sphere chirality has been calculated by Drake and Mason (2) using first-order perturbation theory (PT). One can get the same result, which is furthermore not depending on the sign of the perturbing charge, also if one applies the rules of second-order PT, as has been used in my other lecture already (5). Let for the moment the first excited singlet state be the $\pi_x\pi_x{}^*$-state, and the closest B_{1g}-state, which can bring the magnetic transition moment, the $\pi_x\pi_y{}^*$-state. The electrostatic interaction with both, $\vec{\mu}$ and the quadrupole Q (only the nearer one of the two quadrupoles developped must be considered), for the band at longest wavelengths must then be either always attractive, or always repulsive, and in Fig. 3 the attractive interactions have been chosen. This leads to antiparallel orientation of the two moments, the CD contribution is negative, and we have thus a consignate octant rule. Would instead the $\pi_y\pi_x{}^*$-state be closer in energy to S_1 then a dissignate octant rule would have resulted. As most of the chiral olefins follow a consignate rule (6), the assignment of Drake and Mason (2) for the nearest B_{1g}-state is experimentally proved.

In their work Scott and Wrixon (6) have explained some

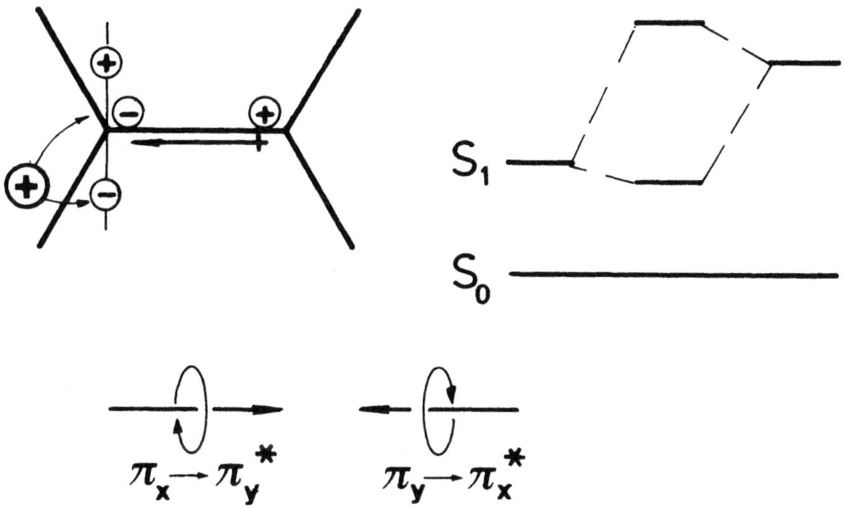

Fig. 3. Mixing of $\pi_x\pi_y^*$ or $\pi_y\pi_x^*$ into the $\pi_x\pi_x^*$-state of an olefin and induction of sign patterns for the electric transition moment and that quadrupole which is nearer to the chirally arranged perturbing positive charge. The direction of the magnetic transition moment corresponding to the sign pattern of the quadrupole depends on the choice of the interacting state.

exceptions to their octant rule by assuming that in these cases the $\pi_x \rightarrow \pi_y^*$-band may appear at longer wavelengths than the $\pi_x \rightarrow \pi_y^*$-band. As can be seen from Fig. 3 it will, however, not matter which of the two states, S_1 or S_i, is of lower energy, the band at longer wavelengths must always be the consignate one, if the missing moment is coming from the mechanism discussed just now.

In a similar way the interchromophoralic mechanism can be discussed, and I will do it only here for the chiral second sphere. The state interacting with the $\pi_x\pi_x^*$-state may e.g. be a $\sigma\sigma^*$-state (C-C, C-H, C-X bond in axial allylic position to the double bond), and Fig. 4 shows the energetically favoured combination of the two respective electric transition moments, which in case of a positive torsion angle (C-)C-C(=C) leads to a positive CD-contribution of the exciton type. The respective dipole strength will be strongest around a torsion angle of 90°, and this is in accordance with the "axial allylic rule" of Mazur et al. (7a) and Burgstahler et al. (7b). It should be noted that agreement with

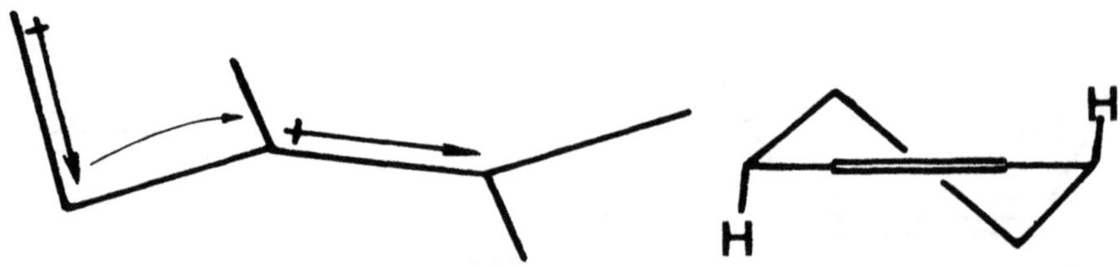

Fig. 4. LEFT: Exciton type interaction between a $\sigma\sigma^*$-transition dipole within an axial allylic bond and the ethylene $\pi_x \to \pi_x^*$-transition, drawn for a positive torsion angle, giving a positive CD. RIGHT: Cyclohexene halfchair with axial allylic C-H - bonds of same absolute conformation as on left side.

experiment is in general obtained only if the axial allylic C-H - bonds are also taken into consideration. Therefore another similar rule has been proposed by Andersen et al. (8) which uses the opposite signs and takes into account in a qualitative way bond polarizabilities.

We may then summarize these results by noting that the several contributions of a group or bond to the CD of an ethylene have not all the same sign and determination of the absolute conformation from the CD should be done rather carefully by comparing empirical data only in very similar systems.

The $n \to \pi^*$-transition of the carbonyl group, which contains also 2 π-electrons, has already been discussed in more detail in the other lecture (5), and the same holds for the 3π-system of the enol ethers.

THE 4π-SYSTEM.

The noncoplanar conjugated diene system was one of the first chromophores where helicity rules were given on theoretical basis (9), and it was proposed that a positive torsion angle around the bond 2-3 should give a positive CD. Later on some exceptions were found and again the axial allylic groups (including hydrogen) were made responsible for these (7). Let's first discuss the coplanar cisoid butadiene (Fig. 5). The lowest energy transition is believed to be from π_2 to $\pi_3^*(B_2)$, although taking into account configurational interaction and doubly excited states might sometimes bring the A_1-state to still lower energy (10). The $A_1 \to B_2$-transition is associated with an electric transition moment in y- and a magnetic one in x-direction. Whereas the direction of the first is determined unequivocally once the signs of the orbital lobes have been stipulated, the direction of the latter depends

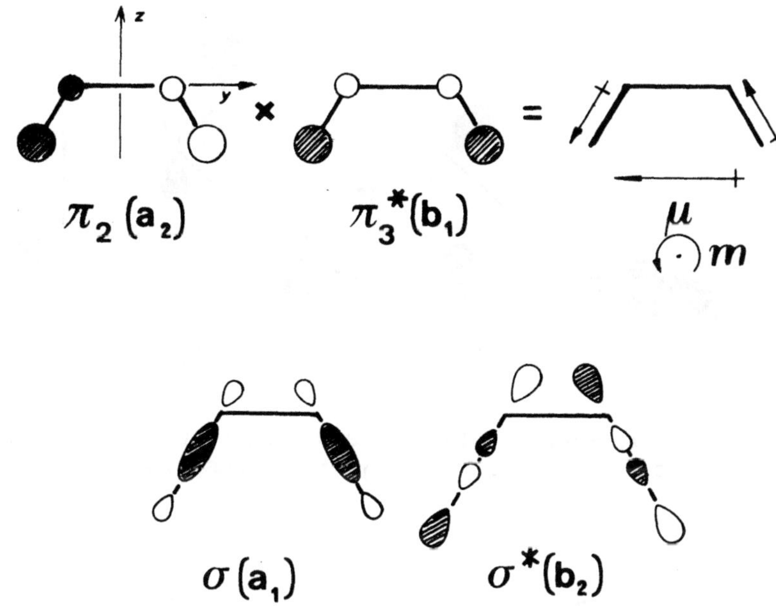

Fig. 5. TOP: HOMO (π_2) and LUMO (π_3*) of a cisoid butadiene system. The transition between them is associated with an electric moment in y-direction and a magnetic one perpendicular to the plane of the system. The direction of the latter depends on the choice of the origin! BOTTOM: Orbitals which may be involved in order to obtain the missing transition moments.

still on the choice of the origin of the coordinate system because it is proportional to the vector cross product $\vec{r}\times\vec{\mu}$ ($\vec{r}$ is the vector from the origin towards $\vec{\mu}$).

In order to get rotational strength one has to steel m_x or/and μ_y. If this should be done intrachromophoralically, then we must mix in some B_1-state which is associated with these two transition moments. Those B_1-states which are nearest in energy can either come from a transition from the HOMO (a_2) into an MO of b_2-symmetry, or from an a_1-orbital into the LUMO (b_1). Both these additional orbitals must be symmetric with respect to the plane of the molecule, so they have to be either σ- or σ*-orbitals (Rydberg bands are not discussed here). Proper MOs are shown in Fig. 5. In Fig. 6 the relative orientation between μ_x and m_y for the $\sigma(a_1)\rightarrow\pi_3^*$-transition is given, and a similar picture is obtained for the transition $\pi_2\rightarrow\sigma^*(b_2)$. That part of $\vec{m}$ which comes from the inherent charge rotation (Fig. 6) is origin-independent, the term $\vec{r}\times\vec{\mu}$ is not.

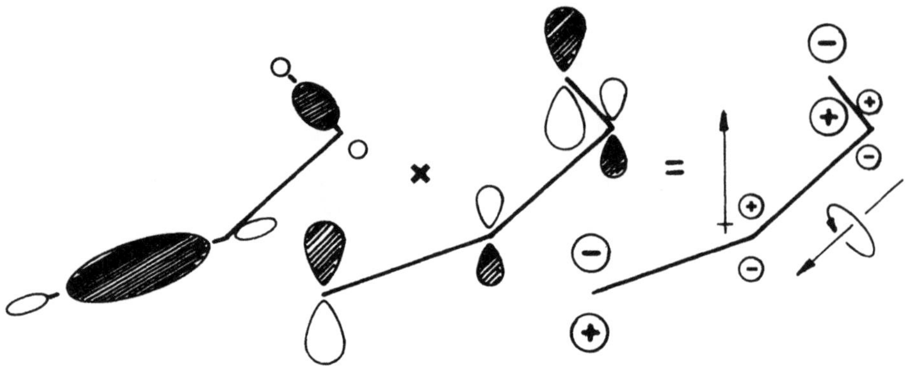

Fig. 6. Direction of electric and (inherent part of) magnetic transition moment for the $\sigma(a_1)\rightarrow\pi_3{}^*$-transition of a planar cisoid butadiene.

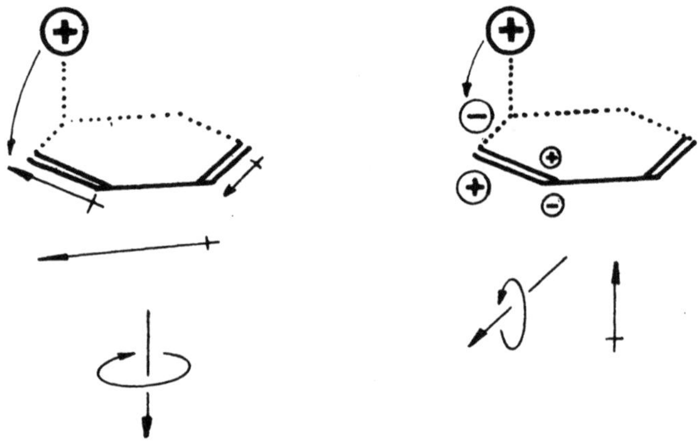

Fig. 7. Induction of the sign pattern for the electric transition moment and the transition quadrupole by a chirally disposed positive perturbing charge (transition of Figs. 5 and 6). Only the interaction with the nearest $\vec{\mu}$ and **Q** is considered.

Mixing of the two excited states B_1 and B_2 by a chiral environment (in Fig. 7 represented by a positive monopole) and taking in each case the attractive interaction for finding the low-energy combination gives parallel orientation for the transition moments in the y-, but antiparallel orientation in the x-direction. The first CD-contribution is thus positive, the second negative, and an unequivocal prediction on this basis is not possible. The same very sensitive dependence of the magnitude and even the sign of the CD from structural details has also been found by detailed calculations (9).

If the diene is not coplanar then these contributions will still play some role, but an additional inherent term will also be non-zero. The situation is, however, not much better than before. The symmetry reduces to C_2 (Fig. 8), and if we follow the

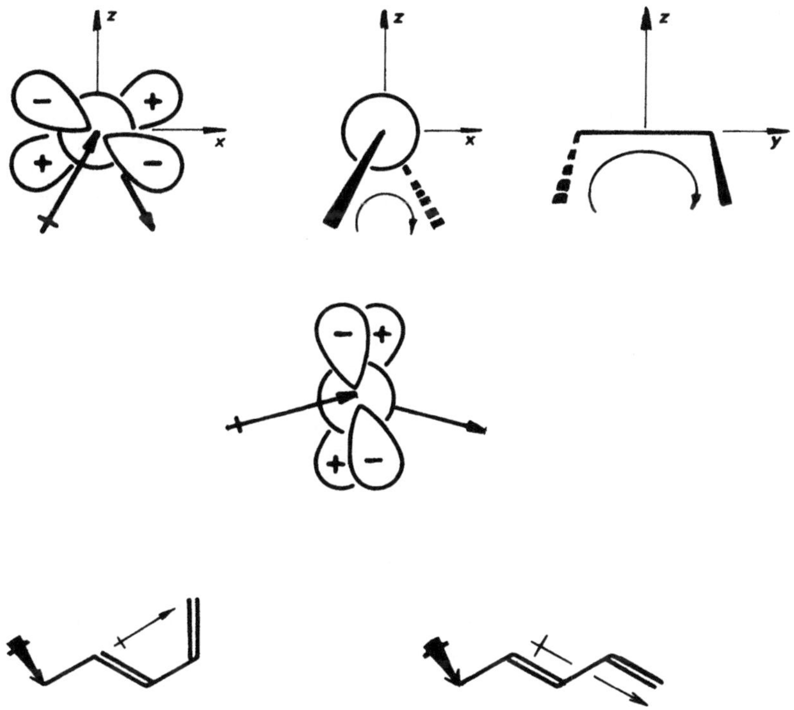

Fig. 8. Noncoplanar butadiene chromophores with negative torsion angles. TOP: Cisoid butadiene: inherent charge rotation and interacting transition dipoles (left); projections in (+y)- and (-x)-direction (middle and right, resp.). The arrows characterize the magnetic transition moments if the origin of the coordinate system is as drawn or on the concave side of the diene. MIDDLE: Transoid butadiene. BOTTOM: Exciton (interchromophoralic) interaction with axial allylic bond $\sigma\to\sigma^*$-transition; for cisoid butadiene it is appr. zero (left), for transoid diene positive (right) for the absolute conformation drawn.

charge translation path from C-1 to C-4, along the y-axis we can easily identify a righthanded screw (μ_y and m_y point both in (-y)-direction), so this term becomes positive. On the other hand, the component in the x-direction of the electric transition dipole for the signs chosen in Fig. 8 is positive, but that of the magnetic one is negative (origin always assumed to be on the concave side of the butadiene); therefore the second term is negative, and from this "helicity" alone no sign-prediction is possible without detailed calculations. As the developed charges are larger, however, at C-1 and C-4 than at C-2 and C-3, some electron charge has to go also through the middle bond C-2/C-3, and for the negative torsion angle chosen this path forms a lefthanded screw (Fig.8). This contribution to the inherent CD of a noncoplanar cisoid butadiene is thus negative.

A still another contribution to the CD comes from the exciton interaction of the two "partial" dipoles along the (formal) double bonds. One has also to add a third dipole which points from C-1 towards C-4 because of the nonequal charge distribution, but if one inserts these vectors into the formula for such exciton interactions between several electric transition moments the third term becomes zero. The combined motion of charge within these two partial dipoles follows a lefthanded screw for a negative torsion angle, so the exciton term is negative.

Altogether we have then more negative than positive contributions to the CD. If their magnitudes are similar then a negative CD is expected for the given helicity of the diene moiety (negative torsion angle), and this is indeed the old experimental rule (9).

For a transoid coplanar butadiene the transition of lowest energy has only an electric transition moment in the plane of the system and parallel to the bonds C-1/C-2 and C-3/C-4. If one steels the missing in-plane magnetic transition moment in analogous way as for the cisoid diene then this is approximately perpendicular to $\vec{\mu}$, any intrachromophoralic mixing of such states by the chiral environment will thus (nearly) not give rise to CD. If the diene is twisted (Fig. 8), then the inherent sense of srewness along the middle bond is righthanded for a negative torsion angle, the exciton interaction term on the other hand remains negative as for the similar cisoid diene. Experimentally one mostly finds a negative CD for this absolute conformation (9), so the exciton term must be bigger than the before mentioned one, in full agreement with the results obtained for twisted ethylenes.

Interchromophoralic interactions *e.g.* with a $\sigma\to\sigma^*$-transition will be very sensitive to small changes of geometry in case of cisoid butadienes, because in the idealized conformation (Fig.8) the "inducing" and the "induced" electric transition moment vectors are nearly coplanar. On the other hand such an interaction is quite insensitive to smaller deviations in case of the transoid dienes, and the absolute conformation shown in Fig. 8 gives rise to a positive CD for the band at longest wavelengths. This is the rule which was experimentally put forward by Burgstahler et al. (7b).

In conclusion we may thus learn from such a qualitative MO-treatment that also for dienes one should be cautious in assigning absolute conformations from CD and should compare only similarly substituted systems.

CONJUGATED ENONES.

For the $\pi\rightarrow\pi^*$-transition the same rules and difficulties are found as for the corresponding butadienes (7b,11), so they must not be discussed again. The $n\rightarrow\pi^*$-transition deserves, however, some consideration. It is electrically forbidden and magnetically allowed ($\vec{m}$ appr. along the C=O - bond), so we have to steel an electric transition moment in the same direction. As in the case of the simple ketones (cf. (5)) admixture of π_2 can afford this. To find the appropriate rule for a coplanar system one can follow the procedure outlined in (5) to find the sign pattern around the oxygen atom. If we then proceed to back-octants we have twice to cross nodal surfaces, so the sign pattern must be opposite to that for saturated ketones (Fig. 9), and this is indeed

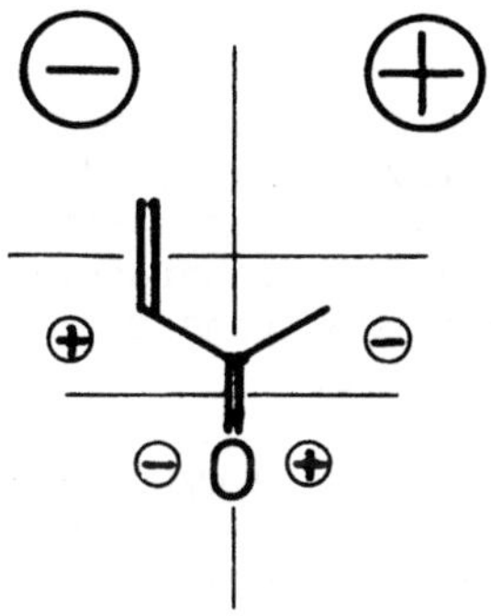

Fig. 9. Sector rule for the $n\rightarrow\pi_3^*$ -transition of a coplanar conjugated enone. The signs are for the contributions in upper sectors.

the experimental result, obtained also by more detailed calculations (12). For noncoplanar conjugated enones the inherent chirality of the system should determine the sign. As the energy of the nonbonding orbital is higher than of π_2 the HOMO consists of the energetically unfavoured combination of the original n- and π_2-orbitals. Fig. 10 shows this for a negative torsion angle (C=)C-C(=O), and whatever phasing in the LUMO (π_3^*) we may use, the angle between $\vec{\mu}$ and $\vec{m}$ will be acute, the CD is thus positive for such an absolute conformation (13). Nearly all of the investigated noncoplanar conjugated enones follow indeed this rule (14).

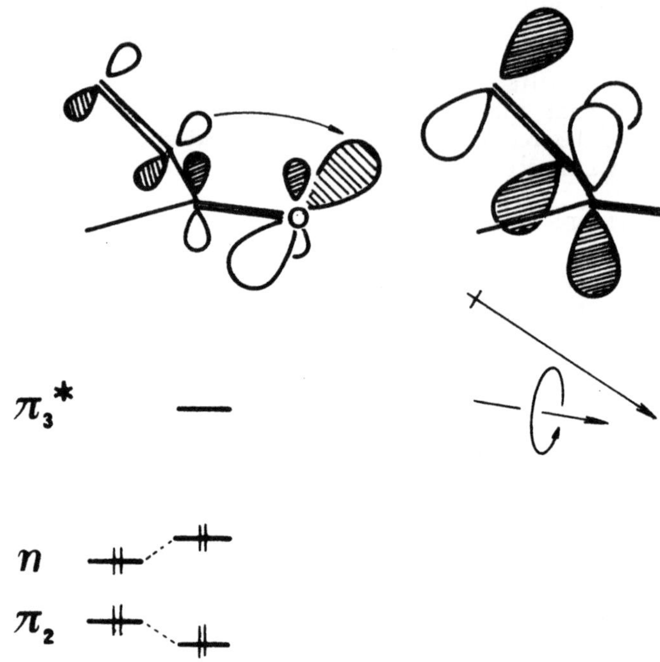

Fig. 10. $n\rightarrow\pi_3{}^*$-Cotton effect of noncoplanar conjugated enones. The transition is magnetically allowed, electrical transition moment can be obtained by admixing some π_2-character to the n-orbital. The HOMO is their energeticaly unfavoured combination, for a negative torsion angle the two transition moments are (nearly) antiparallel, the CD is negative.

Using Walsh-orbitals of a cyclopropane one can derive in similar manner at rules for the $\pi\rightarrow\pi^*$-bands of vinyl cyclopropanes (15) and the $n\rightarrow\pi^*$-bands of cyclopropyl ketones (13). One can also extend this treatment to conjugated polyenes and can by this rationalize even the CD-spectra of all-trans and mono-cis carotenoids (16).

OTHER UNSATURATED SYSTEMS.

Actually most of the investigated chromophores contain one or more π-bonds, but the more complicated the system becomes (crossed conjugated polyenes, aromatic compounds, heteroconjugated systems, etc.) the less such a qualitative MO-treatment will be suitable for the prediction of the correlation between the sign of a Cotton effect and the absolute conformation. On the other hand this same statement is valid also for more detailed calculations, so as a guideline to find out scopes and limitations of rules the treatment given here might still be useful during the next few years for the "practical chiropticists".

Acknowledgement - Financial support from Deutsche Forschungsgemeinschaft and Fonds der Chemischen Industrie is highly appreciated.

REFERENCES

1. For recent reviews on rules in CD cf. a) M.Legrand and M.J. Rougier, in H.B.Kagan, Stereochemistry Vol.2, p.33, Thieme, Stuttgart (1977). - b) F.Snatzke and G.Snatzke, in H.Kienitz, R.Bock, W.Fresenius, W.Huber, and G.Tölg (Eds.), Analytiker Taschenbuch 1979, Springer, Heidelberg (1979).
2. A.F.Drake and S.F.Mason, Tetrahedron 33, 937 (1977).
3. A.Moscowitz and K.Mislow, J.Am.Chem.Soc. 84, 4605 (1962). - M.Yaris, A.Moscowitz, and S.Berry, J.Chem.Phys. 49, 3150 (1968). - C.C.Levin and R.Hoffmann, J.Am.Chem.Soc. 94, 3446 (1972).
4. M.Traetteberg, Acta Chem.Scand. B29, 29 (1975).
5. G.Snatzke, preceding lecture.
6. A.I.Scott and A.D.Wrixon, Tetrahedron 27, 2339, 4787 (1971).
7. a) A.Yogev, D.Amar, and Y.Mazur, J.C.S.Chem.Commun. 339 (1967). b) A.W.Burgstahler and R.C.Barkhurst, J.Am.Chem.Soc. 92, 7601 (1970)
8. N.H.Andersen, C.R.Costin, D.D.Syrdal, and D.P.Svedberg, J.Am. Chem.Soc. 95, 2049 (1973). - N.H.Andersen, C.R.Costin, and J.R.Shaw, ibid. 96, 3692 (1974).
9. A.Moscowitz, E.Charney, U.Weiss, and H.Ziffer, J.Am.Chem.Soc. 83, 4661 (1961). - E.Charney, Tetrahedron 21, 3127 (1965). - J.S.Rosenfield and E.Charney, J.Am.Chem.Soc. 99, 3209 (1977).
10. K.Schulten and M.Karplus, Chem.Phys.Let. 14, 305 (1972).
11. C.Djerassi, R.Records, E.Bunnenberg, K.Mislow, and A.Moscowitz, J.Am.Chem.Soc. 84, 870 (1962). - C.Djerassi and J.E. Gurst, J.Am.Chem.Soc. 86, 1755 (1964).
12. K.Kuriyama, Footnote on p. 461 of W.Nagata and Y.Hayase, J. C.S. (C) 460 (1969).
13. G.Snatzke, Angew.Chem., in press.
14. W.B.Whalley, Chem.& Ind. 1024 (1962). - G.Snatzke, Tetrahedron 21, 421,439 (1965).
15. S.Itō, K.Kuriyama, T.Norin, and G.Snatzke, unpublished results.
16. G.Snatzke, F.Snatzke, and S.Liaaen-Jensen, unpublished results.

CIRCULAR DICHROISM AND FLUORESCENCE DETECTED CIRCULAR DICHROISM OF MACROMOLECULES

I. Tinoco, Jr.

Chemistry Department and Chemical Biodynamics Laboratory
University of California, Berkeley, CA 94720 USA

INTRODUCTION

I will concentrate in this chapter on experimental and theoretical methods which are pertinent mainly to polymers and macromolecules. The optical properties of polymers are determined by the interactions between two or more similar chromophores. Therefore, to calculate the circular dichroism of a polymer we need the properties of the individual chromophores (the monomers) and we need to evaluate their interactions. The crucial assumption that has been made in calculations of circular dichroism of polymers is that electrons are localized in each chromophore. Each chromophore is represnted by a charge distribution which interacts with the other chromophores in the presence of the light. Any electron exchange or electron transfer between chromophores is assumed to have a negligible effect on the circular dichroism. Calculations of circular dichroism of polypeptides and polynucleotides have been reasonably successful using these methods [1].

A unique property of polymers and of macromolecules is their large size. This presents special opportunities and problems in both experimental and theoretical methods. It is relatively easy to orient large molecules either by flow [2] or by electric fields [3]. Thus the circular dichroism can be measured along specific directions. This provides more and different parameters than just the average circular dichroism. However, the large size may require a change in the theory relating circular dichroism and molecular parameters. The usual (Rosenfeld) [4] theory was derived for molecules small compared to the wavelength of light. It is important to know under what conditions this theory is inapplicable.

An experimental technique which is useful for any size molecule, but which has special advantages for large molecules is fluorescence

Stephen F. Mason (ed.), Optical Activity and Chiral Discrimination. 57–85.

detected circular dichroism [5]. The circular dichroism is obtained by measuring the intensity of the emitted fluorescence when the molecules are excited by circularly polarized light. The fluorescence is directly proportional to the absorption, therefore the circular dichroism of transitions which lead to fluorescence is measured. This is a great advantage for a macromolecule which has many chromophores, but only a few fluorophores. Only the circular dichroism of the fluorophores, or of groups which can transfer energy to the fluorophores, is measured. Information about a specific region in a macromolecule can thus be obtained. Fluorescence detection can also provide direction-dependent circular dichroism information. If the fluorophore does not have time to reorient before emission, the circular dichroism along a particular direction in the molecule is obtained. That is, photoselection provides the circular dichroism along specific directions in the molecule, although all the molecules in the sample are randomly oriented [6].

CIRCULAR DICHROISM OF LARGE MOLECULES.

Consider an electronic transition in a molecule from state 0 to A. It has corresponding wavefunctions ψ_0 and ψ_A and corresponding energies E_0 and E_A. The transition probability in the presence of incident light propagating along unit vector $\underset{\sim}{e}_1$ and with its polarization state specified by unit vector $\underset{\sim}{e}$ depends on the transition integral:

$$\underset{\sim}{e}\cdot\underset{\sim}{T}_{0A} = \int \psi_0^* \, \underset{\sim}{e}^{\, i2\pi \underset{\sim}{e}_1 \cdot \underset{\sim}{r}/\lambda_{0A}} \, \underset{\sim}{e}\cdot\underset{\sim}{p}\psi_A \, d\tau \qquad (1)$$

$\underset{\sim}{r},\underset{\sim}{p}$ = the position and momentum vectors

λ_A = the wavelength of the transition, $hc/(E_A-E_0)$

The molar absorptivity, ε, is proportional to the absolute square of the transition integral. It is thus clear that the absorptivity will in general depend on the direction of incidence and state of polarization of the light. For molar absorptivity with units of ℓ/mole cm

$$\varepsilon(\underset{\sim}{e}_1 \text{ incident, } \underset{\sim}{e} \text{ polarized-light}) = 3G\left(\frac{e\lambda_{0A}}{2\pi mc}\right)^2 \underset{\sim}{e}\cdot\underset{\sim}{T}_{0A}\underset{\sim}{T}_{A0}^*\cdot\underset{\sim}{e}^* \qquad (2)$$

$$G = \frac{8\pi^3 N_0 \lambda g(\lambda)}{6909hc}$$

e,m = the charge and mass of an electron; c = the speed of light

$g(\lambda)$ = a shape function which specifies the shape of the absorption band corresponding to the electronic transition; its integral over wavelength is equal to one. * means the complex conjugate and $\underset{\sim}{e}$ can in general be complex.

To obtain the circular dichroism of an oriented collection of molecules we simply use Eq. (2) to calculate the absorptivity for left circularly polarized light with unit vector $\underset{\sim}{e}_L = (1/\sqrt{2})(\underset{\sim}{e}_2 - i\underset{\sim}{e}_3)$, and for right circularly polarized light with unit vector $\underset{\sim}{\varepsilon}_R = (1/\sqrt{2})(\underset{\sim}{e}_2 + i\underset{\sim}{e}_3)$

$$\varepsilon_L - \varepsilon_R(\underset{\sim}{e}_1 \text{ incident}) = 3G\left(\frac{e\lambda_{OA}}{2\pi mc}\right)^2 \text{Im}[\underset{\sim}{e}_3 \cdot \underset{\sim}{T}_{OA}\underset{\sim}{T}_{AO}^{*} \cdot \underset{\sim}{e}_2 - \underset{\sim}{e}_2 \cdot \underset{\sim}{T}_{OA}\underset{\sim}{T}_{AO}^{*} \cdot \underset{\sim}{e}_3] \qquad (3)$$

Here $\underset{\sim}{e}_1$, $\underset{\sim}{e}_2$ and $\underset{\sim}{e}_3$ form a right handed coordinate system and Im means the imaginary part. Writing the molar absorptivity and circular dichroism in terms of transition integrals makes very clear the similarity of the two effects. For oriented systems we must specify the direction of incidence and state of polarization of the light to determine either absorption or circular dichroism.

For a molecule small compared to the wavelength of light the exponential in Eq. (1) is set equal to one and Eq. (2) reduces to the more familiar

$$\varepsilon(\underset{\sim}{e} \text{ polarized}) = 3G\underset{\sim}{e} \cdot \underset{\sim}{\mu}_{OA}\underset{\sim}{\mu}_{AO} \cdot \underset{\sim}{e}^{*} \qquad (4)$$

The electric dipole transition moment is

$$\underset{\sim}{\mu}_{OA} = e \int \psi_O \underset{\sim}{r} \psi_A d\tau$$

Note that the absorption for small molecules depends only on the direction of polarization of light. For a random system we average over all orientations of the molecule and obtain (for small molecules)

$$\varepsilon = G\underset{\sim}{\mu}_{OA} \cdot \underset{\sim}{\mu}_{AO} = G|\mu_{OA}|^2 \qquad (5)$$

For the circular dichroism of a molecule small compared to the wavelength of light the exponential in Eq. (1) is expanded and the first two terms are kept.[7] Eq. (3) reduces to

$$\varepsilon_L - \varepsilon_R(e_1 \text{ incident}) = 6G\left(\frac{e\lambda_{OA}}{2\pi mc}\right)^2 \left(\frac{2\pi}{\lambda_{OA}}\right) [\underset{\sim}{e}_3 \cdot \underset{\sim}{p}_{OA}\underset{\sim}{e}_1 \cdot (\underset{\sim}{r}\underset{\sim}{p})_{AO} \cdot \underset{\sim}{e}_2 - \underset{\sim}{e}_2 \cdot p_{OA}\underset{\sim}{e}_1 \cdot (\underset{\sim}{r}\underset{\sim}{p})_{AO} \cdot \underset{\sim}{e}_3] \qquad (6)$$

For a random system we average over all orientations to obtain (for small molecules)

$$\varepsilon_L - \varepsilon_R = 4G \text{ Im } \underset{\sim}{\mu}_{OA} \cdot \underset{\sim}{m}_{AO} \tag{7}$$

The magnetic dipole transition moment is

$$\underset{\sim}{m}_{AO} = \frac{e}{2mc} \int \psi_A^* (\underset{\sim}{r} \times \underset{\sim}{p}) \psi_O d\tau \tag{8}$$

Eqs. (5) and (7) give the average absorption and circular dichroism in the dipole approximation. They state that the absorption is proportional to the dipole strength ($\underset{\sim}{\mu}_{OA} \cdot \underset{\sim}{\mu}_{AO}$) and that the circular dichroism is proportional to the rotational strength (Im $\underset{\sim}{\mu}_{OA}\ \underset{\sim}{m}_{AO}$). The corresponding equations for the average optical properties based on the transition integrals are quite complicated and will not be given here. They are discussed by Tobias et al.[8]

In the next sections we will investigate when we can and cannot use the small molecule, or dipole, approximation. We can summarize the conclusions here. The critical parameter that determines whether the dipole approximation is valid is

$$\frac{2\pi L_c}{\lambda}$$

L_c is the largest distance for which there is significant interaction (or coupling) in the system. When $(2\pi L_c/\lambda) > 0.2$ the dipole approximation begins to become inadequate. This means that for wavelengths near 2000Å, the interaction length must be less than about 65Å for a valid dipole approximation. It is important to recognize that the length of interaction is critical, not the length of the molecule. For example, in liquid crystals each molecule is small, but the ordered array of molecules gives rise to spectacular optical effects. [9]

A FREE ELECTRON ON A HELIX

A good model whose optical properties can be calculated exactly is an electron constrained to move on a helix (Fig. 1). This model was treated in the dipole approximation in 1964.[10] The wavefunctions and energies are identical to those for a particle in a one dimensional box. For a transition from quantum number n to quantum number m the wavelength of the transition is

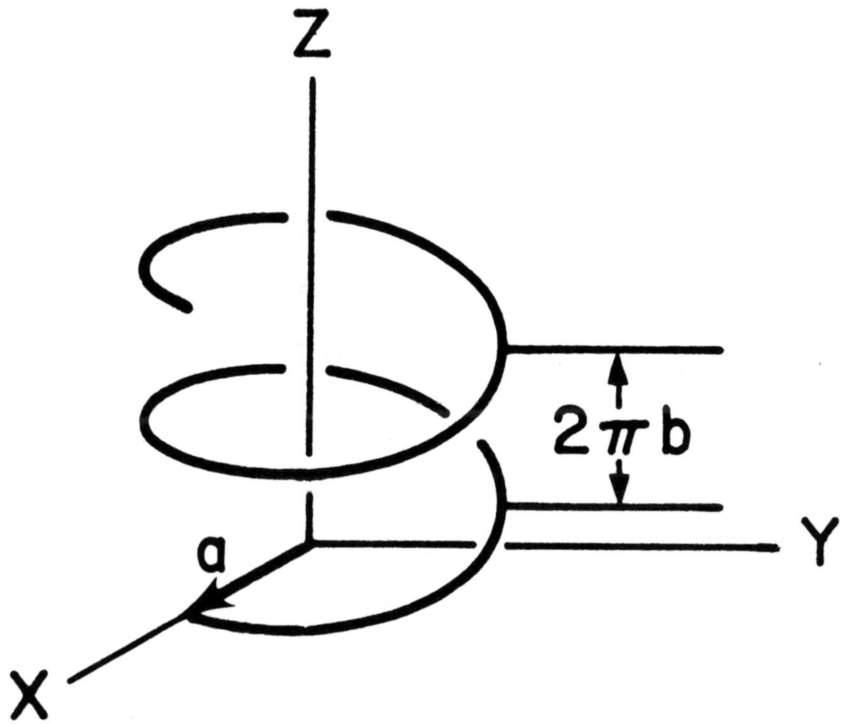

Fig. 1. A helix with pitch (2πb) and radius (a).

$$\lambda_{nm} = \frac{8\mu L^2}{h(m^2-n^2)}$$

μ = mass of electron
L = the contour length along the helix = $2\pi k(a^2+b^2)^{1/2}$
k = number of turns of helix
a = radius of helix
$2\pi b$ = pitch of helix

The transition integrals (Eq. 2) can be obtained in closed form dependent on the radius (a), pitch (2πb) and number of turns (k) of the helix. For an integral number of turns of the helix, selection rules are obtained. Those are given in Table I. We notice that the correct selection rules depend only on the direction of incidence of the light; they do not depend on the polarization direction. It is only in the dipole approximation that the absorption selection rules depend on the polarization direction. For example, one sees that for light incident along

Table I. Selection rules for an electron on a helix of k turns. The directions x, y, z for the helix are shown in Figure 1.

Transition Integral Selection Rules

Light Incidence	Light Polarization	Absorption	Light Incidence	Circular Dichroism
x	y	m-n odd	x	m-n odd
x	z	m-n odd		
y	x	m-n odd, $m \pm n = 2k, 6k, 10k...$	y	m-n odd, $m \pm n = 2k, 6k=10k...$
y	z	m-n odd, $m \pm n = 2k, 6k, 10k...$		
z	x	all transitions	z	all transitions
z	y	all transitions		

Dipole Selection Rules

Light Polarization	Absorption	Light Incidence	Circular Dichroism
x	$m \pm n = 2k$	x	m-n odd
y	m-n odd	y	m-n odd, $m \pm n = 2k$
z	m-n odd	z	m-n odd, $m \pm n = 2k$

the helix axis (z) all transitions interact with the light. In the dipole approximation for z incidence, light polarized along x requires m±n to equal two times the number of turns of the helix, and light polarized along y requires m-n odd.

The magnitudes of the transition integrals (which depend on the pitch and radius of the helix) characterize the absorption and circular dichroism. They can be compared with the dipole approximation to see when the size of the helix invalidates the approximation. For light incident along the helix axis the circular dichroism is proportional to

$$1 \pm \cos\frac{2\pi}{\lambda}(\text{length of helix})$$

The plus sign correspond to (m-n) odd; the minus sign corresponds to (m-n) even. The dipole approximation replaces the cosine by 1, thus only (m-n) odd is allowed. However, as 2π (length of helix)/λ increases from zero, the cosine is no longer equal to 1 and (m-n) even transitions also become allowed. For example, when 2π(length of helix)/λ is equal to 0.25 the cosine is 0.97. For a dipole allowed transition (with the plus sign) the error is 3%, but dipole forbidden transitions (minus sign) now become allowed. When 2π(length of helix)/λ becomes equal to π, cos2π (length of helix)/λ equals -1 and now dipole allowed transitions are forbidden, and dipole forbidden transitions are completely allowed. The dipole approximation is useless. For light incident perpendicular to the helix axis the transition integrals depend on a sum of Bessel functions

$$J_n[\frac{2\pi}{\lambda}\ (\text{radius of helix})]$$

Significant deviations from the dipole approximation occur when 2π(radius of helix)/λ becomes greater than 0.25.

A very general method which allows the calculation of the circular dichroism of a particle constrained to move along any space curve has been published by Balazs et al.[11] Their equations can be applied to any size and shape one-dimensional curve. A good test of the equations for large helices can be made by comparison with measurements of the rotation of linearly polarized microwaves by helices made of copper wire.[12] For right-handed helices containing three turns of radius equal to 0.25 cm and length equal to 1 cm (length of wire in helix is 4.9 cm), the rotation is negative for radiation incident parallel to the helix axis and positive for radiation incident perpendicular to the helix axis. Fig. 2 shows the experimental data for parallel incidence. The data were fit to a three term Drude equation to obtain experimental rotational strengths. The critical wavelengths corresponded to $\lambda_j = 9.44/j$ with j an integer. For perpendicular incidence λ_j for j even had zero rotational strengths and λ_j for j odd were slightly shifted. Table II shows

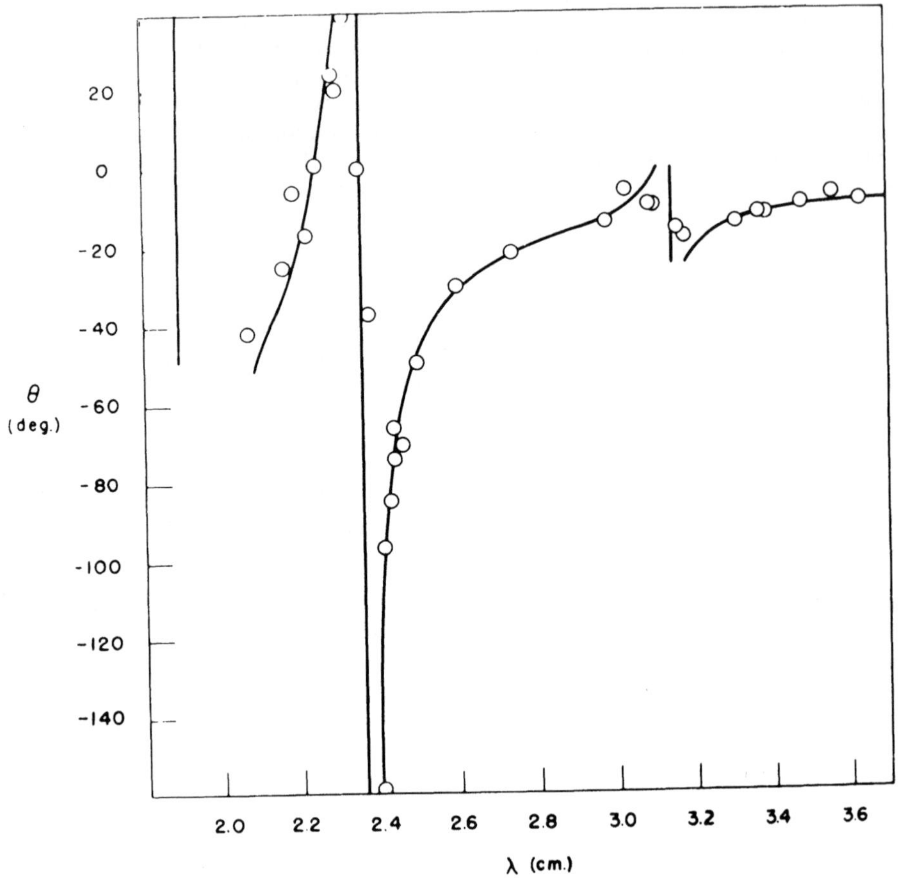

Fig. 2. The measured rotation of linearly polarized microwaves by helices made of copper wire. The helices are right handed with a pitch of 0.33 cm and radius 0.25 cm; they contain three turns. The microwaves are incident parallel to the helix axis. The solid line is calculated from a three term Drude equation with parameters given in Table II. Figure is from reference 12.

the excellent agreement between calculations using the transition integrals and experiment.[13] The calculated signs are correct and the experimental result that the rotational strengths for perpendicular incidence are zero for j even is reproduced. The calculated magnitudes were adjusted to the experimental rotational strength at 2.36 cm. The other values are consistent with the

Table II. Comparison of experimental data with calculated values for the rotation of microwaves by helices of copper wire. The helices are 1 cm long with 3 turns of radius 0.25 cm.

		Rotational Strength Parallel (z incidence)		Rotational Strength Perpendicular (x and y incidence)	
j	λ_j (cm)*	Expt.	Theory	Expt.	Theory
1	9.44	-	- 0.2	+83	+29
2	4.72	-	- 0.4	0	0
3	3.15	- 2.2	- 1.3	+ 4.6	+2.5
4	2.36	-24.2	(-24.2)	0	0
5	1.89	-106	-102	-	-4.2

*These wavelengths correspond to parallel incidence; the wavelength for perpendicular incidence are slightly shifted.

experimental data. In contrast the dipole approximation is totally inadequate for this system; in the dipole approximation the 2.36 cm transition is forbidden.

The absorbance of the helices was also calculated, although it was not experimentally measured. The calculated absorbance illustrates the behavior of light interacting with large molecules. The most significant difference from small molecules is the effect of the direction of incidence on absorbance. Fig. 3 shows the calculated absorbance for the wavelength region where the rotation was measured. The shapes of the curves are arbitrary, but the heights are proportional to the calculated squares of the transition integrals. Although anomalous dispersion of the rotation is found at 3.15 cm (Table II), the calculated absorption is too small to see in Fig. 3. Fig. 3 does show that for x or y polarization, light incident parallel to the helix axis has a different absorption than light incident perpendicular. For light polarized parallel to the helix axis (z), the direction of incidence is unimportant. Fig. 3 also shows that the dipole approximation for these dimensions can give magnitudes less than (x polarization), more than (y polarization) or equal (z polarization) to the transition integrals.

These model calculations for a free electron on a helix were done to show the types of effects to be expected and to illustrate the size of molecules necessary to obtain deviations from dipole-approximation theories. With visible or ultraviolet light these effects may be important for viruses,[14] membranes,[15] chromatin[16] and liquid crystals.[9] Interpretation of rotation of polarized X-rays and X-ray circular dichroism would also require use of transition integrals for all molecules.

CIRCULAR DICHROISM OF POLYNUCLEOTIDES AND NUCLEIC ACIDS

Circular dichroism studies of polypeptides, proteins, polysaccharides, polynucleotides and nucleic acids have been useful in characterizing the structures present in different environments. An excellent comprehensive review [1] has been published recently which covers all these molecules. Here I will concentrate on recent theoretical work on helices and its application to polynucleotides and nucleic acids.

A long helix can be considered as a one dimensional crystal. All unit cells are related by a rotation and a translation along the helix axis. To calculate the optical properties of a nucleic acid we could choose the unit cell to be a base for a single-stranded helix, or a base pair for a double-stranded helix, and so forth. Quantum mechanical theories incorporating helix symmetry and reentrant boundary conditions have been published, [17] but they would be very difficult to apply to real polymers. More practical methods have recently appeared [18,19] which can be applied to nucleic acids. The methods are equivalent and

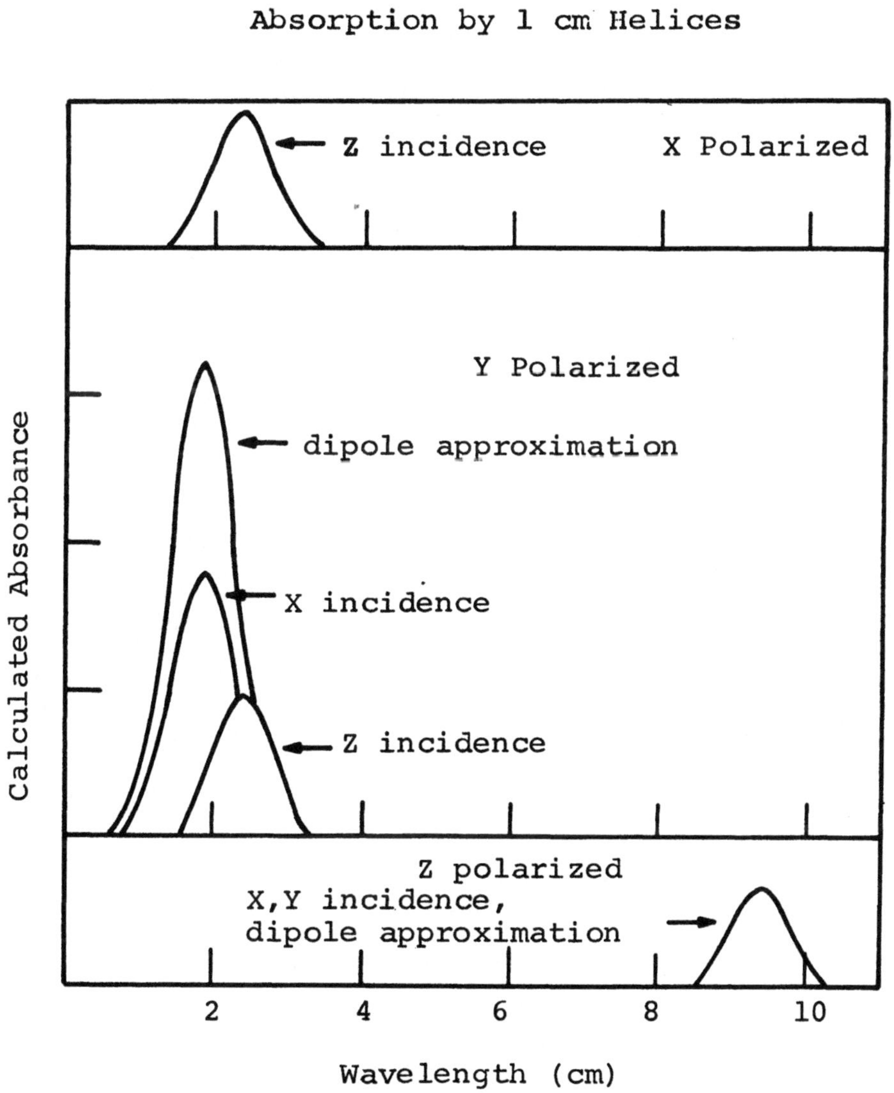

Fig. 3. The calculated absorbance for the helices in Fig. 2. The same scale is used through the figure, but the band shape is arbitrary. For x polarized light and y incidence there is no absorption; the dipole approximation also leads to no absorption.

consider the interaction among monomer units to all orders.

Rhodes and Redmann[18] have used the decorrelation approximation (equivalent to the random phase and time-dependent Hartree approximation) to obtain simple, closed-form expressions for the CD for light incident parallel and perpendicular to the helix axis. An equivalent expression has also been obtained from a classical coupled oscillator method.[19] The practical advance of these methods is the ability to treat interactions to all orders. In the past if one wanted, for example, to include the interactions of 100 transitions to all orders, one had to invert a 100 x 100 matrix. Now, one must at most invert a matrix of dimension equal to the number of transitions in a unit cell.[19] In fact Redmann and Rhodes[18] sum the transitions in a unit cell so that only 3 x 3 matrices must be inverted.

The ability to reduce a very large matrix containing all the transitions in a polymer to a small matrix containing only the transitions in a unit cell rests on the identification of selection rules for a helix. Use of helical symmetry and reentrant boundary conditions requires use of transition integrals (instead of the dipole approximation) to obtain the correct selection rules. However, we can calculate the CD of helical polymers using the dipole approximation, if we do not explicitly use the helical symmetry. That is, we simply calculate the Rosenfeld rotational strength for each transition in the polymer. If the convergence distance for the sum of interaction energies of a transition with all others in the polymer is small compared to the wavelength, then the dipole approximation is valid. For nucleic acids a distance of about 60Å on each side of the transition dipole was required to obtain convergence to 1%. This means that the dipole approximation is reasonable for these systems for wavelengths above 2000Å.

The goal of the theories and methods for calculating CD and absorption is to provide interpretations for the measured properties. We want to be able to deduce structures from the measured circular dichroism. Fig. 4 shows[20] the measured circular dichroism for calf thymus DNA in buffer (B-form) in 80% ethanol (A-form) and in 95% methanol (C-form). The A,B,C labels come from correlations based on measured CD of films of DNA[21] under conditions where X-ray diffraction studies indicate those conformations exist. Although CD calculations for X-ray geometries give similar shapes (Fig. 5)[22] we are not able to draw firm conclusions from the CD calculations alone. If the CD calculations do not agree with experiment we do not know if the calculations are wrong, or if the X-ray geometry is incorrect.

An X-ray fiber diffraction pattern for DNA does not contain as much structural information as one expects from a single crystal. There is a great deal of model building that goes into the final structure. Furthermore, the phosphate groups and the sugars contribute as strongly to the diffraction pattern as the bases do. For circular dichroism the bases are the dominant

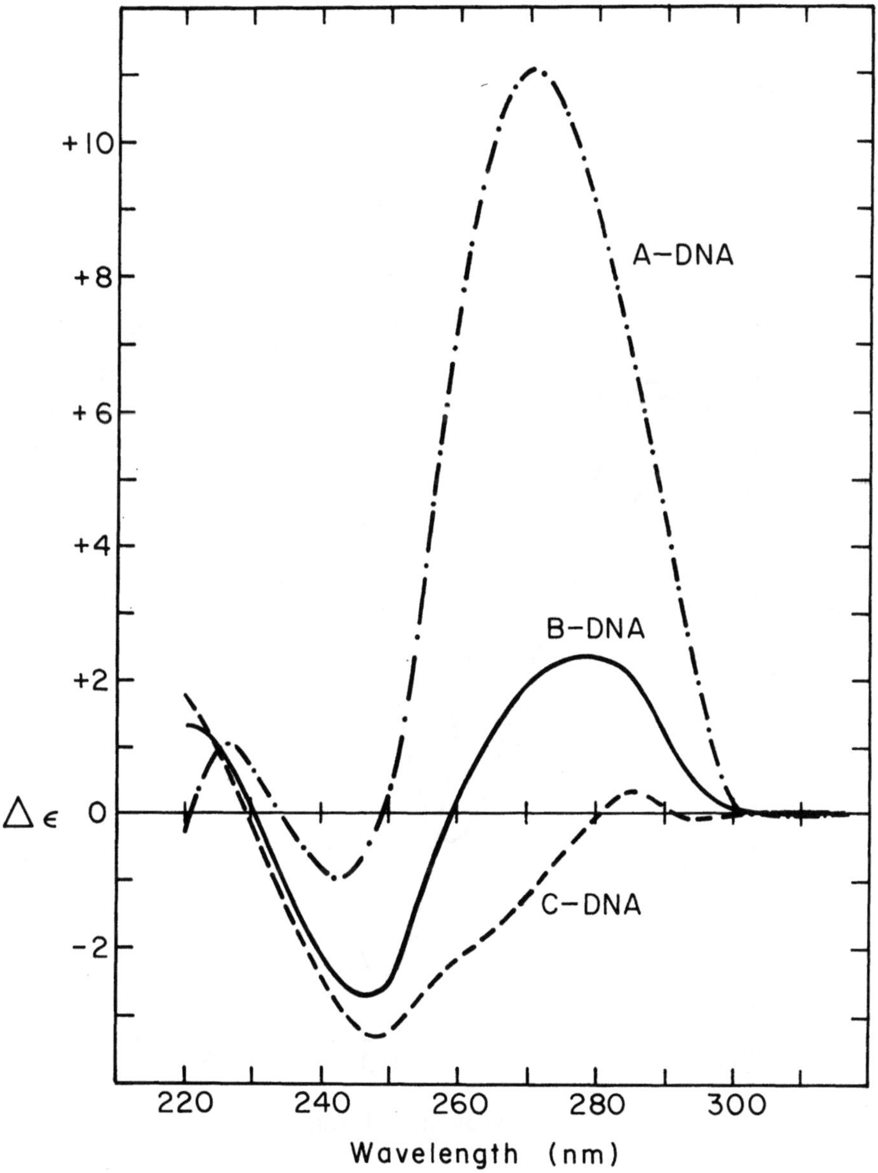

Fig. 4. Experimental circular dichroism measured for calf thymus DNA in pH 7 buffer (B-form), in 80% ethanol plus buffer (A-form) and in 95% methanol plus buffer (C-form). Figure is from Sprecher, Baase and Johnson, reference 20.

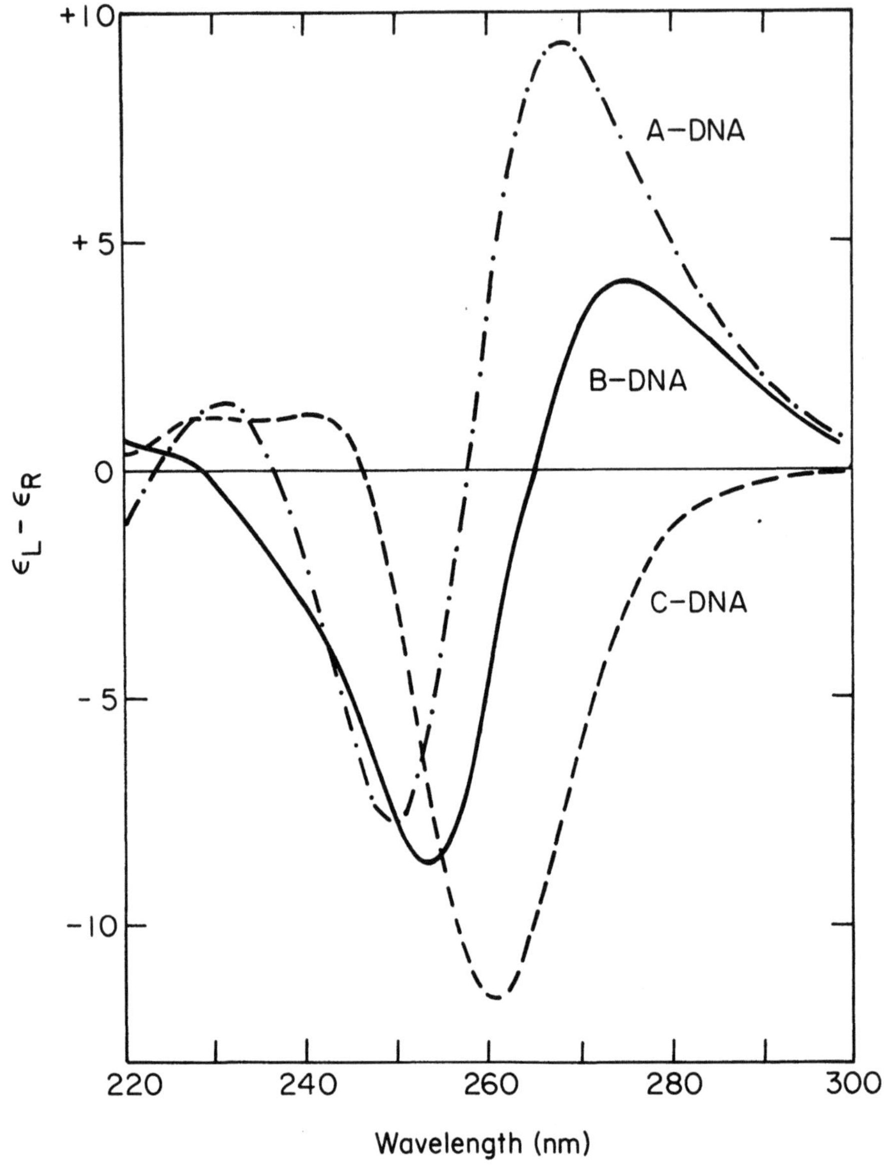

Fig. 5. Calculated circular dichroism for DNA in different conformations. A random sequence of 20 base pairs containing equal amounts of each of the four bases was used. Mononucleotide data are from Cech and Tinoco, Biopolymers 16, 43 (1977); calculations were done by Dr. B. B. Johnson.

contribution. In our CD calculations only the bases are considered; the sugar and phosphate groups are ignored. This may be a poor approximation for the near ultraviolet wavelength region. [20] However, a combination of CD and X-ray data may be a good way to obtain the best structure. Studies of polyinosinic acid provide an example of this. Two separate[23,24] X-ray diffraction studies on polyinosinic acid fibers led to similiar conclusions for the structure. Both groups proposed a four stranded structure of right-handed helices with 11 bases per strand. However, Arnott et al.[23] concluded that each base was tilted, with the perpendicular to the plane of the base tilted 8.9° away from the helix axis. Zimmerman et al.[24] had the bases perpendicular to the helix axis. Circular dichroism calculations Fig. 6[25] allow us to choose the structure with the tilted bases. To test whether the differences in calculated CD were entirely due to the tilt and not to other differences in the two proposed geometries, we used the Arnott et al. geometry except we made the bases perpendicular to the helix axis. The negative CD at 270 nm disappeared and the calculated CD now became similar to that calculated for the Zimmerman et al. structure. We conclude that CD is sensitive to the tilt of the bases and can be used to help establish structures for nucleic acids and polynucleotides.

FLUORESCENCE DETECTED CIRCULAR DICHROISM

Fluorescence intensity is directly proportional to absorption; the proportionality constant is the quantum yield. For chiral molecules in solution the quantum yield is independent of the polarization of the light. Therefore the difference in fluorescence intensity when left and right circularly polarized light is incident is a measure of the circular dichroism. A commercial circular dichroism instrucment with minor modifications[26] can be used to measure the fluorescence detected circular dichroism. Essentially, the photomultiplier is moved to right angles to the incident light as shown in Fig. 7. Fluorescence detection provides advantages over the usual direct detection in certain systems. However, it also provides new information which cannot be obtained by any other method.

The general equation relating tht measured ellipticity to molecular properties is [5]

$$\theta_F^{\circ} = -14.32\left(\frac{\Delta\varepsilon_F}{\varepsilon_F} - R_\ell\right) \qquad (9)$$

$$R_\ell = \frac{\Delta A}{A} - \frac{2.303\Delta A 10^{-A}}{(1-10^{-A})}$$

θ_F° = measured ellipticity in degrees
$\Delta\varepsilon_F$ = circular dichroism of fluorophore
ε_F = molar absorptivity of fluorophore with same units as $\Delta\varepsilon_F$
ΔA = circular dichroism of sample
A = absorbance of sample

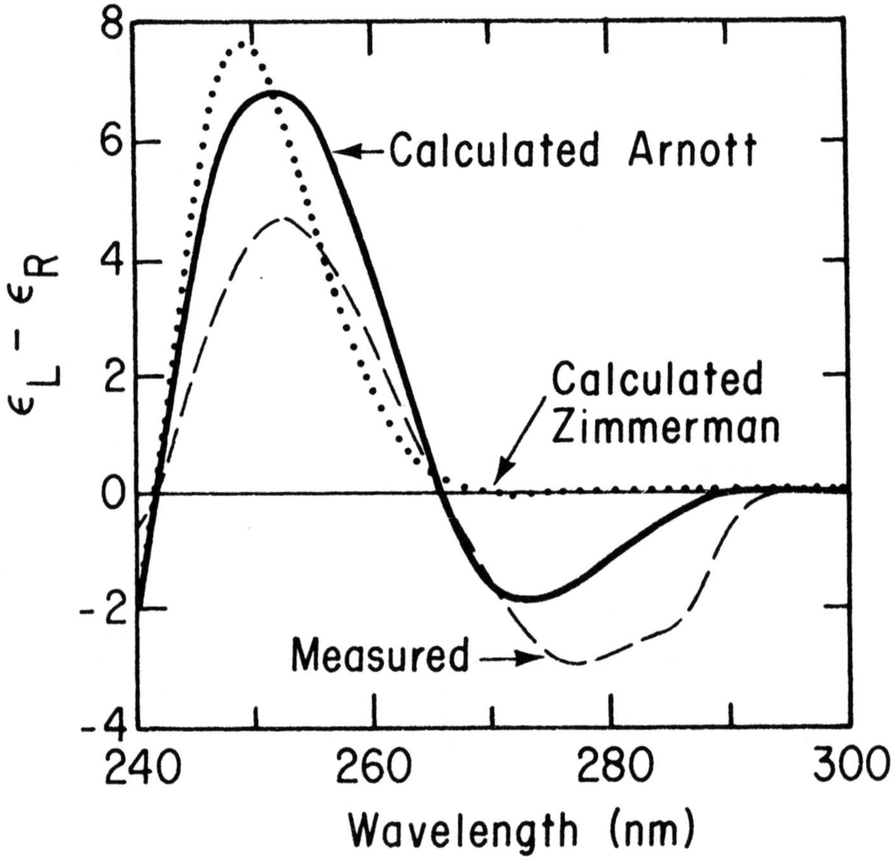

Fig. 6. Calculated circular dichroism for two different four stranded structures for polyinosinic acid compared with measured data. The comparison indicates that the structure with the bases tilted by about 9° to the helix acis (reference 23) is more reasonable.

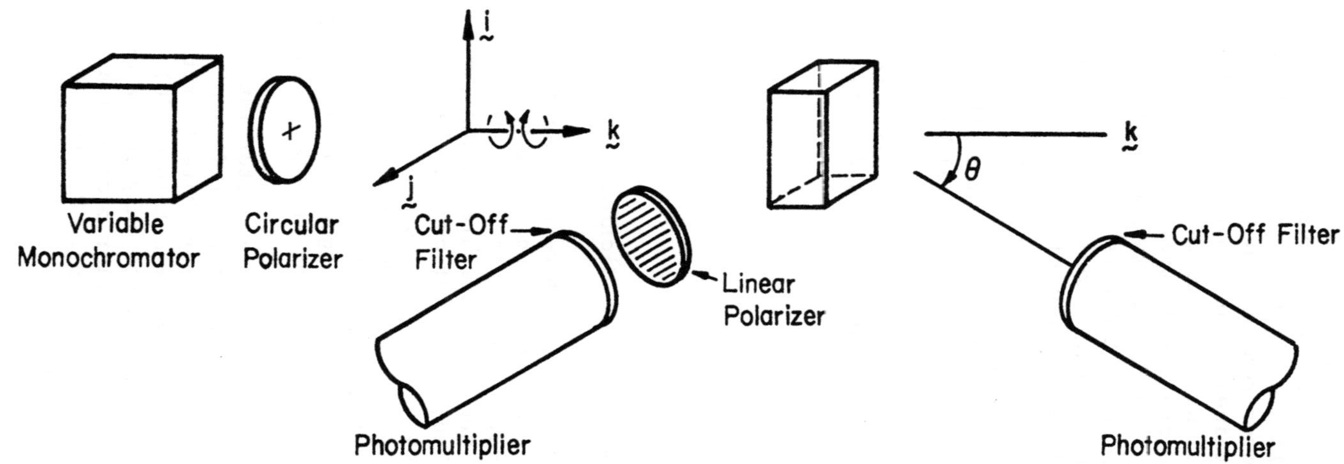

Fig. 7. A diagram of the experimental arrangement for measuring fluorescence detected circular dichroism. The linear polarizer is only used for studying systems in which photoselection is important.

There are two contributions to the measured ellipticity: one is the Kuhn dissymmetry factor of the fluorescent species $g_F = \Delta\varepsilon_F/\varepsilon_F$; the other ($R_\ell$ term) is the circular dichroism of all the species in the sample. In the limit of low concentrations only the Kuhn dissymmetry factor remains and the circular dichroism of the fluorophore can be obtained. For a molecule with a quantum yield greater than 0.1 or 0.2, fluorescence detection can provide an order of magnitude greater sensitivity than direct absorption detection. If the fluorophore is non-chiral and has no induced circular dichroism so that $\Delta\varepsilon_F$ is zero, then only the R_ℓ term contributes. This means that a non-chiral fluorophore can be used as a reporter molecule to detect the circular dichroism of a non-fluorescent, but chiral molecule. This is most useful for study of large, highly scattering particles such as viruses. The scattered light produces artifacts [14] in the usual measurement method which are largely avoided by the fluorescence detection. [27]

When more than one fluorescent species is present in a solution the Kuhn dissymmetry factor becomes a quantum-yield-weighted sum over species.

$$g_F = \frac{\Delta\varepsilon_F}{\varepsilon_F} = \frac{\Sigma\phi_i c_i \Delta\varepsilon_i}{\Sigma\phi_i c_i \varepsilon_i} \tag{10}$$

ϕ = quantum yield
c = concentration
$\varepsilon, \Delta\varepsilon$ = molar absorptivity and circular dichroism

For a non-rigid molecule which has more than one conformation in solution, fluorescence detection can uniquely provide information about the conformations with high quantum yield. A conformation which dominated the usual circular dichroism, but whose fluorescence was quenched, would not contribute to the sum and would thus allow study of the other conformations present. The quantum yield in the equation above also applies to fluorescence due to energy transfer. That is, the circular dichroism of a species which absorbs the light and then transfers it to another species which emits the light will be included in the sum.

Fluorescence detection introduces a new factor into circular dichroism: the fluorescence lifetime. If the lifetime is short compared to the time necessary for the fluorophore to rotate, then a photoselected circular dichroism is obtained. [6] The measured circular dichroism will depend not only on the average circular dichroism, but also on the circular dichroism along the direction of the emission transition moment. For the detector perpendicular to the incident light

$$g_F = \frac{4\Delta\varepsilon_F/\varepsilon_F - 2\Delta\varepsilon_{33,F}/\varepsilon_F}{3 + \varepsilon_{33,F}/\varepsilon_F} \tag{11}$$

$\Delta\varepsilon_{33,F}, \varepsilon_{33,F}$ = the circular dichroism and absorptivity along the direction of the fluorescence emission transition moment

If the detector is placed at an angle of 54.75° (the magic angle) to the incident light, only the average circular dichroism is obtained. By making measurements at different angles of the detector, or equivalently, with a perpendicular detector having a linear polarizer in front of it, one can obtain both the circular dichroism along the emission transition moment and the average CD.

FLUORESCENCE DETECTED CIRCULAR DICHROISM STUDIES OF TRANSFER RNA

Transfer RNA (tRNA) molecules read the genetic message in messenger RNA and translate it into a protein sequence. Each tRNA is specific for one amino acid. Three bases of the tRNA (the anticodon) bind to the three bases (the codon) in the messenger RNA which code for that amino acid. The three bases of the tRNA are in a single stranded loop called the anticodon loop. It is important to know the conformation of this region of the tRNA and to know how it changes upon binding to the codon. As the tRNA molecules contain about 75 nucleotides, it is difficult to learn about such a small fraction of the molecule in solution. However, phenylalanine tRNA contains a fluorescent base right next to the anticodon. This provides the necessary fluorophore for fluorescence detected circular dichroism. Fig. 8 shows the measured Kuhn dissymmetry factor for phenylalanine tRNA in the presence and absence of the codon U-U-C. [28] For a single absorbance band the Kuhn dissymmetry factor should be approximately constant over the band; it should give a horizontal line vs. wavelength. In Fig. 8 we see three such regions (230, 264, 307) corresponding to three absorption bands of the fluorescent base. The measured circular dichroism corresponds to interactions of the transitions of the fluorescent base with neighboring transitions. These interactions for each transition can be approximated by a first order perturbation treatment. For a transition from 0 to a in the fluorophore (i) interacting with other transitions 0 to b in other chromophores (j) [5]

$$g_F = \frac{\Delta\varepsilon_F}{\varepsilon_F} = \frac{4R_{0a}}{D_{0a}} = \frac{2\pi\nu_{0a}}{\mu_{i0a}^2 c} \sum_j \sum_b \frac{\nu_{0b} V_{i0a;j0b} \tilde{R}_{ij} \cdot \tilde{\mu}_{i0a} \times \tilde{\mu}_{j0b}}{h(\nu_{0b}^2 - \nu_{0a}^2)} \tag{12}$$

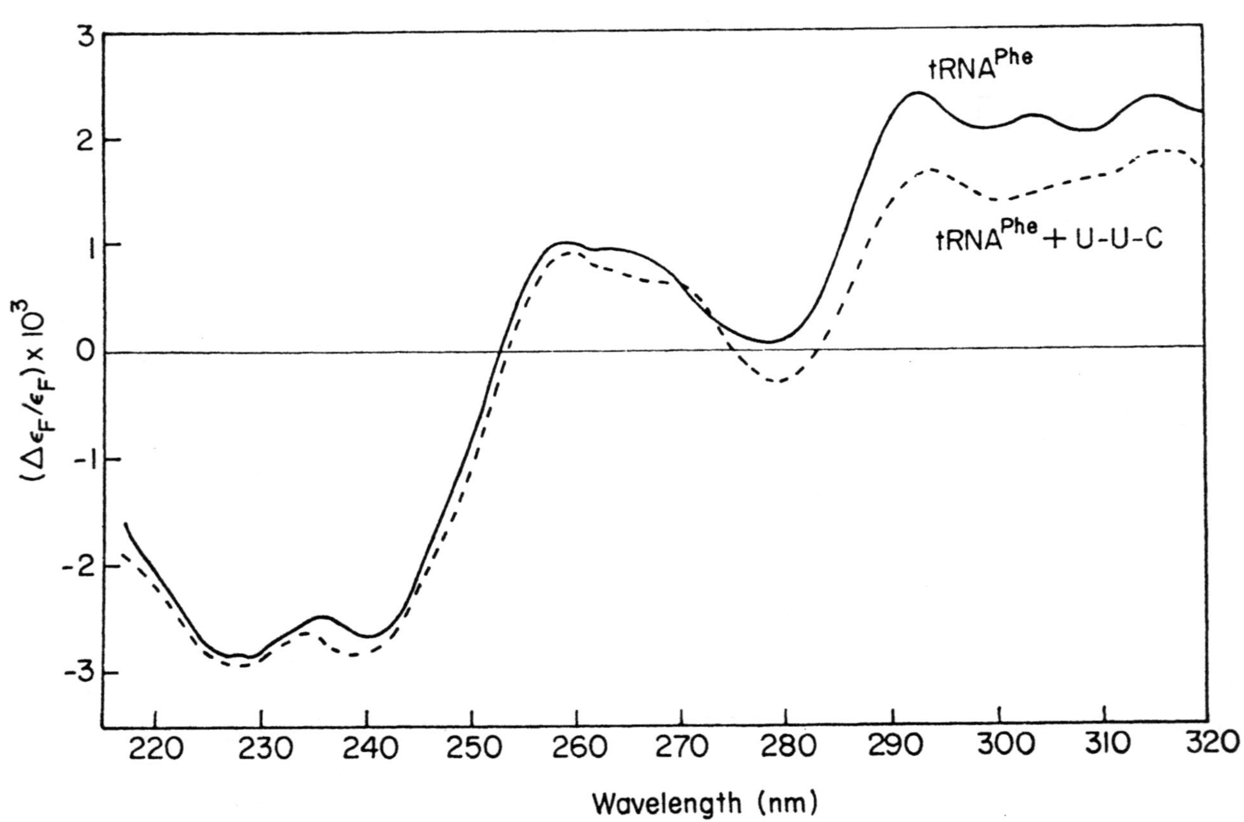

Fig. 8. Measured Kuhn dissymmetry factor for phenylalanine tRNA in the presence and absence of the codon U-U-C. [Reference 28]

ν_{0a}, ν_{0b} = frequency of transition

$\underset{\sim}{R}_{ij} = \underset{\sim}{R}_j - \underset{\sim}{R}_i$, the vector distance from fluorophore i to chromophore j

$\underset{\sim}{\mu}_{i0a}, \underset{\sim}{\mu}_{j0b}$ = electric transition dipoles

$V_{i0a;j0b}$ = interaction energy between groups

h,c = Planck's constant and the speed of light

In order to be able to apply this equation we need to know the properties of the individual chromophores and fluorophore. We also need a geometry; we need to know the positions and orientations of the various groups. Molecular orbital calculations [29] were used to obtain transition monopoles for calculating the interaction energy. The interaction energy is approximated by Coulomb's Law between transition monopoles in different groups. The transition monopoles are also used to determine the directions of the transition dipoles; the magnitudes of the transition dipoles were obtained from absorption spectra of the nucleotides.

Equation (12) was used to calculate the Kuhn dissymmetry factor for three transitions of the Y base fluorophore in the anticodon loop of phenylalanine tRNA. [13] Several groups have published structures deduced from X-ray diffraction of this tRNA. [30] A side view of one proposed [30c] crystal structure of the anticodon loop is shown in Fig. 9. Calculations for the four published crystal structures are compared with the experimental results in Table III. Although the calculated values are of the right order of magnitude the signs are completely inconsistent. Furthermore, the range of values calculated indicate the differences in geometry obtained by the four different groups.

Table III. Fluorescence detected dissymmetry factor for $tRNA^{Phe}$

	$\Delta\varepsilon_F/\varepsilon_F \times 10^3$		
Wavelength (nm)	Expt.	Calculated B,C-form DNA	Calculated tRNA Crystal
307	+2	+2	- 1 to +1
264	+1	+2	- 3 to -7
230	-3	-4	+0.1 to +1

The calculated values in Table III included interactions of the Y-base with all other bases in the anticodon loop and stem-16 bases. However, the same sign pattern and similar magnitudes

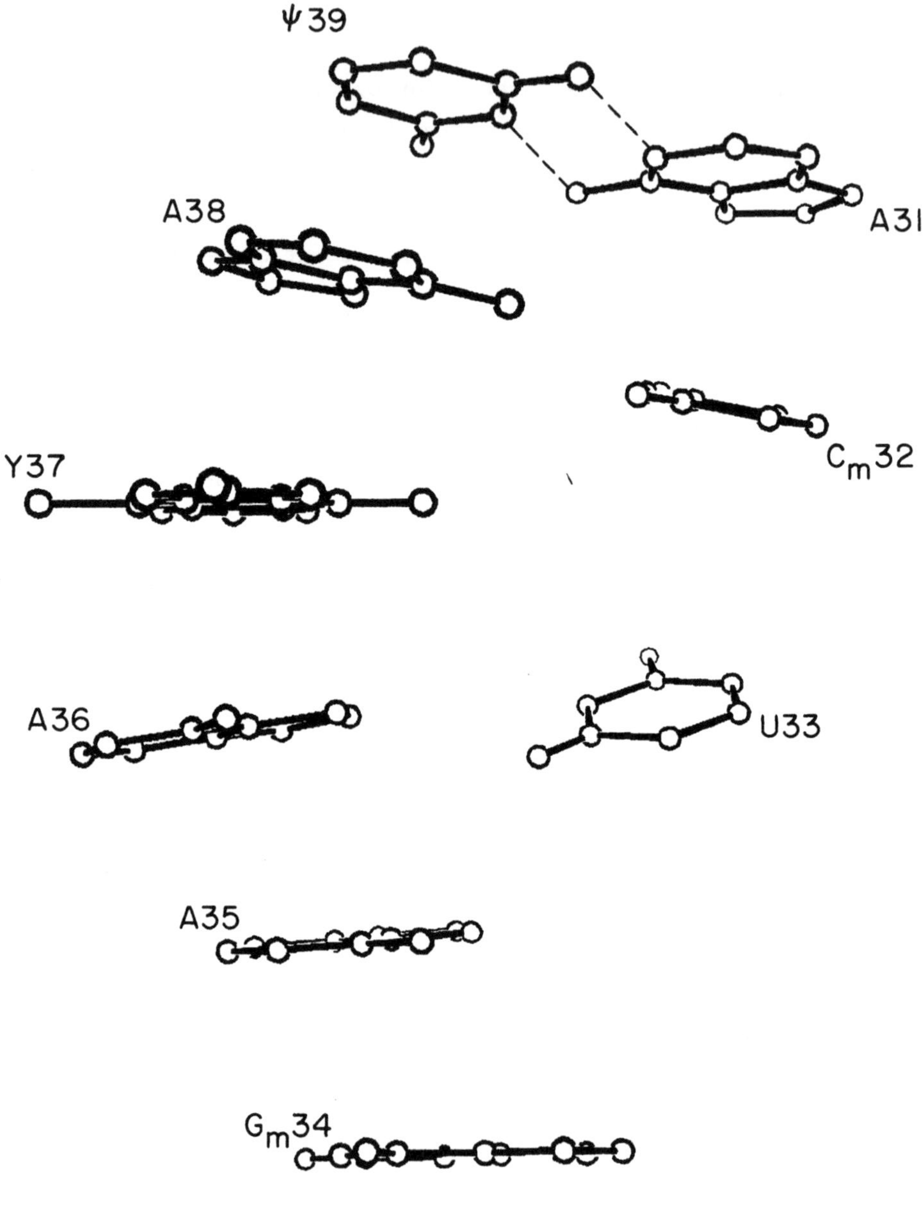

Fig. 9. The conformation of the bases in the anticodon loop of phenylalanine transfer RNA as deduced from X-ray diffraction by Sussman and Kim. [Reference 30c]

were obtained when only the two nearest bases on each side of the Y-base were considered. We [13] therefore asked what orientation of these five bases would give agreement with experiment. A simple helical arrangement of these bases stacked as in one strand of B-form [31] or C-form [32] DNA give good agreement in sign and magnitude with experiment (see Table III). RNA [33] or A-DNA geometry does not agree. We can therefore conclude: (1) the detailed arrangement of bases in the tRNA anticodon loop in a crystal is not well established. (2) The arrangement of bases in the crystal is probably not the same as in solution. The changes in the Kuhn dissymmetry factor seen experimentally on forming the codon-anticodon complex can be satisfactorily explained by the formation of a double-stranded helix without large changes in loop geometry.

FLUORESCENCE DETECTED CIRCULAR DICHROISM STUDIES OF DINUCLEOSIDE PHOSPHATES

Stacking-unstacking equilibria in single-stranded polynucleotides have been extensively studied. Dinucleoside phosphates are of interest in themselves and as a model system for the more complicated polymers. The simplest assumption which can be made about the equilibrium for a dinucleoside phosphate is that there exists only two significant species in solution: a stacked species and an unstacked species. Fluorescence detected circular dichroism can give new information about this two-state hypothesis.

Fluorescent dinucleoside phosphates can be easily made by treating adenine containing compounds with chloroacetaldehyde. [34] Fluorescent derivatives containing 1,N^6 ethenoadenine (εA) are produced. We [35] studied the fluorescent derivatives of adenylyl-3'-5' adenosine (εApεA), adenylyl-3'-5'-cytidine, (εApεC), and adenylyl-3'-5'-uridine (εApU). A nonfluorescent derivative of cytosine (3,N^4 ethenoadenine, εC) is also formed, but uracil does not react with chloroacetaldehyde.

The CD <u>vs.</u> temperature of all three compounds is shown [35] in Fig. 10. Note that there is a uniform decrease of CD with increasing temperature and that several isosbestic points are seen in the spectra. This is consistent with a two-state hypothesis. However, fluorescent detected circular dichroism clearly distinguishes εApεA from εApεC and εApU. The dissymmetry factor, $\Delta\varepsilon_F/\varepsilon_F$, for both εApεC and εApU is essentially independent of temperature (Fig. 11), whereas $\Delta\varepsilon_F/\varepsilon_F$ for εApεA changes by more than a factor of two over the same temperature range (Fig. 12). The fluorescence detected CD does not sense the most strongly stacked state, because this state is quenched in the dinucleoside phosphates. [34] Therefore, the FDCD is only measuring unstacked or partially stacked fluorescent species. The fact that $\Delta\varepsilon_F/\varepsilon_F$ is independent of temperature indicates that only one such species exists for εApεC and εApU. However, for

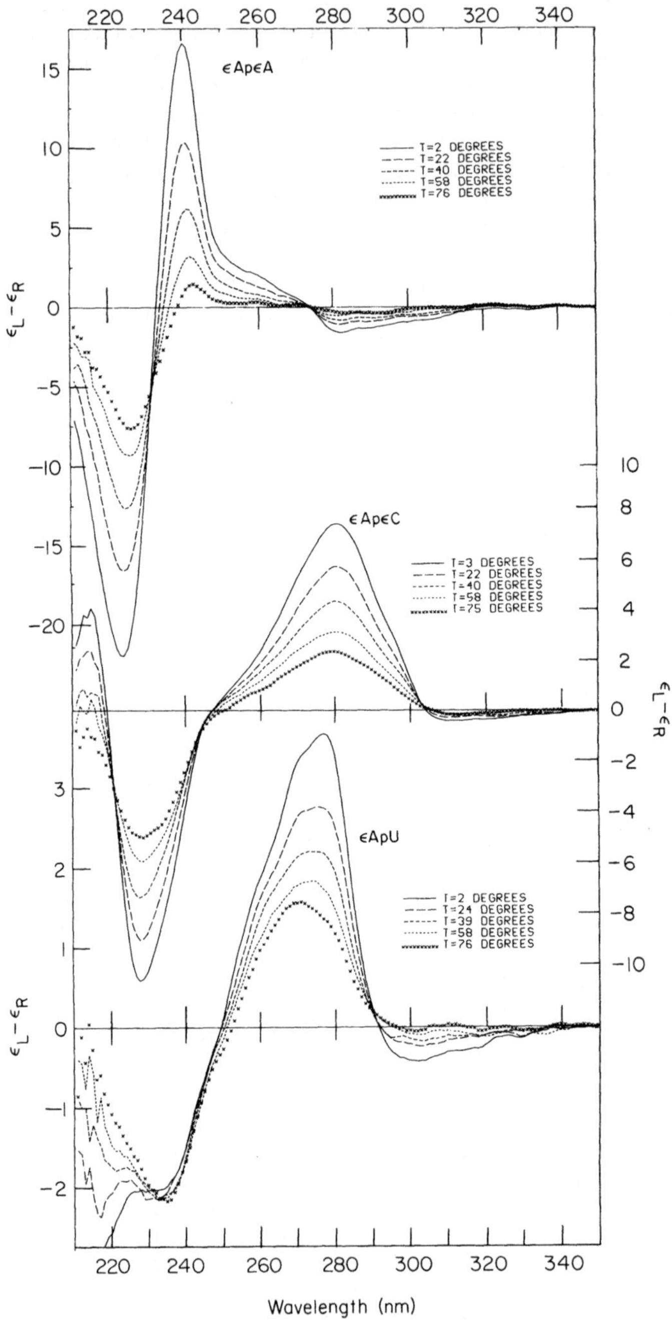

Fig. 10. The CD of three dinucleoside phosphates: εA is 1,N^6 ethenoadenosine; εC is 3,N^4 ethenocytidine; U is uridine. [Ref. 35]

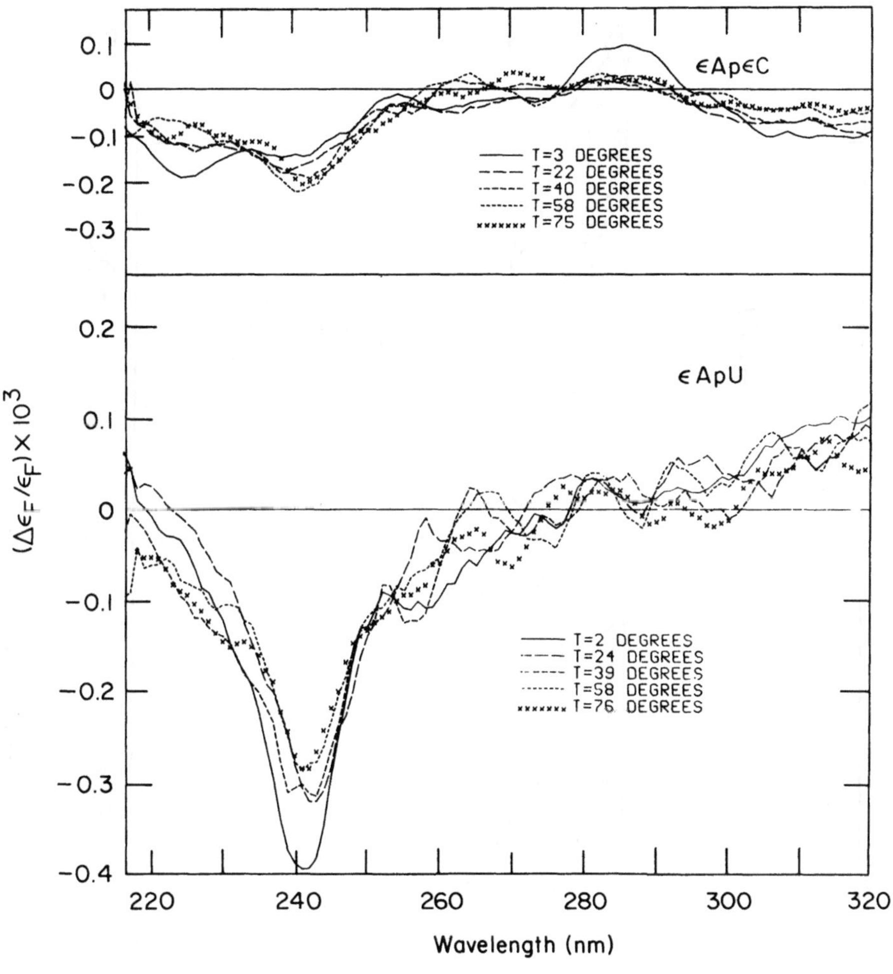

Fig. 11. The Kuhn dissymmetry factor vs. temperature for εApεA. Data are from reference 35.

εApεA there must be at least two such species. This means that in a solution of εApεA at least three species exist: a stacked non-fluorescent species plus two fluorescent species. We conclude that the optical properties of εApεC and εApU are consistent with a two state model for stacking-unstacking, but εApεA requires at least three states. We can estimate the circular dichroism of the fluorescent species from the measured $\Delta\varepsilon_F/\varepsilon_F$ by assuming that the molar absorptivity, ε_F, is equal to that of the mononucleotide.

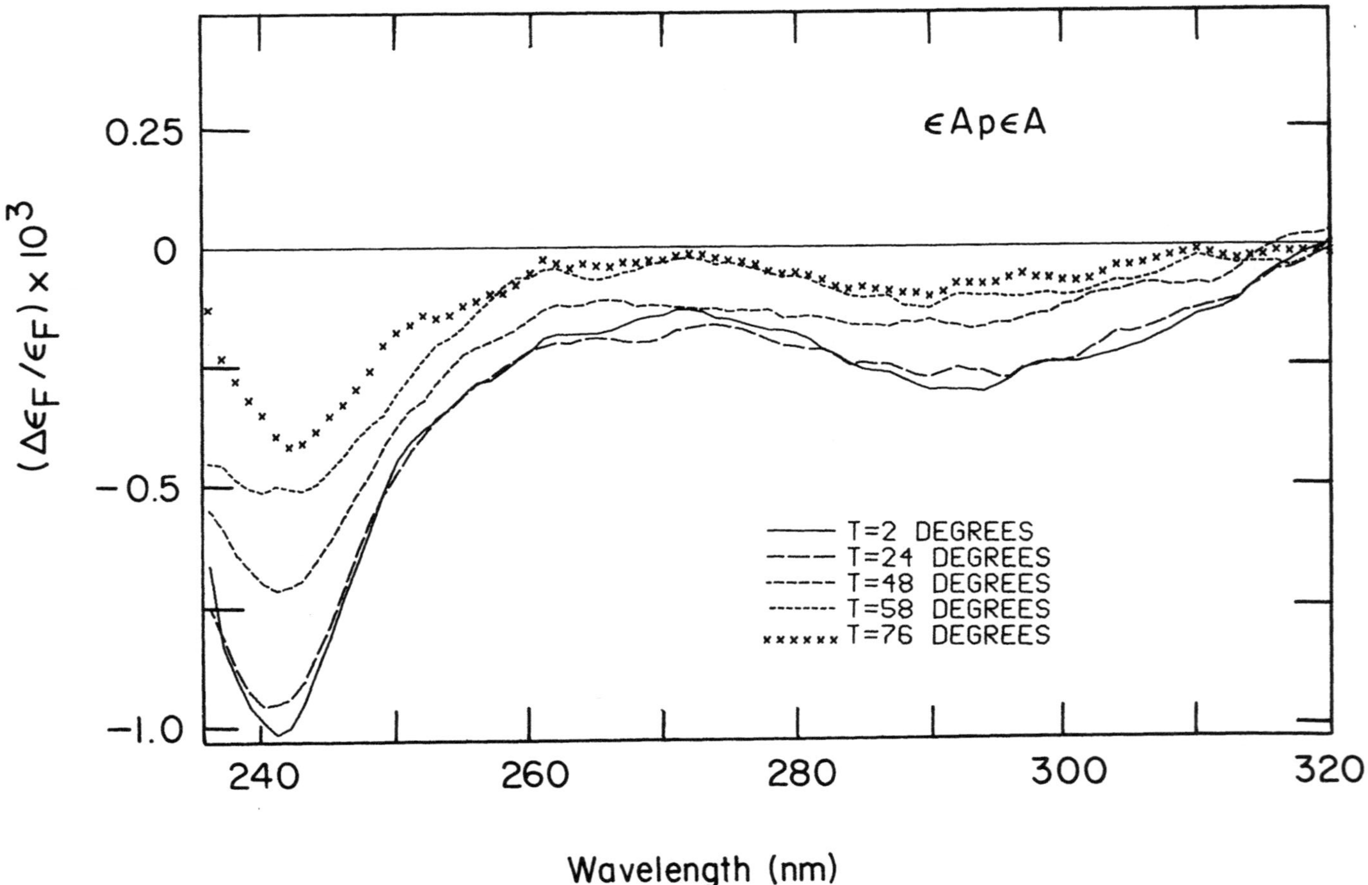

Fig. 12. The Kuhn dissymmetry factor vs. temperature for εApεA. Data are from reference 35.

That is, we assume that the approximately 20% hypochromicity all occurs in the stacked species. The results are shown in Fig. 13 where the CD of the various species present in εApεA are compared. The mononucleotide, pεA, represents the CD of the completely

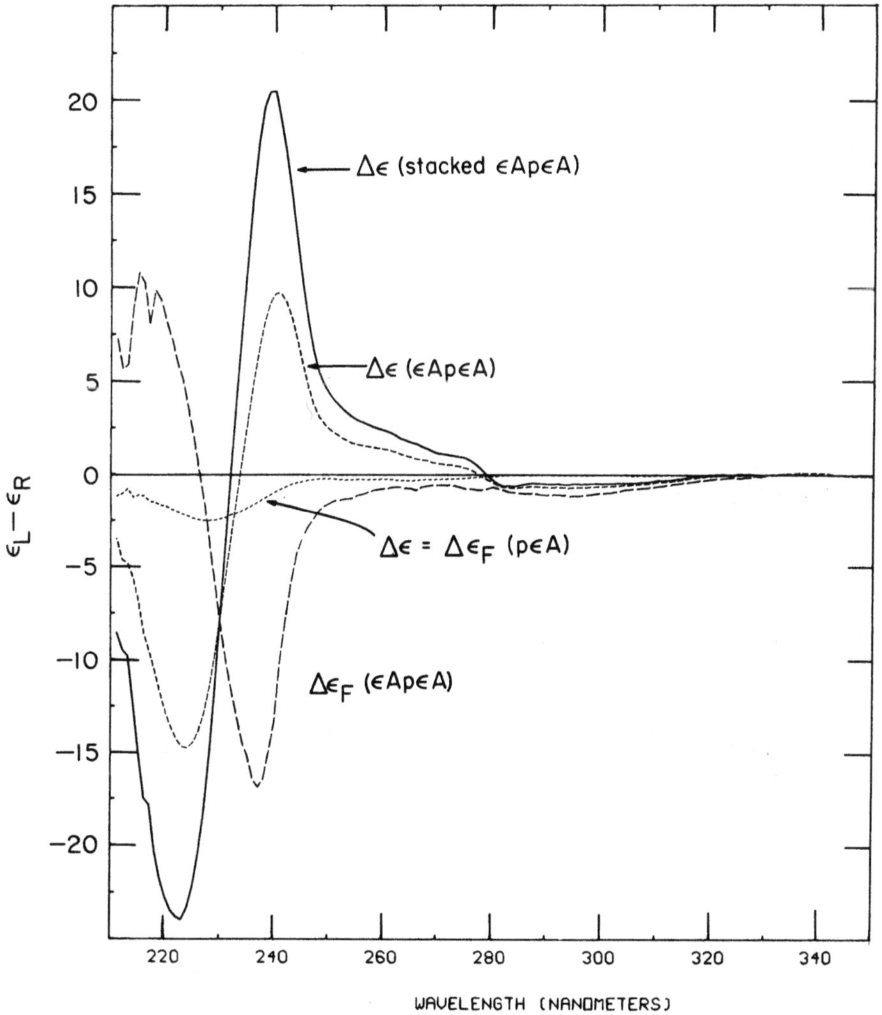

Fig. 13. The circular dichroism of the fluorescent species in εApεA. The CD of completely unstacked species ($\Delta\varepsilon_F$,pεA), the completely stacked species (——) and a partially stacked intermedi ate ($\Delta\varepsilon_F$,εApεA) are compared with the measured CD (Δε,εApεA). Data are from reference 35.

stacked state was obtained by extrapolation from the temperature dependence of the CD. The partially stacked, fluorescent species is seen to have an opposite sign CD relative to the stacked, non-fluorescent species. This indicates a very different conformation from the stacked state, but detailed calculations will be necessary to further characterize this conformation.

Acknowledgment. Many of the results here are based on unpublished work of Dr. B. B. Johnson, Dr. D. S. Moore and Dr. C. Reich. I also want to thank Mr. C. Bustamante, Dr. B. B. Johnson and Dr. M. F. Maestre for many helpful discussions. Prof. W. C. Johnson, Jr. kindly read the manuscript and made detailed comments. His help is greatly appreciated. Support for this work was provided by National Institutes of General Medical Sciences grant GM 10840 and by the Division of Biomedical and Environmental Research of the U.S. Department of Energy.

REFERENCES

1. See R. W. Woody, J. Polymer Sci.: Macromolecular Reviews, 12, 181-321 (1977) for a comprehensive review.
2. S. Y. Chung and G. Holzwarth, J. Mol. Biol. 92, 449-466 (1975).
3. E. Charney and J. B. Milstein, Biopolymers 17, 1629-1655 (1978).
4. L. Rosenfeld, Z. Physik 52, 161 (1928).
5. I. Tinoco, Jr. and D. H. Turner, J. Am. Chem. Soc. 98, 6453-6456 (1976).
6. I. Tinoco, Jr., B. Ehrenburg and I. Z. Steinberg, J. Chem. Phys. 66, 916-920 (1977).
7. I. Tinoco, Jr., Adv. Chem. Phys. 4, 113-159 (1962).
8. I. Tobias, T. R. Brocki and N. L. Balazs, J. Chem. Phys. 62, 4181-4183 (1975).
9. G. Holzwarth and N. A. W. Holzwarth, J. Opt. Soc. Am. 63, 324-331 (1973).
10. I. Tinoco, Jr. and R. W. Woody, J. Chem. Phys. 40, 160-165 (1964).
11. N. L. Balazs, T. R. Brocki and I. Tobias, Chem. Phys. 13, 141-151 (1976).
12. I. Tinoco, Jr. and M. P. Freeman, J. Phys. Chem. 61, 1196-1200 (1957).
13. D. S. Moore and I. Tinoco, Jr., unpublished.
14. B. P. Dorman and M. F. Maestre, Proc. Natl. Acad. Sci., U.S. 70, 255-259 (1973).
15. M. Glaser and S. J. Singer, Biochemistry 10, 1780-1787 (1971).
16. H. J. Li, C. Chang, Z. Evagelinou and M. Weiskopf, Biopolymers 14, 211-226 (1975).
17. C. W. Deutsche, J. Chem. Phys. 52, 3703 (1970); F. M. Loxsom, J. Chem. Phys. 51, 4899 (1969).
18. W. Rhodes and S. M. Redmann, Chem. Phys. 22, 215-220 (1977); S. M. Redmann and W. Rhodes, unpublished.

19. A. I. Levin and I. Tinoco, Jr., J. Chem. Phys. 66, 3491-3497 (1977).
20. C. A. Sprecher, W. A. Baase and W. C. Johnson, Jr., Biopolymers, in press (1978).
21. M. J. Tunis-Schneider and M. F. Maestre, J. Mol. Biol. 52, 521 (1970).
22. B. B. Johnson and I. Tinoco, Jr., unpublished results.
23. S. Arnott, R. Chandrasekharan and C. M. Mertilla, Biochem. J. 141, 537-543 (1974).
24. S. B. Zimmerman, G. H. Cohen and D. R. Davies, J. Mol. Biol. 92, 181-192 (1975).
25. C.L. Cech and I. Tinoco, Jr., Nucleic Acids. Res. 3, 399-404 (1976).
26. D. H. Turner, Methods in Enzymology 49, 199-214 (1978).
27. M. F. Maestre, unpublished data.
28. K. Yoon, D. H. Turner and I. Tinoco, Jr., J. Mol. Biol. 99, 507-518 (1975).
29. W. Hug and I. Tinoco, Jr., J. Am. Chem. Soc. 95, 2803-2813 (1973).
30. (a) Ladner et al., Nucleic Acids Res. 2, 1629-1637 (1975);
(b) Quigley et al., Nucleic Acids Res. 2, 2329-2339 (1975);
(c) Sussman & Kim, Biochem. Biophys. Res. Comm. 68, 89-96 (1976);
(d) Stout et al., Nucleic Acids Res. 3, 1111-1123 (1976)
31. S. Arnott and D. W. L. Hukins, Biochem. Biophys. Res. Comm. 47, 1504-1509 (1974).
32. D. A. Marvin, M. Spencer, M. H. F. Wilkins and L. D. Hamilton, J. Mol. Biol. 3, 547-565 (1961).
33. S. Arnott, D. W. L. Hukins and S. D. Dover, Biochem. Biophys. Res. Comm. 48, 1392-1399 (1972).
34. G. L. Tolman, J. R. Barrio and N. J. Leonard, Biochemistry 13, 4869-4878 (1974).
35. C. Reich and I. Tinoco, Jr., unpublished data.

NATURAL AND MAGNETIC CIRCULAR DICHROISM SPECTROSCOPY IN THE VACUUM ULTRAVIOLET*

Otto Schnepp

Department of Chemistry, University of Southern California

I. INTRODUCTION

Studies of Natural and Magnetic Circular Dichroism were extended into the vacuum ultraviolet spectral region, i.e. to wavelengths shorter than 200nm, beginning about 1970. The purpose of this effort is to make possible the measurement of CD and MCD for chromophores which have limited or no absorption at longer wavelengths, such as ethylene, alcohols, cyclopropane or to extend such measurements to more excited electronic states. Two basic thrusts can be distinguished in the published work on natural CD. One is directed toward the elucidation and determination of conformations, while the second seeks to obtain data additional and often complementary to absorption measurements aimed at the characterization and assignment of excited states. MCD spectroscopy is usually confined to the latter direction.

The first review on CD and MCD spectroscopy in the vacuum ultraviolet was published in 1973 as part of the proceedings of a NATO ASI (Schnepp, 1973). A recent review on natural CD in the vacuum ultraviolet is in press at this time (Johnson, 1978).

An important consideration, which prompted those workers in the field of CD and MCD who are interested in the assignment of excited states is as follows. In the absence of extensive fine structure in the absorption spectrum, it is difficult, if not impossible to characterize the excited state. As a result, it was considered of importance to seek additional observables for this purpose. The CD and MCD spectra seemed attractive sources for such information. However, many basic and interesting chromophores have limited or no absorption at wavelengths longer than 200nm

Stephen F. Mason (ed.), Optical Activity and Chiral Discrimination. 87–106.

and therefore these workers decided to develop instrumentation extending the spectral region of measurement to shorter wavelengths.

CD spectroscopy for the charecterization of states can be carried out in two ways. First, the natural CD of an optically active derivative of the chromophore of interest can be studied. In this case, increased resolution as compared to absorption is obtained between overlapping unstructured bands, due to the fact that CD bands are positive or negative. Also, semiquantitatively, transitions of a centrosymmetric chromophore can be characterized as electric dipole allowed as opposed to magnetic dipole allowed by a measure of the anisotropy factor $g = \Delta \varepsilon/\varepsilon$. It has been shown (Mason, 1962) that a relatively large value of g is evidence for a magnetic dipole allowed transition ($g \rightarrow g$) in the symmetric chromophore whereas a small value indicates an electric dipole allowed transition ($g \rightarrow u$). In some instances of the former case the band is prominent in the CD spectrum but is not observable in the absorption. In addition, the development of good ab-initio theory would allow specific assignments to be made using now not only the energy and absorption intensity but also the rotatory strength which has both characteristic magnitude and sign. As will be described, this has recently been successfully done for a twisted ethylene chromophore, viz. trans-cyclooctene. Since the CD of a substituted optically active derivative must be studied, the assumption is made that the spectrum observed is characteristic of the chromophore itself with only a minor perturbation. This assumption may not always be valid. Also the possibility of the presence of more than one conformer must be investigated since a superposition of the CD spectra of these may exhibit fortuitous maxima. Since CD theory is dealt with extensively in other chapters, no further elucidation will be given here.

Secondly, the magnetic field induced circular dichroism (MCD) can be studied of the symmetric chromophore itself. The information so obtained is therefore free from the ambiguities described for CD and it is also different in nature from that obtained for CD. We shall be interested in two types of observed MCD line shapes. The derivative type (or dispersive type) is observed only in case of a degenerate state and is called an "A-term"; it corresponds to a first order Zeeman effect. In atoms, absorption lines are very narrow and the Zeeman splitting is large compared to the line width. The split components are circularly polarized when viewed parallel to the magnetic field lines (e.g. through the pole pieces of a magnet or along the axis of the coil of a superconducting magnet). For a molecule where the absorption bands are broad (several thousand cm^{-1}) compared to the Zeeman splitting (of the order of $10 cm^{-1}$), the MCD which measures $\Delta \varepsilon = \varepsilon_L - \varepsilon_R$ results in a derivative or S-shape signal with a zero coinciding with the maximum of the absorption. By studying this type of line, and by moment analysis, we obtain a quantity $\mathcal{A}/D$ which is related to the magnetic moment of the degenerate state by a simple proportionality factor. Specifically, (Stephens, Mowery and Schatz, 1971):

$$\mathcal{A}/D = \langle\Delta\ \varepsilon\rangle_1/\langle\varepsilon\rangle_0 \quad (1)$$

or $\mathcal{A}/D$ is given by the ratio of the first moment of $\Delta\ \varepsilon$ over the zeroeth moment of ε; the nth moment of a quantity α is given by:

$$\langle\alpha\rangle_n = \int(\alpha/\nu)\ (\nu-\nu_0)^n d\nu \quad (2)$$

In Eq. (2), ν designates the frequency, ν_0 is the central frequency determined by the condition $\langle\varepsilon\rangle_1 = 0$ and the integral is in principle over all frequencies (zero to infinity). Theoretically, the quantities $\mathcal{A}$ and D for a transition between states $A\rightarrow J$ are given by Eqs. (3) and (4).

$$\mathcal{A}\ (A\rightarrow J) = \tfrac{1}{2} \sum_{\lambda,\lambda'} \mathrm{Im}\ [\langle A|\underset{\sim}{\mu}_e|J_\lambda\rangle \times \langle J_{\lambda'}|\underset{\sim}{\mu}_e|A\rangle \cdot \langle J_\lambda|\underset{\sim}{\mu}_m|J_{\lambda'}\rangle]\ (3)$$

$$D(A\rightarrow J) = \sum_\lambda |\langle A|\underset{\sim}{\mu}_e|J_\lambda\rangle|^2. \quad (4)$$

Here μ_e and μ_m are the electric and magnetic dipole moment operators, respectively, and λ,λ' designate the components of the degenerate excited state J. Note also, that $\mathcal{A}$ is given by a vector triple product. The second type of MCD line shape is identical to that of the absorption and is called a "B-term". It corresponds to a second order Zeeman effect which involves a coupling of the observed state with all other states of the molecule. Such an effect is possible for all states both degenerate and non-degenerate. The corresponding equations defining B/D in terms of measured spectra and B in terms of quantum-mechanical quantities are given as Eqs. (5) and (6).

$$B/D = \langle\Delta\ \varepsilon\rangle_0/\langle\varepsilon\rangle_0 \quad (5)$$

$$B = \sum_\lambda \mathrm{Im}\ [\sum_{K_\kappa \neq J} \langle A|\underset{\sim}{\mu}_e|J_\lambda\rangle \times \langle K_\kappa|\underset{\sim}{\mu}_e|A\rangle \cdot \langle J_\lambda|\underset{\sim}{\mu}_m|K_\kappa\rangle]\ \hbar\omega_{KJ}. \quad (6)$$

By considering Eq.(6) we find for a hypothetical case of two interacting states only, J, K, that the B-term of state J is equal in magnitude and opposite in sign to that of state K since

ω_{KJ} is defined by: $\omega_{KJ} = \omega_K - \omega_J$ or the difference in excitation frequencies of states K and J. As in the case of CD, sign changes are often helpful in resolving overlapping or close-lying unstructured bands. Also as already specified above, the observation of an A-term unambiguously characterizes a degenerate excited state and its analysis gives a measurement of the magnetic dipole moment of the state. The study of B-terms gives information regarding interactions between states.

As already mentioned above, a recent review of CD spectroscopy in the vacuum ultraviolet is in press (Johnson, 1978). This author divides the molecules studied into two broad groups, viz. "Small Molecules" and "Biological Molecules". The latter grouping is subdivided into proteins, nucleic acids and sugars. The present paper will be limited to small molecules and chromophores and for CD of large molecules the reader is referred to Johnson's review article. Also, the present paper will place emphasis on CD and MCD results not contained in a previous summary (Schnepp, 1973).

II. INSTRUMENTATION AND TECHNIQUES

The very first instrument used to successfully extend the range of CD measurements into the vacuum ultraviolet was constructed by Feinleib and Bovey (1968). However, only the spectrum of 3-methyl cyclopentanone was reported by these authors, and the principle underlying the instrument was such as to limit its applicability.

The first effective vacuum ultraviolet CD instrument was built in our laboratory (Schnepp, Allen and Pearson, 1970) and a schematic diagram of the layout of the most recent modification (Gross and Schnepp, 1977) is given in Fig.1. The monochromator is a 1m model

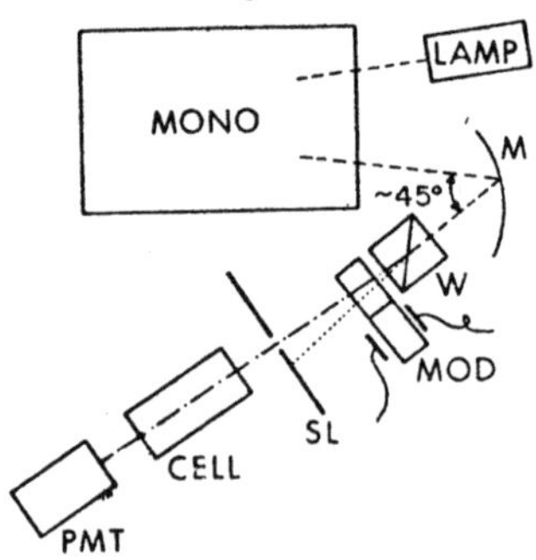

Fig. 1. Schematic of optical system of the circular dichroism instrument. MONO-monochromator, M-Mirror, W-Wollaston, MOD-MODULATOR, SL-separating slit; PMT-photomultiplier tube.

225 McPherson vacuum monochromator and the source is a Hinteregger hydrogen arc modified to give greater power output; typically it is operated at 0.6A with a voltage drop of 2.5KV. A concave mirror M collimates the monochromatic light beam in reflection, thus giving rise to a folded path configuration. A magnesium

fluoride Wollaston prism produces two perpendicularly polarized beams with an angular separation of 1.4°. The selected plane polarized component is converted to circular polarization by a quarter wave retardation element which here is a stress plate modulator made of CaF_2 and driven by a quartz plate. Investigation showed that the polarized components emerging from the MgF_2 Wollaston prism had, for all practical purposes, constant separation at the separating slit SL for all wavelengths out to 135nm. Below about 133nm, the present configuration no longer produces valid CD measurements. The detector is a solar blind photomultiplier tube with an effective cathode diameter of 16mm. This instrument can be modified for MCD measurements by the insertion of a superconducting magnet at the sample cell in Fig. 1. For such measurements, the detector must be adequately shielded and removed from the sample area and magnet by a greater distance to prevent interference.

Following the construction of the CD instrument at the University of Southern California as described, Johnson built another CD instrument for the vacuum ultraviolet at Oregon State University (Johnson, 1971). The principle is similar except that the Wollaston polarizer is placed inside the monochromator and the exit slit serves as the separating slit to select the chosen beam. The stress plate modulator is attached to the exit slit followed by the sample cell compartment and the detector. The advantage of this instrument is compactness but it would not be suitable for MCD measurements. Stevens (formerly Pysh) at Brown University built another instrument similar to that at USC (Pysh, 1976), except that they replaced the Wollaston prism by a biotite reflection polarizer (Robin et al., 1966). Further instruments have been reported by Brahms and co-workers (Brahms, et al., 1977), at the University of Paris, and by Gedanken at Bar-Ilan University in Israel (Gedanken and Levy, 1977). Both these groups also use the biotite polarizer whose advantage lies in the fact that it does not contribute to the wavelength transmission limitation as does the Wollaston prism made of about 1cm thick CaF_2. Mason at Kings College, London (Drake and Mason, 1977) has built an instrument good to about 160nm by flushing the cell compartment with nitrogen, thus eliminating the need for vacuum in this part of the apparatus which greatly facilitates sample handling and interchange. Also, the instrument at Paris has this feature. McGlynn at Louisiana State University, Baton Rouge has built an instrument designed for MCD measurements (Scott et al., 1978) which has a layout very similar to that at USC.

The signal processing electronics centers around a lock-in amplifier to measure the AC signal which appears superimposed on the DC and which is generated due to the difference in transmission of the optically active sample to left and to right circularly polarized light. This modulation is produced by the stress plate modulator which alternates the light polarization between these two modes. The desired quantity, $\Delta\varepsilon = \varepsilon_L - \varepsilon_R$, is obtained to a very good approximation as the ratio of the AC to the DC signal. The sensitivity is of the order of 10^{-5} in this ratio and this is also the ratio of $\Delta\varepsilon/\varepsilon$.

As mentioned earlier, the instrumental configuration with the MgF_2 Wollaston polarizer and using reflection optics for collimation allows measurements to be made to about 135nm. With a biotite reflection polarizer the range can be extended somewhat but it will be limited by the CaF_2 modulator transmission which decreases rapidly beyond 130nm.

Measurements have been made in the gas phase, since most solvents have limited transmission in the vacuum ultraviolet. However, for various reasons it is often desirable to make solution studies and techniques for this have been developed. For example, perfluoro-n-hexane makes measurements possible out to 165nm in a 1mm cell. Also, Johnson has developed a thin cell technique to study aqueous solutions of proteins (Johnson and Tinoco, 1972). Stevens and co-workers on the other hand, has worked with films of β-forming oligopeptides (Baleerski et al., 1976; Toniolo et al., 1976).

III. NATURAL CIRCULAR DICHROISM RESULTS

1. Ethylene Chromophore

A. Trans-Cyclooctene. All ethylene chromophore molecules studied, including straight chain monoolefins but excluding trans-cyclooctene contain four major CD bands in the region of measurement (Mason and Schnepp, 1973; Gross and Schnepp, 1975; Drake and Mason, 1977). For this reason, the CD spectrum of trans-cyclooctene will be taken up first. This molecule is intrinsically optically active due to the twisted configuration of the double bond. The special importance of this CD spectrum lies in the fact that an ab-initio theoretical calculation has recently been reported (Liskow and Segal, 1978) which examined the excited state assignments of this molecule and of its CD spectrum. In Fig.2 the absorption and CD spectra are presented (Mason and Schnepp, 1973). The CD is intense in the N→V region, i.e. the $\pi_x \rightarrow \pi_x^*$ ($1b_{3u} \rightarrow 1b_{2g}$, $A_g \rightarrow B_{1u}$) transition (the symmetry designations for the orbitals and states are for ethylene). In the same spectral region another transition is expected designated variously as $\pi_x \rightarrow \pi_y^*$ or $\pi \rightarrow CH^*$ or $\pi - 3p_t$ ($1b_{3u} \rightarrow 2b_{2u}$, $A_g \rightarrow B_{1g}$) for ethylene and in the twisted configuration this state would be expected to mix strongly with the N→V, $\pi_x \rightarrow \pi_x^*$ transition. The calculation of Liskow and Segal does bear out this expectation in the sense that the rotatory strength of the N→V transition is predicted to be high and the magnitude and sign agree well with the experimental results. However, these authors find that the major component of the state previously designated $\pi_x \rightarrow \pi^*_y$ ($\pi \rightarrow CH^*$) is a 3p Rydberg orbital in agreement with previous theoretical results (Buenker et al., 1971). This latter state does not contribute significantly to the rotation according to the recent calculation. The intense negative CD band peaking at 156.5nm is again confirmed to represent the transition $\sigma \rightarrow \pi^*$ and again sign and magnitude of the rotatory strength agree well with experiment. The same assignment has been made originally (Mason and Schnepp, 1973) based on two theoretical papers (Yaris et al., 1968; Robin et al., 1968). The ab-initio calculation also predicts two Rydberg transitions ($\pi \rightarrow 3s$ and $\pi \rightarrow 3d_\sigma$) with optical rotation in good agreement with the measurements.

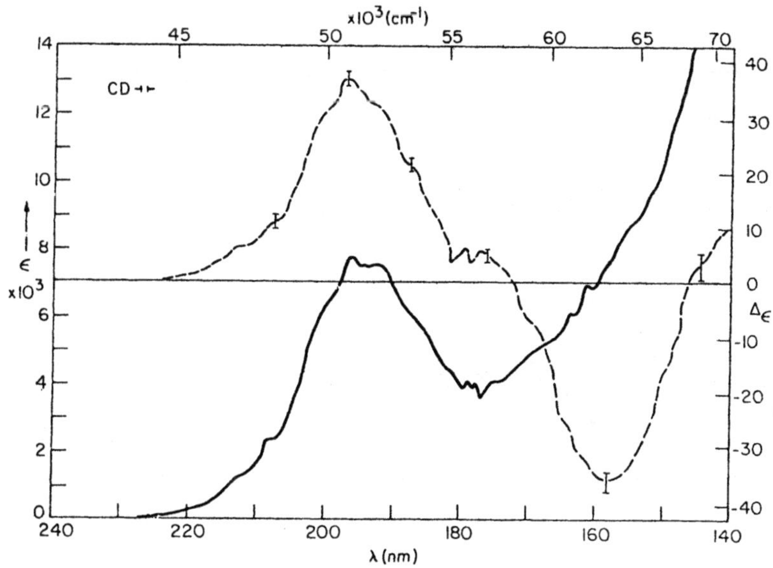

Fig. 2. Absorption and CD spectra of trans-cyclooctene in the gas phase. CD resolution is 0.8nm.

B. Cyclic Monoolefins. Although described previously, we shall briefly recall α-pinene whose absorption and CD spectra are presented in Fig. 3 (Mason and Schnepp, 1973). The CD is typical of the ethylene chromophore in having four major well differentiated bands as already mentioned. In order of increasing frequency, the first is assigned as $\pi \rightarrow 3s$ Rydberg and the second negative CD band is $\pi_x \rightarrow \pi^*_x$, N→V. The third large positive CD band occurs in a region of low absorption and is therefore characterized as magnetic dipole allowed in the symmetric parent chromophore (large g value). It is assigned as $\pi_x \rightarrow p_y$ which, according to all theoretical calculations has an upper orbital that is largely of 3p character. The position of this transition is not in agreement with theory for ethylene (Buenker et al., 1971), which puts it much closer to the $\pi_x \rightarrow \pi_x^*$ valence transition, and even below it in energy. However, Liskow and Segal have shown that on methyl substitution the excitation energy of the $\pi_x \rightarrow \pi_x^*$ $^1B_{1u}$ state is decreased by more than that of the $\pi_x \rightarrow 3p_y$ $^1B_{1g}$ state. This effect results in a different order for these states in trans-2-butene than that in ethylene. As a result, the theory and experiment can be reconciled and the assignment confirmed. The fourth and negative CD band which does not have a distinct peak in α-pinene is again assigned to $\sigma \rightarrow \pi^*$. The narrow positive CD band at 169nm was previously assigned as $\pi_x \rightarrow 4s$ but may well be $\pi \rightarrow 3d_\sigma$ as the theoretical work for trans-cyclooctene indicates.

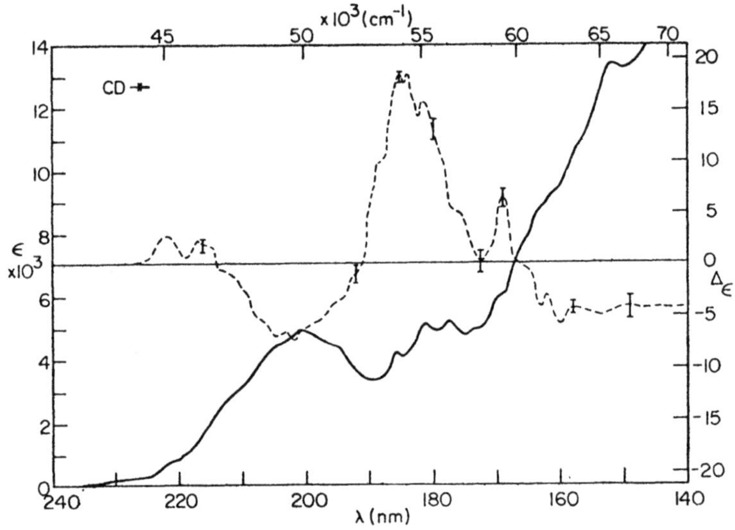

Fig. 3. Absorption and CD spectra of α-pinene in the gas phase. CD resolution is 0.4nm.

The nature of the intense α-pinene CD band at 184nm has been further investigated by measuring the CD spectrum in perfluorohexane solution (Gross, 1976). This spectrum is presented in Fig. 4 and shows that this band does not disappear in the liquid phase but it is shifted to higher energy by $3000 cm^{-1}$, indicating that it has, indeed, considerable Rydberg character.

The nature of the intense α-pinene CD band at 184nm has been further investigated by measuring the CD spectrum in perfluoro hexane solution (Gross, 1976). This spectrum is presented in Fig. 4 and shows that this band does not disapper in the liquid phase but it is shifted to higher energy by $3000cm^{-1}$, indicating that it has, indeed, considerable Rydberg character.

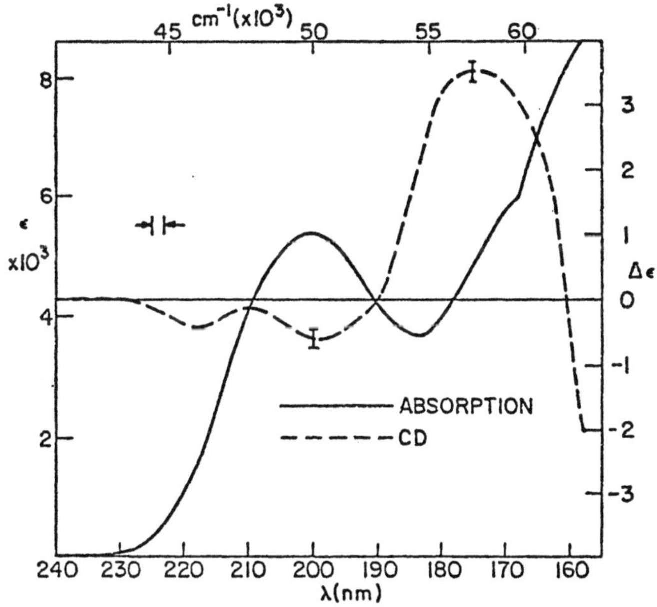

Fig. 4. Absorption and CD spectra of α-pinene in perfluoro hexane solution.

Drake and Mason (1977, 1978) and Drake (1976) have studied the absorption and CD spectra of 17 optically active olefins in the vacuum ultraviolet to about 165nm, in the gas and solution phases. These authors have also made measurements in cooled solutions down to $-190^{\circ}C$, and demonstrated the expected blue shift for the lowest frequency ethylenic transition which had been previously assigned by these authors as $\pi \rightarrow 3s$ Rydberg (Drake and Mason, 1973). This shift is increasingly dramatic as the internal pressure of the solvent is increased on cooling. Fig. 5 shows this effect for Δ^5-cholestene where the positive CD of this transition lies eventually in the region of the $\pi \rightarrow \pi^*$ transition at the lowest temperatures used and has the negative CD of the $N \rightarrow V$ transition on both sides of it.

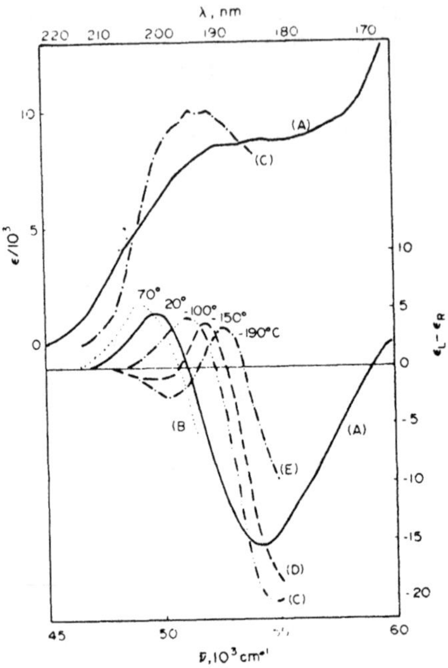

Fig. 5. Absorption and CD spectra (lower curves) of Δ^5-cholestene, (A) n-pentane at +20°, (B) in iso-octane at +70° and 3-methyl pentane, (C) at -100°C, (D) at -150°C and (E) at -190°C.

Drake and Mason (1977) have discussed at length the origin and interrelation of the CD of the $\pi \rightarrow \pi^*$ and the $\pi \rightarrow p_y$ transitions. They find that in most cases the CD bands have opposite signs and propose that this pair represents a "couplet" or a pair of states whose interaction usually gives rise to equal magnitude CD bands of opposite sign. This description looks attractive for many systems but there are exceptions; asymmetry is observed for example in α-pinene and in S, 3-methyl-1-pentene (Fig.6) the CD bands have the same sign (Gross and Schnepp, 1975).

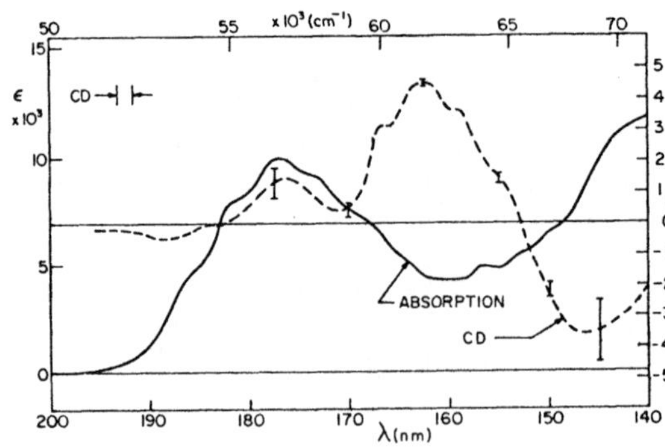

Fig. 6. Absorption and CD spectra of S, 3-methyl-1-pentene, in the gas phase.

C. Straight Chain Mono-Olefins. Gross and Schnepp (1975) have studied the CD spectra of three simple straight chain mono-olefins, viz. S, 3-methyl-1-pentene, (Fig.6), S, 4-methyl-1-hexene and S, 5-methyl-1-heptene in the region 200-140nm in the gas phase. Again the bands characteristic of the ethylene chromophore have been observed. In S, 5-methyl-1-heptene, however, there is no prominent CD peak which coincides with the $\pi\rightarrow\pi^*$ absorption maximum, but only a shoulder appears on the π-p_y band. The cause for this may lie in the fact that these two bands have again the same sign or we may be dealing with the superposition of two or more conformations. It is also suggested that the transition $\pi\rightarrow\sigma^*$ (CC) contributes to the long wavelength CD bands observed in all three molecules studied.

2. Butadiene Chromophore

The cis- and trans-butadiene chromophores α- and β-phellandrene have been studied by Gross and Schnepp (1978). Absorption and CD spectra have been recorded of the gas and solution phases in the spectral region 300-135nm for the former and 300-160nm for the latter. Overlapping valence and Rydberg states were unambiguously differentiated and characterized by comparison of the spectra in the two phases. The principal valence transition of β-phellandrene (trans-butadiene) $\pi_2\rightarrow\pi_3^*$('Ag$\rightarrow$1'Bu) at 227.5nm exhibits an absorption and a CD peak, both of which are also observed in the solution. A weak valence transition at about 195nm also persists as a weak CD shoulder in the liquid phase. The suggested assignment is $\sigma(7a_g)\rightarrow\pi_3^*$ $(2a_u)$, $'A_g\rightarrow{}'A_u$ based on theoretical predictions. The most prominent feature in the CD spectrum of β-phellandrene is a broad, negative band in the region 185-150nm, centered at about 170nm in the gas phase. In the solution phase the CD contour persists to the limit of observation at 160nm. The anisotropy factor, g, is at least ten times larger than that for the 1^1B_u transition, supporting an assignment as a magnetic dipole allowed valence transition in the parent chromophore. The most likely candidate is the transition $\pi\rightarrow\pi^*$, $A_g\rightarrow 3'A_g$, designated $^1A_g^+$ by Schulten et al., (1976), and also predicted in this spectral region by ab-initio calculations (Shih et al., 1972; Buenker et al., 1976). Two Rydberg transitions, one electric dipole allowed and the other one magnetic dipole allowed were also characterized at 205-190nm ($\pi_3\rightarrow$3p) and 180-160nm ($\pi_2\rightarrow$3d), respectively. No evidence was found for the low lying $2^1A_g^-$ state but because of the (-) nature of this state its magnetic dipole transition moment from the ground state may also be expected to be weak, thus precluding its observation in the CD spectrum. Therefore, also, the present results do not give evidence against its existence.

3. Ether Chromophore

Three straight chain ethers have been studied by Sharman (1976). The CD and absorption spectra of (+)-S-sec-Butyl Ethyl Ether in the gas phase are presented in Fig. 7. The CD bands correspond to the three or perhaps four distinguishable absorption regions between 195nm and 140nm. The spectra of the other two ethers investigated (+)-S-2-Methylbutyl Ethyl ether and (+)-S-3-Methylpentyl Ethyl Ether, have similar spectra except that the second CD band is negative in both. The lowest frequency absorption band can be assigned as $n \rightarrow \sigma^*_{CO}$, $^1A_1 \rightarrow {}^1B_1$, in C_{2v} symmetry, in agreement with previous assignments (Tsubomura et al., 1964), or in accordance with Robin (1975) who has assigned this transition as pure Rydberg $n \rightarrow 3s$, giving the same symmetry as above. Actually, the CD spectrum gives an anisotropy factor which is quite high for the first ether ($g = 7.3 \times 10^{-3}$) but smaller for the other ethers ($g = 0.11 \times 10^{-3}$ and 1.6×10^{-3}). Such g-factors seem appropriate for a state of B_1 symmetry for which both the electric dipole and the magnetic dipole transition are allowed. However, it is not possible to be more definitive since no solution spectra have been measured to distinguish between valence and Rydberg transitions. In any case, mixing between the two states is probable. The higher energy transitions have at least a six times lower g-value for the first ether but for the other two compounds the g value is about the same as for the first transition. It is therefore not possible to obtain guidance for the assignments which remain uncertain.

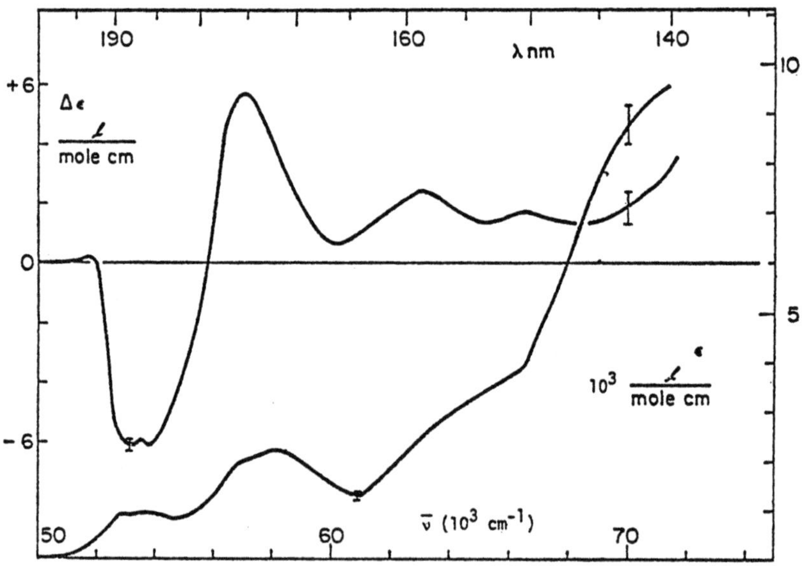

Fig. 7. Absorption and CD spectra of (+)-S-sec-Butyl Ethyl Ether.

4. Alcohol Chromophore

Snyder and Johnson (1973, 1978) have investigated the CD spectra of (+)-2-butanol and 1-borneol out to 135nm. They assigned the first two transitions as $n \to \sigma^*_{OH}$ and $n \to \sigma^*_{CO}$, and excluded the assignment as $n \to 3s$ by using theoretical calculations by the "independent systems" method.

5. Cyclopropane Chromophore

Gedanken and Schnepp (1976a) have studied the CD and absorption spectra of trans 1, 2-dimethyl cyclopropane in the region 210-140nm. The observation of a low frequency structured absorption without a corresponding CD band followed by an intense CD band without corresponding absorption maximum, made possible the ordering of the states expected from the first excited valence state configuration. The lowest excited state is then A_1' (electric and magnetic dipole forbidden) and the second state is assigned as A_2' (electric dipole forbidden and magnetic dipole allowed).

6. Benzene Chromophore

For completeness, the benzene chromophore should be mentioned. However, the work of Allen and Schnepp (1973, 1974) has already been reviewed earlier in some detail (Schnepp, 1973), and therefore we make only brief mention of it here. These authors studied 1-Methylindan and Sec Butyl-benzene and observed transitions in addition to the usual benzene valence transitions. They assigned the lower energy state at 210nm as E_{1g}, $\pi \to \sigma^*$ but it is now believed to be more likely a Rydberg transition of character $\pi \to 3s$, in accordance with theory (Buenker et al., 1968; Peyerimhoff and Buenker, 1970; Hay and Shavitt, 1974). Also this state has recently been observed by multiphoton ionization in benzene (Johnson, 1976). In addition, Allen and Schnepp (1973, 1974) observed a state assigned as the valence state of symmetry E_{2g}.

7. Miscellaneous Compounds

The compound 3-methyl cyclopentanone was used by several groups as criterion for CD resolution capability while developing the vacuum ultraviolet instrument. As a result, a number of CD spectra are in the literature (Feinleib and Bovey, 1968; Schnepp et al., 1970; Johnson, 1971) but they have not been interpreted in detail, although some work has been done on the spectrum (Sharman, 1976).

The CD spectra of 3-methyl cyclopentene and d-camphor have been reported by Gedanken and Levy (1977).

IV. MAGNETIC CIRCULAR DICHROISM RESULTS

1. Acetylene

The MCD spectrum of acetylene gas near 150nm was studied by Gedanken and Schnepp (1976b) and the MCD absorption spectra are presented in Fig. 8. The transition is known to be the first

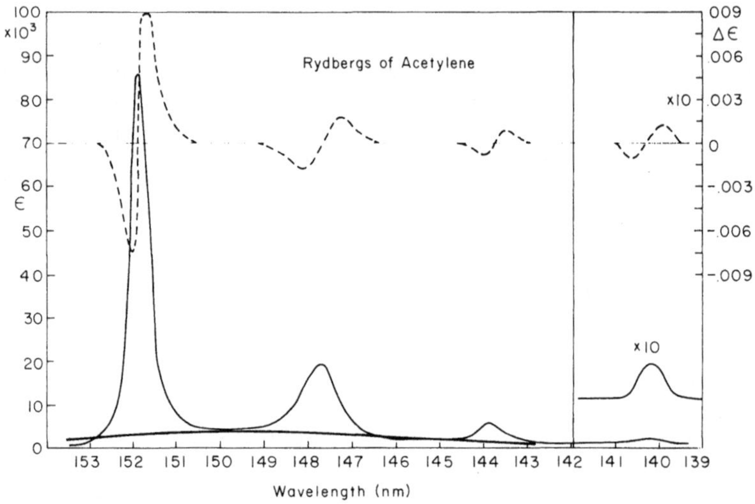

Fig.8. Absorption and MCD spectra of acetylene gas, 154-139nm. (Δ ε per gauss).

member in the Rydberg series $\pi \rightarrow ns$, with n=3. As such the excited state symmetry would be $^1\Pi_u$ for the linear molecular configuration. The vibrational series is due to the C-C stretch and the spacing is $1849cm^{-1}$. The transition is diffuse in absorption and therefore no detailed rotational analysis has been possible. The MCD spectrum exhibits a clear series of completely resolved A-terms which unambiguously characterize the transition as degenerate. Since such a degenerate state can only exist in the linear conformation, it is also determined that the molecule is linear in this state. Furthermore, the magnetic moment of the Π state being proportional to the angular momentum, is known since the angular momentum is here a quantum number of value unity. As a result, the method for determining the magnetic moment or $\mathcal{A}/D$ by moment analysis of the spectrum, as described in the introduction, can be tested. The expected value of $\mathcal{A}/D$ is -0.50 in units of bohr magnetons, and the value obtained by moment analysis of the 0-0 band is -0.51 ± 0.10 with somewhat lower values for the other vibronic bands. However, it must be pointed out that a background absorption continuum had to be corrected for to obtain these values. Without correction, the $\mathcal{A}/D$ value for the 0-0 band was -0.42 ± 0.08 but appreciably lower for the less intense vibronic bands. The subtracted absorption background is indicated in the figure. Its origin is not known and only conjectures are given.

2. Allene

The MCD spectrum of allene was studied by Fuke and Schnepp (1978) in the gas phase and in solution. Valence and Rydberg states have been differentiated by comparison of the spectra of the two phases. In the low frequency region four valence states are expected from the lowest excited configuration, of which only one is electric dipole allowed, and which is assumed to correspond to the intense absorption near 170nm. In agreement with the assignment, this absorption persists in the solution. The MCD spectrum in the low frequency region is weak which also is expected since the four valence states are not coupled by the magnetic field on grounds of symmetry. However, an intense negative B-term is observed in the 193-180nm region which is only present in the gas phase, supporting its assignment as corresponding to the lowest Rydberg transition. Robin (1975) has, in fact, made the assignment as $\pi 2e \rightarrow 3s$. The MCD band is a clear B-term, as already mentioned and the A-term arising from the degeneracy of this state is calculated to be of considerable intensity. Therefore, the MCD results require the conclusion that the degeneracy is removed by some mechanism. This result is in agreement with photoelectron spectroscopic results which indicate a Jahn-Teller type mechanism which causes a splitting of the ^2E ion core by about 4900cm^{-1}, resulting from a coupling with a torsional vibration. The second component of the $\pi 2e \rightarrow 3s$ state can then be expected at about 170nm, also exhibiting a B-term. In fact, there is discrete structure in absorption superimposed on the broad valence band in this region and the MCD spectrum also contains several narrow bands which resemble asymmetric A-terms. However, the correspondence of absorption and MCD peaks indicate B-terms which are interpreted to represent the second component of the $\pi 2e \rightarrow 3s$ transition and a component of the $\pi 2e \rightarrow 3p$ Rydberg transition. The sharp MCD structure disappears in solution.

3. Methyl Halides and Carbon Tetrahalides

Gedanken and Rowe (1975) have studied the MCD spectra of the methyl halides (CH_3I, CH_3Br and CH_3Cl) in the region of the $n \rightarrow \sigma^*$ continuum. The absorption and MCD spectra of CH_3I are shown in Fig. 9. Obviously, the absorption does not contain evidence for more than one transition, whereas the MCD spectrum contains four lobes of alternating signs. In the MCD of CH_3Br and CH_3Cl

only two peaks of opposite sign are observed and these are concluded to correspond to the higher energy bands of CH_3I. Using theory to calculate the magnitude and sign of the expected A-term for a $^1Q(^1\Pi)$ state, this is ruled out and the three lower energy lobes are assigned as B-terms. These represent the three components of the Q-complex whose optically allowed states are $^3Q_1$, $^3Q_0$ and ^1Q. The highest energy MCD lobe of CH_3I is interpreted as part of the underlying A-term of the ^1Q state. This work is a good example of the power of MCD spectroscopy in decomposing structureless overlapping continua. As a result of these measurements the absorption of CH_3I could be resolved into its component transitions.

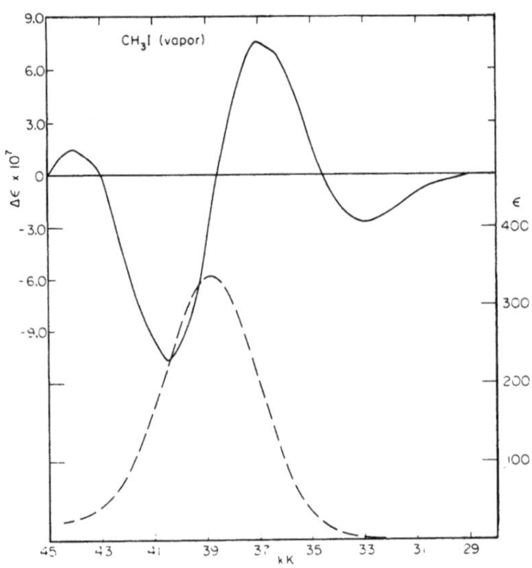

Fig. 9. Absorption and MCD spectra of CH_3I in the gas phase (Δ ε per gauss).

Scott et al.,(1978) have reported the MCD spectrum of CH_3I in the region of the 5p→6s Rydberg absorption bands. Series of A-terms and some B-terms were observed. Vibronic bands belonging to the same electronic state could be identified by the measured magnetic moment from the A-terms and origin assignments of the states were confirmed. Also, vibronically induced transitions were studied and their symmetry assignments confirmed. Four electronic transitions were studied in this manner, which correspond to the expected components of the 5p→6s configuration.

Rowe and Gedanken (1975) have reported the MCD spectra of the carbon tetrahalides CI_4, CBr_4, CCL_4 in the region of the $n_x \rightarrow \sigma^*$ valence transitions where X=Cl, Br, I. Magnetic moments were measured for four states and the MCD spectrum was correspondingly decomposed into four A-terms, assuming gaussian line shapes. Calculations were carried out to determine the magnetic moments theoretically and consistent assignments of the spectra were achieved for the three molecules, account being taken of the changes of spin-orbit coupling in the series.

4. Cyclopropane

Gedanken and Schnepp (1976a) studied the MCD of cyclopropane which is presented in Fig. 10. As is seen in the figure, the

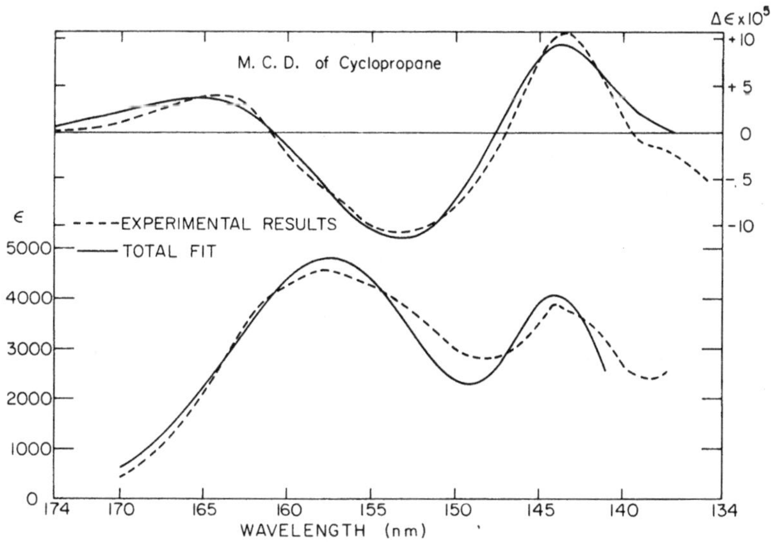

Fig. 10. Absorption and MCD spectra of cyclopropane (Δ ε per gauss).

experimental MCD spectrum (broken line) was reproduced by a curve fit assuming two transitions corresponding to the absorption peaks. These absorptions have previously been assigned to allowed in-plane transitions to degenerate E' states. When results of calculations of the magnetic moments based on simple LCAO wavefunctions were compared with experimentally determined values, it became clear that configuration interaction is important in this molecule and the states cannot be represented by the simple LCAO MO model. It should be pointed out that the absorption spectrum is relatively poor in structre and as a result is only poorly understood.

Most recently, Goldstein and Segal (1978) have carried out an ab-initio calculation of the electronic states of cyclopropane and have included the matrix elements necessary for the evaluation of the magnetic moments and A-terms for the degenerate states and of several B-terms. As a result, it has been possible to match the experimental MCD spectrum and, using it as the major test for the theory, to assign the excited states.

5. Benzene and Derivatives

Kaito et al. (1974) have reported the MCD spectra of symmetric trisubstituted and hexasubstituted chloro- and bromo-benzenes in solution in the $^1E_{1u}$ state region near 200nm. These authors observed A-terms and measured $\mathcal{A}/D$ values in the range -0.2 to -0.3β (bohr magnetons). Allen et al. (1975) measured the MCD of gas phase benzene in order to determine the magnetic moment corresponding to the $^1E_{1u}$ absorption at 180nm. However, a complex spectrum was observed which could not be interpreted. Presumably, the complexity is due to the superposition of Rydberg transitions on the intense valence band. These authors also measured the gas phase MCD spectrum of toluene where it is known that the Rydberg states are appreciably broadened due to the presence of the methyl group. A clear A-term with some vibrational structure was observed which gave value for $\mathcal{A}/D$ of -0.088β. This value was accepted as valid for the magnetic moment of benzene since the splitting of the benzene $^1E_{1u}$ state in the presence of the methyl group substituent is clearly much smaller than the band width and therefore the moment analysis is assumed to be valid. An LCAO calculation for $\mathcal{A}/D$ yielded the value -0.14β, in satisfactory agreement.

More recently, Fuke and Schnepp (1978) have carried out measurements of the MCD spectra of benzene in solution (in perfluorohexane) and of a number of methyl substituted benzenes in the gas phase or in solution. This work gave the surprising result that, whereas $\mathcal{A}/D$ for benzene in solution has the value -0.21β, that for mesitylene (symmetric trimethylbenzene) is only one quarter of the benzene value and hexamethyl benzene has one-half the benzene value. The solution value of toluene is -0.11β. Such large variations of the magnetic moment with methyl substitution are surprising and indicate strong interactions between the $^1E_{1u}$ state with other states. However, the magnetic moment of the $^1E_{1u}$ state represents the difference of the contributions of two orbitals, e_{1g} and e_{2u}. As a result, changes in the orbital contributions of the order of 5% are sufficient to account for the observed variations.

REFERENCES

Allen, S.D. and O. Schnepp (1973). J. Chem. Phys. 59, 4547.
Allen, S.D. and O. Schnepp (1974). Chem Phys. Lett. 29, 210.
Allen, S.D., M.G. Mason, O. Schnepp and P.J. Stephens (1975). Chem Phys. Lett. 30, 140.
Balcerski, J.S., E.S. Pysh, G.M. Bonora and C. Toniolo (1976). J. Am. Chem. Soc. 98, 3470.
Brahms, S., J. Brahms, G. Spach and A. Brack (1977). Proc. Nat. Acad. Sci. USA, 74, 3208.
Buenker, R.J., J.L. Whitten and J.D. Petke (1968). J. Chem. Phys., 49, 2261.
Buenker, R.J., S.D. Peyerimhoff and W.E. Kammer (1971). J. Chem. Phys. 55, 814
Buenker, R.J., S. Shih and S.D. Peyerimhoff (1976). Chem. Phys. Lett. 44, 385(1976).
Drake, A.F. (1976). Chem. Comm., 515
Drake, A.F. and S.F. Mason (1973). Chem. Comm., 253
Drake, A.F. and S.F. Mason (1977). Tetrahedron, 33,937.
Drake, A.F. and S.F. Mason (1978). J. de Phys., in press.
Feinleib, S. and F.A. Bovey (1968). Chem. Comm. 978.
Fuke, K. and O. Schnepp (1978a). Unpublished.
Fuke, K. and O. Schnepp (1978b). Unpublished.
Gedanken, A. and M. Levy (1977). Rev. Sci. Instrum. 48, 104
Gedanken, A. and M.D. Rowe (1975). Chem. Phys. Lett. 34, 39.
Gedanken, A. and O. Schnepp (1976a) Chem. Phys. 12, 341
Gedanken, A. and O. Schnepp (1976b). Chem. Phys. Lett. 37, 373.
Goldstein, E. and G.A. Segal (1978). Unpublished
Gross, K.P. (1976). Ph.D. Thesis, University of Southern California, Unpublished.
Gross, K.P. and O. Schnepp (1975). Chem. Phys. Lett., 36, 531.
Gross, K.P. and O. Schnepp (1977). Rev. Sci. Instrum., 48, 362.
Gross, K.P. and O. Schnepp (1978). J. Chem. Phys. 68, 2647
Hay, P.J. and I. Shavitt (1974). J. Chem. Phys. 60, 2865
Johnson, P.M. (1976). J. Chem. Phys. 64, 4143.
Johnson, W.C.,Jr. (1971). Rev. Sci. Instrum. 42, 1283
Johnson, W.C., Jr. (1978), Ann. Rev. Phys. Chem., 29.
Johnson, W.C., Jr. and I. Tinoco, Jr. (1972). J. Am. Chem. Soc. 94, 4389.
Kaito, A., A. Tajiri and M. Hatano (1974). Chem. Phys. Lett. 25, 548.
Liskow, D.H. and G.E. Segal (1978). J. Am. Chem. Soc. 100, 2945
Mason, M.G. and O. Schnepp (1973). J. Chem. Phys. 59, 1092.
Mason, S.F. (1962). Mol. Phys. 5, 343; J. Chem. Soc. 3285.
Peyerimhoff, S.D. and R.J. Buenker (1970). Theor. Chim. Acta (Berl.), 19, 1.
Pysh, E.S. (1976). Ann. Rev. Biophys. Bioenger. 5,63.
Robin, M.B. (1975). Higher Excited States of Polyatomic Molecules, Academic Press, New York.

Robin, M.B., N.A. Kuebler and Y.H. Pao (1966). Rev. Sci. Instrum. 37, 922.
Robin, M.B., H. Basch, N.A. Kuebler, B.E. Kaplan and J. Meinwald (1968). J. Chem. Phys. 48, 5037.
Rowe, M.D. and A. Gedanken (1975). Chem. Phys. 10, 1.
Schnepp, O. (1973). Chapter in Chemical Spectroscopy and Photochemistry in the Vacuum Ultraviolet, Eds. C. Sandorfy, P.J. Ausloos and M. Robin, D. Reidel, (NATO Advanced Study Institutes Series, Series C, Vol. 8, p. 211).
Schnepp, O., S. Allen and E.F. Pearson (1970). Rev. Sci. Instrum. 41, 1136.
Schulten, K., I. Ohmine and M. Karplus (1976). J. Chem. Phys. 64,4422.
Scott, J.D., W.S. Felps and S.P. McGlynn (1978a). Nucl. Instr. and Methods, 152, 231.
Scott, J.D., W.S. Felps, G.L. Findley and S.P. McGlynn (1978b). J. Chem. Phys. 68, 4678
Sharman, E.H. (1976). Ph.D. Thesis, University of Southern California. Unpublished.
Shih, S., R.J. Buenker and S.D. Peyerimhoff, (1972). Chem. Phys. Lett. 16,244.
Snyder, P.A. and W.C. Johnson, Jr., (1973). J. Chem. Phys. 59, 2618.
Snyder, P.A. and W.C. Johnson, Jr.(1978). J. Amer. Chem. Soc., 101
Stephens, P.J., R.L. Mowery and P.N. Schatz (1971). J. Chem Phys. 55, 224.
Toniolo, C., G.M. Bonora, M. Palumbo and E.S. Pysh (1976). Peptides, 597.
Tsubomura, H., K. Kimura, K. Kaya, J. Tanaka and S. Nagakura (1964). Bull. Chem. Soc. Japan, 37, 417.
Yaris, M., A. Moscowitz and R.S. Berry (1968). J. Chem. Phys. 49, 3150.

Addendum

S.D. Allen and M. Brith-Lindner, (Chem. Phys. Letters, (1977) 47, 32), in a study of the MCD spectrum of ethylene, find evidence for an additional state in the region of N → V transition and discuss its assignment.

*Work supported by a grant from the National Institutes of Health.

THE OPTICAL ACTIVITY OF LANTHANIDE AND TRANSITION METAL COMPLEXES

F.S. Richardson

Department of Chemistry, University of Virginia
Charlottesville, Virginia, USA

1. INTRODUCTION

The optical activity of chiral transition metal complexes has attracted the attention of coordination chemists and molecular spectroscopists for many years. The optical rotatory properties of coordination compounds have played a prominent and venerable role in the development of inorganic stereochemical analysis and structure elucidation, and experimental and theoretical research in this area has been particularly intense over the past 20 years. The first observation of circular dichroism (CD) in the visible absorption bands of transition metal complexes was reported by Cotton in 1895.[1] Over the period 1911-1919, Werner[2] resolved into optical isomers a wide variety of bis- and tris-chelated complexes containing achiral ligands and transition metal ions from each of the three transition metal series of the periodic table. This work established the octahedral structure of hexa-coordinated complexes and posed the problems of molecular stereochemistry and absolute configuration of metal coordination compounds. The optical rotatory properties of the Werner complexes were the subject of considerable study in the 1930's, most notably by Jaeger,[3] Mathieu,[4] and Kuhn.[5-7] The first purely theoretical examination of the origins of optical activity in chiral transition metal complexes was made by Kuhn and Bein.[5,6] Their treatment of these systems was based on the general coupled-oscillator model of molecular optical activity proposed earlier by Kuhn.[8-10] This model was purely classical and it was used to relate the absolute configuration to the sign and the form of the visible-region Cotton effects observed for tris-chelated complexes.

Stephen F. Mason (ed.), Optical Activity and Chiral Discrimination. 107–160.

The first definitive determination of the absolute configuration of a chiral metal complex was reported by Saito and coworkers[11] in 1955, using the anamalous x-ray scattering method. Saito and coworkers[11] found that the tris (ethylenediamine) cobalt (III) isomer which is dextrorotatory at the sodium D-line, (+)-$[Co(en)_3]^{3+}$, has the Λ-configuration.[12] This finding was contrary to the configurational assignment predicted according to the Kuhn and Bein coupled oscillator model for tris-chelated complexes.[5,6]

Moffitt[13] introduced the first quantum mechanical theory of optical activity in chiral transition metal complexes. He adopted a crystal-field model on which to represent the spectroscopic states of the metal ion d-electrons, and used the "one-electron" theory of molecular optical activity proposed by Condon, Altar, and Eyring[14] to develop expressions for the rotatory strengths of the metal d-d transitions. Moffitt applied his theory specifically to tris-chelated complexes of Co(III) and Cr(III) which have exact trigonal dihedral (D_3) symmetry. An error in sign in the d-d transition matrix of the angular momentum operator led Moffitt to incorrect conclusions regarding the optical rotatory properties of these systems, and Sugano[15] subsequently demonstrated that Moffitt's model could not account for the net optical activity observed for the ${}^1A_{1g} \rightarrow {}^1T_{1g}$ transition in trigonal dihedral complexes of Co(III) and for the ${}^4A_{2g} \rightarrow {}^4T_{2g}$ transition in Cr(III) complexes of trigonal dihedral symmetry. However, the general aspects of Moffitt's model remained intact and his work provided an important stimulus for much of the subsequent theoretical effort in this area.

The availability of improved optical rotatory dispersion (ORD) and circular dichroism (CD) instrumentation in the early 1960's led to a great resurgence in the study of molecular optical activity and in the use of optical rotatory properties as probes of molecular stereochemistry and electronic structure. Biopolymer systems, organic carbonyl compounds, and chiral transition metal complexes received the most attention (experimentally and theoretically) during this time. The rather highly developed state of ligand-field theory, in its applications to sorting out the optical spectra associated with the d-d transitions of transition metal complexes, was especially important to providing a theoretical framework (albeit, approximate) within which d-d optical activity could be interpreted or rationalized. Spectra-structure relationship were, and remain, somewhat less well-defined for chiral transition metal complexes than for the $n \rightarrow \pi^*$ optical activity of carbonyl compounds (for example), but they are sufficiently good to provide valuable working hypotheses and guidelines for researchers in the field.

Developments in the theory, understanding, and applications of the optical activity of chiral transition metal complexes have been enormously aided by the very large quantity of experimental data reported over the past 15 years on a wide variety of metal complex structural types (classified with respect to metal ion, ligand donor atoms, ligand chelation modes, chelate bridging groups and sidechains, and general symmetry characteristics). Furthermore, the dramatic increase in the number of metal complexes whose detailed structural features have been elucidated by single-crystal x-ray diffraction techniques has been important to developing reliable spectra-structure correlations.[16] Furthermore, improvements and refinements in the theoretical descriptions and empirical characterizations of the d-d, ligand-metal charge-transfer, and ligand-ligand spectroscopic states of a wide variety of metal complexes have, over the past 15 years, led to considerable refinements in the theoretical models and interpretive schemes employed in analyzing the chiroptical spectra of optically active metal complexes.

Optically active lanthanide complexes have received considerably less attention than transition metal complexes. The optically active lanthanide complexes which have been studied derive their optical activity entirely from the coordination of chiral ligands. There have been no reports of the preparation and isolation of lanthanide ion complexes comprised of achiral ligands dissymmetrically arrayed about the lanthanide ion. The ε and $\Delta\varepsilon$ values associated with the lanthanide f-f transitions are generally quite small making CD/absorption experiments extremely difficult to perform except on samples with very high concentrations of metal complex. Circularly polarized luminescence (CPL) has proved to be the preferred technique for studying lanthanide ion f-f optical activity in a number of systems.[17,18]

The enormous scope of the recent and current literature on experimental studies of optically active metal complexes precludes our reviewing, or even summarizing, the field here. We choose rather to briefly review and describe the major developments in the theory of optical activity of metal complexes, to discuss the current theories and models as they are related to spectra interpretation and structure elucidation, and to briefly discuss special effects arising, for example, from vibronic interactions, solvent perturbations, and crystalline environments. The principal focus will be on the metal localized d-d (or f-f) transitions of the complexes.

2. MODELS FOR d-d OPTICAL ACTIVITY

2.1. Classical Coupled Oscillator Model

Kuhn and Bein[5,6] proposed the first theoretical model for the optical activity of chiral transition metal complexes. Selecting the tris complexes, $Co(en)_3^{3+}$ and $Co(ox)_3^{3-}$, as model systems, Kuhn and Bein[5,6] postulated that electronic transitions localized on the metal ion gained optical activity by coupling with electric dipole oscillators localized on the three bidentate ligands. More specifically, they represented the optical electron on the metal ion as a 3-dimensional isotropic harmonic oscillator with a characteristic frequency (say, ν_M) and the ligands by three linear oscillators of frequency $\nu_L (\nu_L > \nu_M)$ directed along the edges of the octahedron spanned by the chelate rings. On this model, the spatial arrangement of the three ligand oscillators is dissymmetric (chiral) and their coupled motions give rise to a dissymmetric force field. If the metal ion (3-dimensional) oscillator is, in turn, coupled to the ligand oscillators via this dissymmetric force field, then optical activity will be observed at the frequencies of the perturbed metal oscillator (i.e., in the metal ion absorption bands near ν_M), as well as at the frequencies of the perturbed ligand oscillators (i.e., in the ligand absorption bands near ν_L). In the Kuhn and Bein model, the metal ion oscillator represented linear displacements of the optical electron and the metal-ligand coupling mechanism was assumed, therefore, to be electric dipole-electric dipole. This model was entirely classical in its details and did not take into account the detailed nature of the electronic transitions responsible for the visible absorption bands of the metal complexes. Subsequent characterization of these transitions as essentially parity-forbidden d-d excitations pointed to the necessity of treating the optical electrons as circular, rather than linear, oscillators. In quantum mechanical language, the d-d transitions may be characterized in general as having substantial magnetic dipole character and only very weak electric dipole character. The fundamental representation of the Kuhn and Bein coupled-oscillator model is, therefore, not appropriate for treating the optical activity of the d-d transitions in transition metal complexes.

2.2. Ionic Model

Moffitt[13] introduced the first theory of optical activity in transition metal complexes based on quantum mechanical principles. This theory was introduced in 1956 and it made use of the Rosenfeld expression for electronic rotatory strengths,[19] the "one-electron" theory of molecular optical activity proposed by Condon, Altar, and Eyring,[14] and the crystal-field theory of transition metal ion d-d absorption spectra.[20] Moffitt's model has become known as the "ionic model" owing to its representation of the ligand environment as an array of point charge distributions (in the spirit of the crystal-field model for metal complexes). The central problem was to calculate the rotatory strength, R_{ij}, associated with a specific d-d transition, $i \rightarrow j$, of a transition metal complex. The Rosenfeld expression for rotatory strength is given by

$$R_{ij} = \mathrm{Im} < \psi_i | \hat{\mu} | \psi_j > \cdot < \psi_j | \hat{m} | \psi_i > \quad , \tag{1}$$

where $\hat{\mu}$ and $\hat{m}$ are the electric dipole and magnetic dipole operators, respectively. Considering only hexacoordinate, tris-chelated complexes of exact trigonal dihedral (D_3) symmetry, Moffitt adopted a crystal-field perturbation model in which, to zeroth-order, the d-electrons of the metal ion were subjected to a crystal-field potential of exact octahedral (O_h) symmetry. The magnetic dipole transition moments of eq. (1) were then calculated using the d-electron eigenstates of the zeroth-order (octahedral) crystal-field Hamiltonian. Using first-order perturbation theory, he then introduced the ungerade components of the trigonal dihedral (D_3) crystal-field and permitted mixing of the metal ion d- and p-orbitals. This admixture of metal ion d- and p-orbitals resulted in non-vanishing values for the electric dipole transition moments (calculated to first-order) of eq. (1). The d-d optical activity was assumed to arise, then, from (dissymmetric) crystal-field-induced mixing of the metal ion d- and p-orbitals.

Moffitt applied his model to tris-chelated complexes of Co(III) and Cr(III), and predicted strong optical activity within the $^1A_{1g} \rightarrow {}^1T_{1g}$ transition of Co(III) and within the $^4A_{2g} \rightarrow {}^4T_{2g}$ transition of Cr(III). These transitions are both magnetic dipole allowed under octahedral (O_h) selection rules. The magnetic dipole forbidden transitions, $^1A_{1g} \rightarrow {}^1T_{2g}$ for Co(III) and $^4A_{2g} \rightarrow {}^4T_{1g}$ for Cr(III), were predicted to be optically inactive within the approximations of the Moffitt model. These predictions were in good qualitative agreement with experimental observation since the magnetic dipole allowed transitions (referring to selection rules based on the octahedral parentage of the ground and excited states) do in fact exhibit substantially stronger optical activity than do the magnetic dipole forbidden transitions in the tris-oxalato and tris-ethylenediamine complexes of Cr(III) and Co(III). However, Moffitt's predictions were based on calculations which included an error in sign in the d-d transition matrix of the angular momentum operator (and, consequently, the magnetic dipole operator). Sugano[15] showed that when this error is corrected, Moffitt's model leads to zero net rotatory strength within the $^1A_{1g} \rightarrow {}^1T_{1g}$ and $^4A_{2g} \rightarrow {}^4T_{2g}$ transitions of Co(III) and Cr(III), respectively. Sugano[15] further showed that in order for Moffitt's model to yield a net rotatory strength in these transitions, the crystal-field expansion would have to include at least one ungerade term with $\ell \geq 9$. More specifically, the chiral (D_3) trigonal crystal-field perturbation potential must include at least one term which transforms as the pseudoscalar A_{1u} irreducible representation of the O_h point group. The lowest-order term transforming as A_{1u} in O_h has $\ell = 9$. In his calculations, Moffitt truncated the (D_3) trigonal crystal-field expansion after the leading ungerade term ($\ell = 3$) which transforms as T_{2u} in the O_h point group. Sugano's conclusions were based on the uniquely deduced arguments of group theory and therefore depended on the physical model only insofar as the zeroth-order basis states of the perturbation model were taken to be eigenstates of an octahedral (O_h) crystal-field Hamiltonian. A crystal-field potential term with $\ell = 9$ can mix d-orbitals ($\ell = 2$) only with metal orbitals of aximuthal quantum numbers $\ell = 7, 9$, or 11, a most unrealistic condition. Moffitt's ungerade trigonal potential ($\ell = 3$) can mix the metal d-orbitals with p- and f-orbitals, but to first-order the rotatory strengths associated with the two trigonal (D_3) components of the $^1A_{1g} \rightarrow {}^1T_{1g}$ transition of Co(III) will be equal in magnitude and opposite in sign. The net $^1A_{1g} \rightarrow {}^1T_{1g}$ rotatory strength will, therefore, vanish. The same result is obtained for the $^4A_{2g} \rightarrow {}^4T_{2g}$ transition of Cr(III).

Hamer[21] extended Sugano's analysis[15] of Moffitt's one-electron crystal-field perturbation model to show that for any metal complex assumed to be centrosymmetric to zeroth-order, a first-order net d-d rotatory strength is generated only by a ligand-field potential possessing at least one component which transforms under the pseudoscalar irreducible representation of the point group of the zeroth-order complex.

Despite the initial failure of the one-electron crystal-field model (often referred to as the "ionic" model) to account for the net optical activity observed in the d-d transitions of tris-chelated Cr(III) and Co(III) complexes, the elegance and simplicity of this model inspired numerous additional efforts to adapt its basic physical and symmetry-determined features to the problem at hand.[22-33] Poulet[22] modified Moffitt's treatment of tris-chelated Co(III) and Cr(III) complexes by allowing some trigonal splitting to occur within the triply degenerate octahedral excited states. It was postulated that a gerade component of the D_3 crystal-field potential lifted the degeneracy of the zeroth-order octahedral states and, as in Moffitt's treatment,[13] the lowest-order ($\ell = 3$) ungerade component of the

D_3 potential induced the optical activity. This, of course, resulted in a predicted CD spectrum for the Cr(III) ${}^4A_{2g} \rightarrow {}^4T_{2g}$ transition with two closely spaced components of identical intensities and opposite signs. The net CD intensity was calculated to be zero (as required by Sugano's analysis). Piper and Karipedes[23] extended Moffitt's treatment to include 3d, 4p, and 4f orbitals in the zeroth-order metal orbital basis set for calculations on tris-chelated Co(III) and Cr(III) complexes. Restricting their perturbation treatment to first-order in the $\ell = 3$ component of the D_3 crystal-field, they too found the net d-d rotatory strength to be zero.

The first extension of Moffitt's ionic model which led to a nonvanishing net d-d rotatory strength for tris-chelated transition metal complexes was reported by Sugano and Shinada.[24,34] In their physical model, Sugano and Shinada[24,34] represented the ligand environment by two sets of point-dipoles located on the ligand atoms coordinated directly to the metal ion. One set produced a trigonal (D_3) potential of odd (ungerade) parity (and with $\ell = 3$ transformation properties), and the other set produced a trigonal (D_3) potential of even (gerade) parity (and with $\ell = 2$ transformation properties). The gerade potential is effective in mixing the metal d-orbitals amongst themselves and in lifting the degeneracy of the (D_3) trigonal components of the octahedral (zeroth-order) parent states. The ungerade potential is effective in mixing zeroth-order octahedral states of opposite parities (or in promoting d-p and d-f orbital mixing on the metal ion). The gerade and ungerade perturbation potentials employed by Sugano and Shinada transformed as T_{2g} and T_{2u}, respectively, under the symmetry operations of the O_h point group. The direct product representation spanned by products of the ungerade (T_{2u}) and gerade (T_{2g}) potentials includes in it one component which transforms as the pseudoscalar A_{1u} irreducible representation of the O_h point group. Carrying out their perturbation calculations to second-order in the rotatory strength, Sugano and Shinada found that the net second-order rotatory strength contributions were nonvanishing and that the signs of these contributions were dependent upon the sign of the trigonal splitting parameter (as determined by the gerade perturbation potential) and upon the detailed nature of the ungerade perturbation potential. The dominance of one trigonal field component rotatory strength over the other was predicted to be entirely determined by the sign and magnitude of the trigonal field splitting parameter.

Sugano and Shinada's extension[24,34] of the one-electron crystal-field model to second-order in the rotatory strength quantities was successful in accounting for the net d-d rotatory strengths observed for tris-chelated metal complexes, and it also accounted for the non-zero (but small) optical activity observed in the regions of magnetic dipole forbidden transitions such as the $^1A_{1g} \rightarrow {}^1T_{2g}$ transition of Co(III) complexes and the $^4A_{2g} \rightarrow {}^4T_{1g}$ transition of Cr(III) complexes. The latter feature of Sugano and Shinada's treatment originated with interactions (and mixings) between different d-d excited states under the influence of the gerade trigonal potential. The simultaneous consideration of the lowest-order gerade and ungerade components of the trigonal perturbation potential required only d-d and d-p metal orbital mixings to achieve net optical activity, and avoided the need for including $\ell \geq 7$ metal orbitals in the zeroth-order basis set as was required by the original Moffitt model. The representation of the ligand environment by sets of point-dipoles located on the donor atoms of the ligands provided only a crude physical approximation to the real structural features of the complexes, but it remained within the spirit of applied crystal-field theory.

Richardson and co-workers[26,30,32,33] have carried out perhaps the most extensive theoretical investigations on the one-electron crystal-field model of optical activity in chiral transition metal complexes. These investigations have included studies of a wide variety of structural types, of various sources of dissymmetry in the ligand environment (e.g., configurational, conformational, and vicinal), and of the convergent nature of the perturbation and crystal-field expansions employed in the model. In this work, a pure crystal-field representation of the ligand environment was employed in which all atoms or groups (molecular fragments) of the ligands were treated as point charge distributions. The signs and magnitudes of the (partial) charges assigned to each perturber site in the ligand environment were assumed to be derivable from local bond moments and formal charges (of ionized groups) characteristic of the ligand ground state charge distributions. Each term in the crystal-field perturbation potential expansion was taken as a sum over all ligand perturber sites in the complex. This latter feature of the model permitted a partitioning of effects attributable to configurational dissymmetry (reflecting the distribution of chelate rings about the metal ion), conformational dissymmetry (reflecting conformational features and preferences within chelate rings), the presence of asymmetrically substituted sites in the ligands, and local (dissymmetric) distortions within the metal-donor atom cluster of the complex.

In most (but not all) cases examined by Richardson and co-workers, a centrosymmetric microsymmetry was assumed to zeroth-order for the metal-donor atom chromophoric cluster and rotatory strength expressions were developed to second-order in a chiral perturbation potential. The chiral perturbation potential was chosen such that only the metal d-, p-, and f-orbitals were required in constructing the zeroth-order basis states of the perturbation model. The principal focus of these studies was on how the net CD intensity might be expected to distribute itself among the various d-d transitions of a given metal complex, and on how the sign and intensity observables of a given CD spectrum may be related to specific stereochemical and electronic structural features of a metal complex. The latter problem was addressed through the formulation of detailed "sector" or "regional" rules for metal complex CD spectra.

Sector (or regional) rules in chiroptical spectroscopy are used to correlate the sign and, in some cases, the intensity observables of a CD spectrum with the relative positions or spatial arrangements of the chromophoric and extrachromophoric atoms (or groups) of a chiral molecular system. They are useful, therefore, in deducing stereochemical information concerning optically active systems. In most applications to simple optically active organic compounds, sector rules are applied to just one electronic transition at a time (e.g., the $n \rightarrow \pi^*$ transition of chiral carbonyl compounds in the famous octant rule[35]) and are based on simple pairwise (chromophore-perturber) interaction mechanisms.[36] For transition metal complexes, however, Richardson[26-29] proposed that the most appropriate (and reliable) sector rules be based on the net optical activity associated with all the d-d transitions in a given complex (that is, the algebraic sum of all d-d CD intensities), and that the sector rules must include consideration of three-way interactions involving the metal ion (chromophore) and two different ligand perturber groups. The latter emphasis of three-way interactions (leading to so-called "mixed" sector rules) is a consequence of the second-order nature of the perturbation model employed by Richardson, and the necessity for considering total or net d-d optical activity (rather than the optical activity of individual d-d transitions) arises from the assumed strong coupling (and intensity borrowing) between the relatively closely-spaced d-d transitions of most metal complexes. Detailed accounts of how Richardson's sector rules may be applied to the prediction and interpretation of metal complex CD spectra have been given,[26-29,37] and some use of them has been reported in the experimental literature. The general validity of the sector rules proposed by Richardson rests on the validity of the one-electron crystal-field model from which they were derived. Although the symmetry-determined aspects of this model are straightforward and valid, its completeness and its reliability in modeling the optical activity of transition metal complexes remain in question.

Strickland and Richardson[30] employed the one-electron crystal-field perturbation model in performing detailed calculations on the d-d optical activity of Ni(II) in crystalline $NiSO_4 \cdot 6H_2O$, and Richardson and Hilmes[32] performed similar calculations on the d-d optical activity of Cu(II) in crystalline $ZnSeO_4 \cdot 6H_2O$. The site symmetry at the metal ion in both of these systems in C_2, while the microsymmetry of the MO_6 clusters is very nearly octahedral (O_h). Second-order perturbation theory was used in each case to calculate the d-d rotatory strengths. In the case of

Cu(II) in $ZnSeO_4 \cdot 6H_2O$, vibronic interactions of the pseudo Jahn-Teller type were also taken into account in calculating the d-d CD spectrum.[32] Kato[38] has performed a theoretical analysis of the d-d optical activity observed for a series of divalent transition metal ions doped into single crystals of $ZnSeO_4 \cdot 6H_2O$. Her analysis was based on Sugano and Shinada's[24,34] model for d-d optical activity and included consideration of Cu(II), Ni(II), Co(II), Fe(II), and Mn(II).

The deficiencies of the one-electron crystal-field perturbation model in properly representing the physical aspects of the ligand environment and its interactions with the metal ion are readily apparent and widely recognized. However, the more subtle problems with this model such as dealing with the proper convergence criteria for the perturbation potential expansion and the order to which the perturbation calculation (of rotatory strengths) should be carried are not so widely appreciated. For six-coordinate metal complexes of near octahedral (O_h) microsymmetry, it seems clear that a first-order treatment of the model is inappropriate due to the rather "unphysical" requirement of including metal orbitals of $\ell \geq 7$ angular momentum in the zeroth-order basis set and of using a perturbation potential of extremely short distance dependence (R^{-10}). As shown by Sugano and Shinada[24,34] and by Richardson,[26] a second-order treatment will generate net d-d optical activity when the metal orbital basis set is restricted to d-, p-, and f-orbitals and the perturbation potential is limited to just two low-order terms, a gerade term with $\ell = 2$ and an ungerade term with $\ell = 3$. Richardson[33] has also pointed out that if the model is carried to third-order (in the rotatory strengths) a single ungerade term of $\ell = 3$ in the perturbation potential will lead to net d-d optical activity. Recently, Hilmes and Richardson[33] reported calculations of d-d optical activity in trigonal dihedral (D_3) complexes based on the one-electron crystal-field ("ionic") perturbation model carried to "all-orders" in the leading non-cubic (trigonal) terms of the crystal-field interaction potential. This was essentially a variational calculation in which the metal orbital basis set was restricted to 3d-, 4p-, and 4f-orbitals, and the crystal-field perturbation potential was restricted to the Y_2^0 and $(Y_3^{-3} - Y_3^3)$ trigonally symmetric terms. The "perturbed" wave functions were obtained by diagonalizing the trigonal Hamiltonian matrix and these wave functions were then used to calculate the rotatory strengths associated with trigonally perturbed d-d transitions.

The "ionic" or one-electron crystal-field model of d-d optical activity has had a very strong and useful influence on the interpretation and understanding of the chiroptical spectra associated with dissymmetric transition metal complexes. Whatever the merits, or otherwise, of this very simple model, it has provided a convenient focus from which many aspects of the problem can be discussed and has served as a point of departure for more refined theoretical treatments. However, it seems clear at this point that the ionic model has little, if any, quantitative usefulness and that many of the qualitative deductions drawn from it should be considered with some circumspection. It has provided valuable working hypotheses for experimentalists in the field and its formalism has been important in analyzing the symmetry-determined aspects of d-d optical activity in a wide range of metal complex structural types. The shortcomings of the model lie deeper than truncation of the perturbation treatment after first- or second-order and inclusion or exclusion of higher-order terms (large values of ℓ) in the expansion of the chiral parts of the ligand field. Modifications of the basic model to include consideration of an "extended" chromophoric unit comprised of the metal ion and the ligand donor atoms, and the inclusion of ungerade metal-ligand charge-transfer states in the zeroth-order basis set,[39] improve the physical representation of the problem but do not enhance its quantitative utility.

In 1963, Bürer[40] suggested a ligand-field approach to the problem of calculating d-d rotatory strengths in tris-chelated complexes of the type studied by Moffitt[13] and by Piper and Karipedes.[23] In this approach it was presumed that explicit consideration of metal-ligand bonding (or antibonding) would lead to a trigonal potential of a sign opposite to that used by Piper and Karipedes[23] in their crystal-field treatment, and would also yield a trigonal splitting energy with a sign opposite that assumed in the crystal-field calculation. Bürer never reported rotatory strength calculations based on his proposed ligand-field model and this approach has not been pursued further.

2.3. Molecular Orbital Models

The first theoretical studies of optical activity in transition metal complexes to employ a molecular orbital representation of the spectroscopic states were reported by Liehr[41,42] and by Karipedes and Piper.[43] In the latter study, the molecular orbitals were constructed in the LCAO (linear combination of atomic orbitals) approximation, the ligand orbital basis set was restricted to the 2s and 2pσ atomic orbitals on the ligand donor atoms, and only the 3d and 4p orbitals of the metal ion were included. Metal-ligand π-bonding was neglected. Optical activity was generated by chiral distortions within the metal-donor atom cluster of the complex. Dissymmetry in the non-donor atom parts of the ligand environment was presumed to contribute only indirectly to the d-d metal ion optical activity through ligand-induced distortions of the metal-donor atom cluster away from an achiral microsymmetry. Karipedes and Piper[43] applied their molecular orbital model to the tris-ethylenediamine and tris-oxalato complexes of Co(III) and Cr(III), and made correlations between the signs of the various CD bands associated with metal ion d-d transitions and specific structural distortions within the metal-donor atom chromophoric cluster. However, since precise relationships between these distortions in the ML_6 cluster and the distribution and structural features of the chelate rings could not be formulated generally or unambiguously, assignments of absolute configuration based on this model remained tentative. The Karipedes and Piper[43] molecular orbital model leads to the prediction that two complexes (of a given metal ion) with different absolute configurations will exhibit identically signed rotatory strengths (and CD bands) if the chelate rings in the two systems induce similar (chiral) distortions within the ML_6 cluster. This result follows from the neglect of direct interactions between the non-donor atoms of the chelate systems and the chromophoric electrons of the metal ion. The non-donor atoms and groups in the ligand environment merely serve to "mechanically" distort the ML_6 cluster.

Liehr constructed molecular orbital models for the optical activity of six-coordinate complexes of trigonal dihedral symmetry[41] and four-coordinate complexes of digonal dihedral symmetry.[42] The essential feature of Liehr's models is a "significant angle of mis-match" between the directions of maximum charge density for the metal d-orbitals and the donor orbitals of the ligand atoms. In the trigonal dihedral (six-coordinate) systems, the ligand donor atoms were assumed to be situated at the vertices of a regular octahedron but the donor atom σ-orbitals were taken to be "canted" with respect to the metal-donor atom internuclear axis. This angle of cant (denoted α by Liehr[41]) was assumed to reflect the detailed structural features of the chelated rings, and all the dissymmetry in the ligand environment was assumed to be communicated to the metal ion via this deviation of each primary metal-donor atom linkage from axial symmetry (about the M-L internuclear axis). In this treatment the rotatory strengths turn out to be proportional to $\sin\alpha$. Liehr's model for digonal dihedral (four-coordinate) systems depended similarly on distortions within M-L σ-bonds of the complex.

Liehr did not carry out detailed calculations based on his models and the conceptual basis of his work is not easily transformed into working hypotheses which can be tested by experiment. Piper and Karipedes[44] calculated the d-d dipole strengths of the $Cu(en)_3^{3+}$ system using Liehr's bent-bond model and concluded that it underestimates the electric dipole transition integrals by at least an order of magnitude.

In 1973, Strickland and Richardson[45] reported molecular orbital calculations on the d-d rotatory strengths of trigonally distorted six-coordinate complexes of Co(III) and Cr(III) with nitrogen and oxygen donor atoms. The electronic states of spectroscopic interest were constructed from molecular orbitals calculated on a modified Wolfsberg-Helmholz model. Calculations were carried out on ML_6 clusters (M = metal atom, L = donor or ligator atom) in which either the nuclear geometry of the ML_6 system had trigonal dihedral (D_3) symmetry or the donor orbitals of the ligand atoms (L) were trigonally disposed about the metal atom. That is, chirality was introduced into the ML_6 cluster either by a Piper representation (trigonal nuclear geometry)[43] or by a Liehr representation (ligand donor orbitals canted from the M-L axes in an octahedral ML_6 cluster).[41,42] Additional calculations were carried out in which both a trigonally distorted ML_6 cluster and (dissymmetrically) canted donor orbitals were present simultaneously. Trigonal distortion parameters for donor atom displacements (from octahedral geometry) and for donor orbital directions (with respect to the M-L axes) were varied to simulate the various structural features known to occur in a variety of tris-chelated complexes of Co(III) and Cr(III). The atomic orbital basis set used in these molecular orbital calculations consisted of the 4s, 4p, and double-ζ 3d metal orbitals and the 2s and 2p orbitals on each ligand atom (either oxygen or nitrogen). M-L σ- and π- interactions were treated separately in the molecular orbital parameterization scheme.

The results reported by Strickland and Richardson[45] suggest that neither the Piper model[43] nor the Liehr model[41] provides an adequate representation of the source of d-d optical activity in trigonal dihedral metal complexes. However, the more refined and comprehensive molecular orbital model employed by Strickland and Richardson[45] did prove to be useful in deducing qualitative information about the sensitivity of d-d rotatory strengths to various kinds of distortions within the ML_6 cluster of these systems. Furthermore, since the computational parameters of the model were defined to reflect specific structural features of the ligand environment, rather specific spectra-structure correlations could be deduced from the calculated results.

In 1974, Evans, Schreiner, and Hauser[46] reported molecular orbital calculations on the d-d optical activity of the tris-ethylenediamine complexes of Co(III) and Cr(III). The level of approximation and general features of the molecular orbital model employed in this study[46] were similar (but not identical) to those of Strickland and Richardson's model.[45] However, whereas Strickland and Richardson considered only distorted ML_6 clusters, Evans, Schreiner, and Hauser (ESH) included all atoms (M,C,N, and H) in their calculations. ESH used structure parameters derived from the x-ray crystallographic study of D-$Co(en)_3Br_3 \cdot H_2O$ reported by Nakatsu.[47] The only structure variations considered in the ESH study were "lel" $\rightarrow$ "ob" geometry changes within the metal-ethylenediamine chelate rings. ESH did, however, thoroughly investigate the sensitivity of their rotatory strength calculations to various computational features of their molecular orbital model and to various operator representations of the rotatory strength quantity. They also carried out comparative rotatory strength calculations using complete multicenter operator matrices (angular momentum, dipole velocity, and dipole length) and successively approximate ones. The ESH calculations proved successful in relating the absolute configurations of $Co(en)_3^{3+}$ and $Cr(en)_3^{3+}$ to the natural CD observed for the magnetic dipole allowed ${}^1A_{1g} \rightarrow {}^1T_{1g}$ and ${}^4A_{2g} \rightarrow {}^4T_{2g}$ transitions in the Co(III) and Cr(III) complexes, respectively. Furthermore, the calculations correctly accounted for the relative signs of the trigonal components of the rotatory strengths (associated with the magnetic dipole allowed transitions), as well as the net ${}^1A_{1g} \rightarrow {}^1T_{1g}$ and ${}^4A_{2g} \rightarrow {}^4T_{2g}$ rotatory strengths. The model also showed sensitivity of d-d rotatory strength to ligand (chelate ring) conformation.

Richardson and coworkers[48,49] have reported additional "all-atom" molecular orbital calculations of d-d rotatory strengths for a number of four-coordinate and six-coordinate Cu(II) complexes of variable ligand type and coordination geometry. These calculations were based on a modified Wolfsberg-Helmholz molecular orbital model and they included consideration of all atoms in the metal complex system. The rotatory strengths associated with ligand-to-metal charge-transfer transitions and with ligand-ligand transitions were also calculated and reported in these studies.[48,49]

C. Schafer[50] reported a theoretical analysis of optical activity in chiral transition metal complexes based essentially on the angular overlap model[51,52] of metal-ligand interactions. This model focused primarily on chiral distortions within the metal-donor atom cluster of tris-chelated and cis-bis-chelated complexes of d^3 and low-spin d^6 transition metal ions.

2.4. Independent Systems/Perturbation Model

Perhaps the most generally applicable and useful theoretical treatments of d-d optical activity in transition metal complexes are those based on the so-called independent systems/perturbation (ISP) model as developed by Mason and coworkers[31,37,53-55] and by Richardson and coworkers.[27,56] The basic concepts of this model derive primarily from the work of Tinoco[57] (on polymer systems) and of Hohn and Weigang[58] (on small organic molecules). According to this model, the metal complex is partitioned into a chromophoric unit (either the metal ion or the metal-donor atom cluster) and a set of extrachromophoric perturber atoms (or groups) situated in the ligand environment. To zeroth-order in the perturbation representation, the chromophoric and extrachromophoric groups are treated as non-interacting (independent) sub-systems. Interactions between the sub-systems are then treated by first- or second-order perturbation techniques. The interaction potential is generally expressed in terms of electrostatic interactions between non-overlapping charge distributions localized on the various sub-systems, and is commonly represented by a bicentric electrostatic multipolar expansion. This model subsumes the crystal-field representation of metal complexes insofar as it includes all interactions arising from point-charge distributions in the ligand environment interacting with the various multipolar components of metal localized charge distributions and transition densities. This aspect of the ISP model leads to the so-called static-coupling (SC) interaction terms in the final spectroscopic expressions which arise from ligand ground state point-charge interactions with multipolar components of the metal ion (or chromophoric) electronic transition densities. The d-d optical activity generated in the one-electron crystal-field (or "ionic") model described previously can be associated entirely with this static-coupling mechanism.

In addition to the SC interaction terms, the general ISP model also includes so-called dynamic-coupling (DC) interaction terms arising from multipole-multipole interactions between transition densities located on the various sub-systems of the metal complex. The DC terms arise from the correlated (or "complementary") motions of electrons on different sub-systems undergoing (virtual) transitions induced by the electrostatic interaction potential operating between the sub-systems. In most applications of the ISP model reported to date, expansions of the extra-chromophoric (perturber) group charge distributions have been truncated after the (electric) dipolar term and only the multipole (metal)-dipole (ligand) DC terms have been retained. To first-order in perturbation theory, these multipole (metal)-dipole (ligand) DC terms lead to rotatory strength expressions (for the metal ion d-d transitions) which depend upon ligand (perturber) group polarizabilities (and polarizability anisotropies), perturber group positions relative to the metal ion, and multipolar transition matrix elements characteristic of the metal ion chromophoric electrons.

Sector rules based on the DC mechanism for generating d-d optical activity within the ISP model have been worked out and proposed by Mason[31,37] and by Richardson.[27,56] In general, these sector rules are different (qualitatively and quantitatively) from those based on the SC mechanism leading to considerable ambiguity in the use of sector rules for sorting out spectra-structure correlations in metal complex CD spectra. Calculations and spectral interpretations based solely on the DC aspects of the ISP model have enjoyed considerable success as reported in several recent studies,[53-56] and it would appear that the DC mechanism is perhaps dominant over the SC mechanism in complexes possessing no charged groups.

Although the ISP model neglects all exchange interactions between the chromophoric and extrachromophoric sub-systems of the complex, this neglect is not so serious when the chromophoric sub-system is defined to include the metal ion and the ligand donor atoms. However, consideration remains restricted to through-space versus through-bond interactions and thus the physical representation of the model must be considered highly approximate.

Very recently, Schipper[59] has reported a theory of d-d optical activity based on what he calls the associate-induced circular dichroism (AICD) theory. This theory is based essentially on the general ISP model carried to second-order in perturbation theory.[60] The derived rotatory strength expressions include SC, DC, and "mixed" SC/DC contributions, and the theory was applied in a general way to a wide variety of metal complex types.

3. DESCRIPTION OF INDEPENDENT SYSTEMS/PERTURBATION MODEL

3.1. General Aspects

The basic assumption of the independent systems/perturbation model is that the metal complex may be partitioned into a chromophoric sub-system (the metal ion or the metal-donor atom cluster) and a set of extrachromophoric sub-systems. To zeroth-order the spectroscopic properties of the chromophoric unit are assumed to be independent of the remainder of the complex. Interactions between the chromophoric and extrachromophoric sub-systems are then treated by perturbation techniques and the spectroscopic properties of the perturbed chromophore are calculated from the resultant (perturbed) wave functions of the chromophoric sub-system. Pairwise interactions between the chromophoric and extra-chromophoric (perturber) sub-systems are generally expressed in terms of bicentric multipolar expansions representing the electro-static interactions between charge distributions localized on the respective (interacting) sub-systems. Exchange interactions (due to overlap) are neglected. Terms in the interaction poten-tial which influence the energies of the d-d spectroscopic states but which do not contribute to d-d rotatory strengths are generally neglected or are absorbed into the zeroth-order representation of the system.

In our present treatment of the general ISP model, we shall find it convenient to express the total electronic Hamiltonian of the metal complex as,

$$H = H_A + H_B + V_A + V_{AB} \quad , \tag{2}$$

where H_A denotes the electronic Hamiltonian of the achiral chromophoric sub-system (A), H_B denotes the total electronic Hamiltonian of the collection of extrachromophoric sub-systems (B) in the ligand environment, V_A allows for a local (chiral) distortion within the chromophoric sub-system (A), and V_{AB} represents all pairwise A-B interactions within the overall complex. In general, the chromophoric sub-system (A) will be taken as a metal-donor atom cluster of an idealized geometry (with, for example, O_h, D_{4h}, or D_{2h} symmetry). Deviations from this idealized (zeroth-order) geometry of A are taken into account by the interaction term, V_A, in eq. (2). The B sub-systems will be taken as atoms, groups of atoms, or chemical bonds in the ligand environment. The Hamiltonian H_B may be further partitioned as,

$$H_B = \sum_r h_r \tag{3}$$

where r labels individual perturber sub-systems.

Defining $H^\circ = H_A + H_B$ as the zeroth-order Hamiltonian operator in our general ISP model, the zeroth-order basis states of the model may be expressed as product functions of the type:

Ground State

$$|A_oB_o) = |A_o \prod_r b_{ro}) \quad , \tag{4}$$

Excited States (singly-excited)

$$|A_mB_o) = |A_m \prod_r b_{ro}) \quad , \tag{5a}$$

$$|A_oB_{ri}) = |A_ob_{ri} \prod_{s \neq r} b_{so}) \quad , \tag{5b}$$

Excited States (doubly-excited)

$$|A_mB_{ri}) = |A_mb_{ri} \prod_{s \neq r} b_{so}) \quad , \tag{6a}$$

$$|A_oB_{ri,sj}) = |A_ob_{ri}b_{sj} \prod_{t \neq r,s} b_{to}) \quad , \tag{6b}$$

where (r,s,t) label individual B sub-systems, (i,j) label excited states localized on the B sub-systems, m denotes an excited state of the chromophore (A), and o denotes ground state. We note that $H_A|A_\alpha) = E_\alpha|A_\alpha)$ and $h_r|b_{rj}) = E_{rj}|b_{rj})$. The inclusion of doubly-excited states in the zeroth-order basis set is necessary only if the perturbation expansion of the wave functions is carried to second-order.

Applying non-degenerate perturbation theory to second-order in $H' = V_A + V_{AB}$, we obtain for the $|A_mB_o\rangle$ state:

$$|A_mB_o\rangle = |A_mB_o) + \sum_{\alpha\neq m}\sum_{\beta} \frac{(A_\alpha B_\alpha|H'|A_mB_o)}{E_m-(E_\alpha+E_\beta)}|A_\alpha B_\beta)$$

$$+ \sum_{\alpha'\neq m}\sum_{\beta'}\left[\sum_{\alpha\neq m}\sum_{\beta} \frac{(A_{\alpha'}B_{\beta'}|H'|A_\alpha B_\beta)(A_\alpha B_\beta|H'|A_mB_o)}{(E_m-E_{\alpha'}-E_{\beta'})(E_m-E_\alpha-E_\beta)}\right.$$

$$\left. - \frac{(A_mB_o|H'|A_mB_o)(A_{\alpha'}B_{\beta'}|H'|A_mB_o)}{(E_m-E_{\alpha'}-E_{\beta'})^2}\right]|A_{\alpha'}B_{\beta'}) \ , \qquad (7)$$

where $\alpha(\alpha')$ and $\beta(\beta')$ are used as generic indices for zeroth-order states on the A and B sub-systems, respectively. Rounded kets denote zeroth-order state functions and pointed kets denote perturbed wave functions. An expression similar to (7) may be written for the perturbed ground state, $|A_oB_o\rangle$, of the metal complex. Eq. (7) is appropriate only for non-degenerate systems. However, if we redefine our state indices (such as m,n,i,j,α,β, etc.) to reflect components of degenerate states and further choose zeroth-order degenerate states with components diagonal in H', then we may use eq. (7) with the understanding that the summations are taken over states and (degenerate) state components.

3.2. Electronic Rotatory Strengths (General Expressions)

The quantity of primary interest is the electronic rotatory strength. For the $o\to m$ transition localized on the chromophore (A):

$$R_{om} = \mathrm{Im}\,\langle A_oB_o|\hat{\mu}|A_mB_o\rangle \cdot \langle A_mB_o|\hat{m}|A_oB_o\rangle \qquad (8a)$$

$$= \mathrm{Im}(P_{om}\cdot M_{mo}) \ , \qquad (8b)$$

where P_{om} and M_{mo} denote the electric and magnetic dipole transition moments, respectively. To second-order in perturbation theory, the rotatory strength may be re-expressed as:

$$R_{om} = R_{om}^{(0)} + R_{om}^{(1)} + R_{om}^{(2)} \tag{9}$$

where,

$$R_{om}^{(0)} = \mathrm{Im}(P_{om}^{(0)} \cdot M_{mo}^{(0)}) , \tag{10a}$$

$$R_{om}^{(1)} = \mathrm{Im}\,(P_{om}^{(0)} \cdot M_{mo}^{(1)} + P_{om}^{(1)} \cdot M_{mo}^{(0)}) , \tag{10b}$$

and,

$$R_{om}^{(2)} = \mathrm{Im}\,(P_{om}^{(0)} \cdot M_{mo}^{(2)} + P_{om}^{(1)} \cdot M_{mo}^{(1)} + P_{om}^{(2)} \cdot M_{mo}^{(0)}) . \tag{10c}$$

In eqs. (9), (10a), (10b), and (10c), the superscripts refer to zeroth (0), first (1), and second (2) order contributions to the rotatory strengths and/or transition moments.

Restricting our treatment to systems with centrosymmetric chromophoric units (A), both $P_{om}^{(0)}$ and $R_{om}^{(0)}$ will vanish for all d-d transitions. The appropriate rotatory strength expression is, therefore, given (to second-order) by:

$$R_{om} = R_{om}^{(1)} + R_{om}^{(2)} \tag{11a}$$

$$= \mathrm{Im}(P_{om}^{(1)} \cdot M_{mo}^{(0)}) + \mathrm{Im}(P_{om}^{(1)} \cdot M_{mo}^{(1)} + P_{om}^{(2)} \cdot M_{mo}^{(0)}) . \tag{11b}$$

To calculate the rotatory strength, eq. (11), it is first necessary to evaluate $P_{om}^{(1)}$, $P_{om}^{(2)}$, $M_{mo}^{(0)}$, and $M_{mo}^{(1)}$ using the perturbation model outlined in Section 3.1. This requires explicit consideration of the perturbation Hamiltonian, $H' = V_A + V_{AB}$, as well as specification of the zeroth-order basis states to be included in the perturbation expansion.

3.3. Interaction Hamiltonian

We have defined the interaction (perturbation) Hamiltonian, H', as comprised of two parts: V_A and V_{AB}. The operator V_{AB} represents electrostatic interactions between charge distributions localized in the extra-chromophoric ligand environment (B) and transition densities associated with (virtual) electronic excitations localized on the chromophore (A). The operator V_A represents all chiral components of the metal-ligand interactions within the metal-donor atom chromophoric cluster (A). Assuming non-overlap between the A and B charge distributions and representing V_{AB} in terms of a sum of bicentric multipolar expansions, the V_{AB} interaction potential may be expressed as:

$$V_{AB} = \sum_r \sum_{\ell_A=0}^{\infty} \sum_{\ell_r=0}^{\infty} V_{Ar}(\ell_A, \ell_r) \; , \tag{12}$$

where,

$$V_{Ar}(\ell_A, \ell_r) = \sum_{m_A} \sum_{m_r} T_r(\ell_A, \ell_r; m_A, m_r)\, D(\ell_A; m_A) D_r(\ell_r; m_r) \; . \tag{13}$$

In eqs. (12) and (13), the summation $\sum_r$ is taken over all ligand perturber sites, m_A runs from $-\ell_A$ to $+\ell_A$, and m_r runs from $-\ell_r$ to $+\ell_r$. The quantity $T(\ell_A, \ell_r; m_A, m_r)$ is the (m_A, m_r) component of a $(\ell_A + \ell_r)$-th rank tensor which describes the orientational dependence of the interaction between a 2^{ℓ_A} multipole on A and a 2^{ℓ_r} multipole on perturber r. The multipole components of the charge distributions on A and r are denoted, respectively, by $D(\ell_A; m_A)$ and $D_r(\ell_r; m_r)$. The general form of the tensor operator, $T(\ell_A, \ell_r; m_A, m_r)$, is given by:

$$T(\ell_A, \ell_r; m_A, m_r) = \frac{(-1)^{\ell_r + m_r + m_A}}{R_r^{\ell_A + \ell_r + 1}} \left[\frac{(\ell_A + \ell_r + m_A + m_r)!\,(\ell_A + \ell_r - m_A - m_r)!}{(\ell_A + m_A)!\,(\ell_A - m_A)!\,(\ell_r + m_r)!\,(\ell_r - m_r)!} \right]^{\frac{1}{2}}$$

$$\times \; C_{-m_A - m_r}^{(\ell_A + \ell_r)} (\Theta_r, \Phi_r) \; . \tag{14}$$

The multipole moment operators, $D(\ell_A; m_A)$ and $D_r(\ell_r; m_r)$, may be expressed as:

$$D(\ell_A; m_A) = -\sum_{\varepsilon} e r_{\varepsilon}^{\ell_A} C(\ell_A; m_A)(\theta_{\varepsilon}, \phi_{\varepsilon}) \; , \tag{15}$$

and,

$$D_r(\ell_r; m_r) = \sum_{\delta} e Z_{\delta} r_{\delta}^{\ell_r} C(\ell_r; m_r)(\theta_{\delta}, \phi_{\delta}) \; . \tag{16}$$

The electrons on A are labeled by ε with coordinates $(r_\varepsilon, \theta_\varepsilon, \phi_\varepsilon)$. The charged particles on perturber r are labeled by δ with coordinates $(r_\delta, \theta_\delta, \phi_\delta)$ and charge $Z_\delta e$ (where e is the magnitude of the electron charge). The position of ligand perturber group r with respect to the chromophoric center (A) is defined by the set of coordinates (R_r, Θ_r, Φ_r). The general form of the $C(\ell;m)(\theta,\phi)$ operator is:

$$C(\ell;m)(\theta,\phi) = [4\pi/(2\ell+1)]^{\frac{1}{2}}\, Y_{\ell,m}(\theta,\phi) \ , \qquad (17)$$

where $Y_{\ell,m}(\theta,\phi)$ is a spherical harmonic function of rank ℓ. In eq. (13), $D(\ell_A;m_A)$ and $D_r(\ell_r;m_r)$ are pure electronic operators while $T_r(\ell_A,\ell_r;m_A,m_r)$ is a factor determined entirely by the position of perturber r relative to the chromophoric group A.

In dealing with chiral distortions within the metal-donor atom cluster, we shall assume it possible to express V_A in precisely the same form as V_{AB}. In this case, however, the extra-chromophoric perturber sites r are replaced by ligand donor atoms displaced with respect to the centrosymmetric reference geometry assumed in our zeroth-order representation of the metal complex. This construct makes less clear the detailed nature of the zeroth-order basis states of the chromophoric sub-system (A), but it should preserve the essential symmetry-determined aspects of the general ISP model.

It is convenient now to partition the interaction potential V_{AB} into its so-called static-coupling (SC) and dynamic-coupling (DC) components. We define the static-coupling component of V_{AB} to be:

$$\mathcal{V}_{AB} = \sum_r \sum_{\ell_A} V_{Ar}(\ell_A, o) \ , \qquad (18)$$

or,

$$\mathcal{V}_{AB} = \sum_r \sum_{\ell_A} \sum_{m_A} T_r(\ell_A,o;m_A,o)\, D(\ell_A;m_A)\, D_r(o;o). \qquad (19)$$

The operator $\mathcal{V}_{AB}$ represents the electrostatic interactions between the chromophore (A) multipoles, $D(\ell_A;m_A)$, and the net charges (monopoles), $D_r(o;o)$, of the perturber groups. $\mathcal{V}_{AB}$ thus contains the familiar "crystal-field" potential and is the operator employed in the original one-electron crystal-field models of d-d optical activity. The dynamic-coupling component of V_{AB} is defined by:

$$\mathcal{U}_{AB} = \sum_r \sum_{\ell_A} \sum_{\ell_r \geq 1} V_{Ar}(\ell_A, \ell_r) \qquad (20)$$

In our subsequent treatment, we shall restrict ℓ_r to $\ell_r = 1$ (dipole components) so that:

$$\mathcal{U}_{AB} = \sum_r \sum_{\ell_A} \sum_{m_A} \sum_{m_r} T_r(\ell_A,1;m_A,m_r)\ D(\ell_A;m_A)\ D_r(1;m_r)\ . \tag{21}$$

The total interaction potential, V_{AB}, is given by:

$$V_{AB} = \mathcal{V}_{AB} + \mathcal{U}_{AB}\ . \tag{22}$$

The interaction operator $\mathcal{U}_{AB}$ describes the (coulombically) correlated motions of electrons in the A and B sub-systems; thus the designation "dynamic-coupling".

Expressions similar to (18)-(22) may be written down for the distortion operator V_A. The operator V_A may also be written as a sum of a static-coupling component ($\mathcal{V}_A$) and a dynamic-coupling component($\mathcal{U}_A$).

The interaction operator V_{AB} (as well as $\mathcal{V}_{AB}$ and $\mathcal{U}_{AB}$) must transform as the totally symmetric irreducible representation of the point group describing the full metal complex. Thus the functions,

$$\underline{F} = \sum_{\ell_A} \sum_{m_A} \underline{F}(\ell_A,m_A) = \sum_r \sum_{\ell_A} \sum_{m_A} T_r(\ell_A,o;m_A,o)\ , \tag{23}$$

and,

$$\underline{\bar{F}} = \sum_{\ell_A} \sum_{m_A} \sum_{m_r} \underline{\bar{F}}(\ell_A,m_A,m_r) = \sum_r \sum_{\ell_A} \sum_{m_A} \sum_{m_r} T_r(\ell_A,1;m_A,m_r), \tag{24}$$

must also tranform as the totally symmetric irreducible representation of this point group. Referring back to the Schellman's general symmetry analysis of the one-electron theory of optical activity (carried to first-order in perturbation theory),[36] optical activity can be induced in the electronic transitions of the symmetric (achiral) chromophore (A) only if the perturbation potential contains at least one component which transforms as the pseudoscalar irreducible representation in the point group of A. In the present context, this implies that the function $\underline{F}$ must contain at least one pseudoscalar term (with respect to the symmetry operations contained in the point group of A) in order for the electronic transitions of A to exhibit optical activity to first-order. Precisely the same requirement obtains for $\underline{\bar{F}}$ if the dynamic-coupling model is carried to just first-order.[27,37] More succinctly, the <u>net</u> first-order rotatory strength associated with any electronic transition of the achiral chromophore (A) will vanish unless either $\underline{F}$ or $\bar{\underline{F}}$ contains at least one term which transforms as a pseudoscalar under the symmetry operations of the A point group (which we shall denote by G_A).

Extension of the ISP model to second-order (in the wave functions and rotatory strengths) leads to interaction terms whose symmetry transformation properties may be evaluated from product functions of the type:

$$\underline{F} \times \underline{F}' = [\sum_{\ell_A} \sum_{m_A} \underline{F}(\ell_A, m_A)][\sum_{\ell'_A} \sum_{m'_A} \underline{F}'(\ell'_A, m'_A)] \ , \tag{25a}$$

$$\underline{F} \times \bar{\underline{F}} = [\sum_{\ell_A} \sum_{m_A} \underline{F}(\ell_A, m_A)][\sum_{\ell_A} \sum_{m_A} \sum_{m_r} \bar{\underline{F}}(\ell_A, m_A, m_r)] \ , \tag{25b}$$

$$\bar{\underline{F}} \times \bar{\underline{F}}' = [\sum_{\ell_A} \sum_{m_A} \sum_{m_r} \bar{\underline{F}}(\ell_A, m_A, m_r)][\sum_{\ell_A} \sum_{m'_A} \sum_{m'_r} \bar{\underline{F}}'(\ell'_A, m'_A, m'_r)] \ . \tag{25c}$$

Second-order rotatory strength contributions will result only if the direct product representations generated by $\underline{F} \times \underline{F}'$, $\underline{F} \times \bar{\underline{F}}$, or $\bar{\underline{F}} \times \bar{\underline{F}}'$ include a pseudoscalar representation (defined within the G_A point group).[26,27,37,59] To second-order, then, neither $\underline{F}$ nor $\bar{\underline{F}}$ need transform as a pseudoscalar within the point group of the achiral chromophore. Note that eq. (25a) pertains to the static-coupling mechanism carried to second-order, eq. (25c) pertains to the dynamic-coupling mechanism carried to second-order, and eq. (25b) pertains to a (simultaneous) static-coupling/dynamic-coupling combination mechanism.

Just as the functions $\underline{F}$ and $\bar{\underline{F}}$ reflect the symmetry properties of V_{AB} and U_{AB}, respectively, so do the functions,

$$\underline{f} = \sum_{\ell_A} \sum_{m_A} \underline{f}(\ell_A, m_A) = \sum_{r} \sum_{\ell_A} \sum_{m_A} D(\ell_A; m_A)\ D_r(o;o) \ , \tag{26}$$

$$\bar{\underline{f}} = \sum_{\ell_A} \sum_{m_A} \sum_{m_r} \bar{\underline{f}}(\ell_A, m_A, m_r) = \sum_{r} \sum_{\ell_A} \sum_{m_A} \sum_{m_r} D(\ell_A; m_A)\ D_r(1; m_r) . \tag{27}$$

The symmetry restrictions on $\underline{f}$ and $\bar{\underline{f}}$ (and on products of these functions) in the generation of optical activity in the electronic transitions of A are precisely the same as those discussed above for $\underline{F}$ and $\bar{\underline{F}}$. Whereas $\underline{F}$ and $\bar{\underline{F}}$ are defined in terms of nuclear positional coordinates, $\underline{f}$ and $\bar{\underline{f}}$ are defined in terms of functions of electronic coordinates. For our "fixed-nuclei" model, the electronic symmetry of the complex must at all times be identical to the symmetry of the nuclear framework.

Sector (or "regional") rules for metal complexes may be derived directly from the $\underline{F}$ and $\bar{\underline{F}}$ functions upon application of the appropriate symmetry restrictions, whereas the details of the chiral metal-ligand electronic interactions responsible for observed optical activity are best analyzed using the $\underline{f}$ and $\bar{\underline{f}}$ functions (and the appropriate symmetry restrictions).

3.4. First-Order Rotatory Strengths

To first-order on the ISP model, the rotatory strength of the o→m electronic transition is given by:

$$R^{(1)}_{om} = \mathrm{Im}\,(P^{(1)}_{om} \cdot M^{(0)}_{mo}) \;, \tag{28}$$

where, $M^{(0)}_{mo} = (A_m \mid \hat{m} \mid A_o)$.

The electric-dipole transition moment, $P^{(1)}_{om}$, may be expressed in terms of a static-coupling (SC) part and a dynamic-coupling (DC) part as follows:

$$P^{(1)}_{om} = P^{(1)}_{om}(SC) + P^{(1)}_{om}(DC) \;, \tag{29}$$

where, in the contracted notation of Table 1,

$$P^{(1)}_{om}(SC) = -\sum_{\alpha \neq o} P^{(0)}_{\alpha m}\, V(\alpha,o)E^{-1}_{\alpha} - \sum_{\alpha \neq m} P^{(0)}_{o\alpha}\, V(\alpha,m)\Delta E^{-1}_{\alpha m} \;, \tag{30}$$

and,

$$P^{(1)}_{om}(DC) = \sum_{\beta \neq o} \mu_{o\beta} U(o\beta,mo)\;[2E_{\beta}/(E^2_m - E^2_{\beta})] \;. \tag{31}$$

The set of states (A_α) are taken from the set of <u>ungerade</u> states localized on the achiral (centrosymmetric) chromophore. These may be described in terms of metal ion excitations (such as, for

Table 1. Contracted Notation

A. Interaction Matrix Elements

1. General

$$V(\alpha\beta,\alpha'\beta') \equiv (A_\alpha B_\beta | \mathcal{V}_A + \mathcal{V}_{AB} + \mathcal{U}_A + \mathcal{U}_{AB} | A_{\alpha'} B_{\beta'})$$

2. Static Coupling

$$\mathcal{V}(\alpha,\alpha') \equiv (A_\alpha | \mathcal{V}_A + \mathcal{V}_{AB} | A_{\alpha'}) \equiv (A_\alpha B_o | \mathcal{V}_A + \mathcal{V}_{AB} | A_{\alpha'} B_o)$$

3. Dynamic Coupling

$$\mathcal{U}(\alpha\beta,\alpha'\beta') \equiv (A_\alpha B_\beta | \mathcal{U}_A + \mathcal{U}_{AB} | A_{\alpha'} B_{\beta'})$$

B. Transition Energy Sums and Differences

$$E_{\alpha\beta\gamma} = E_\alpha + E_\beta + E_\gamma$$

$$\Delta E_{\alpha\beta} = E_\alpha - E_\beta$$

C. Transition Dipole Matrix Elements

1. Metal Ion Chromophore

$$P^{(0)}_{\alpha\gamma} \equiv (A_\alpha | \hat{\mu}(A) | A_\gamma); \quad M^{(0)}_{\alpha\gamma} \equiv (A_\alpha | \hat{m}(A) | A_\gamma)$$

2. Ligand (Perturber) Group

$$\mu_{o\beta} \equiv (B_o | \hat{\mu}(B) | B_\beta); \quad m'_{o\beta} \equiv (B_o | \hat{m}'(B) | B_\beta)$$

Summations over all perturber sites $\underline{r}$ are implicit in the operators $\hat{\mu}(B)$ and $\hat{m}'(B)$.

example, d-p and d-f transitions), ligand-metal charge-transfer states within the metal-donor atom chromophoric cluster, or (composite) ligand-ligand excitation localized on the donor atoms of the chromophoric cluster. The set of states (B_β) represent excited states localized in the perturbing (extra-chromophoric) ligand environment. The first-order rotatory strength may now be expressed as:

$$R_{om}^{(1)} = R_{om}^{(1)}(SC) + R_{om}^{(1)}(DC) \quad , \tag{32}$$

where,

$$R_{om}^{(1)}(SC) = \mathrm{Im}[P_{om}^{(1)}(SC) \cdot M_{om}^{(0)}] \quad , \tag{33}$$

and,

$$R_{om}^{(1)}(DC) = \mathrm{Im}\,[\,P_{om}^{(1)}(DC) \cdot M_{om}^{(0)}] \quad , \tag{34}$$

To first-order, the SC and DC contributions to the rotatory strength are additive and may be treated separately.

As noted previously, in writing $o \rightarrow m$ we take the ground state (o) to be non-degenerate and we define m to include all components of any excited state which may be degenerate in the symmetry group G_A of the unperturbed achiral chromophore A. For example, if we specify $o \rightarrow m$ to be the ${}^1A_{1g} \rightarrow {}^1T_{1g}$ transition of an octahedral (O_h) Co^{3+} chromophoric unit, then R_{om} corresponds to the total (net) rotatory strength associated with this transition.

In order for $R_{om}^{(1)}(SC)$ to be non-vanishing, the static-coupling operator $V = V_A + V_{AB}$ must contain at least one component which transforms as a pseudoscalar function under the symmetry operations of the G_A point group. In order for $R_{om}^{(1)}(DC)$ to be non-vanishing, the dynamic-coupling operator $U = U_A + U_{AB}$ must contain at least one component which transforms as a pseudoscalar function (in G_A).

The static-coupling contribution, $R_{om}^{(1)}(SC)$, arises from the dissymmetric components of the "Stark field" created at the metal ion by the ligand environment (represented as a distribution of point charges). Symmetry considerations dictate that only those components which transform as a pseudoscalar function under the symmetry operations of the G_A point group will make non-vanishing contributions to $R_{om}^{(1)}(SC)$. Referring back to eq. (23), the lowest-order $\underline{F}(\ell_A, m_A)$ functions contributing to $R_{om}^{(1)}(SC)$ are listed below for several centrosymmetric G_A groups:

G_A	$\underline{F}(\ell_A, m_A)$	ℓ_α
O_h	$\underline{F}(9, m_A)$	7,9,11
D_{6h}	$\underline{F}(7, m_A)$	5,7,9
D_{4h}	$\underline{F}(5, m_A)$	3,5,7
D_{2h}	$\underline{F}(3, m_A)$	1,3,5
C_{2h}	$\underline{F}(1, m_A)$	1,3

where ℓ_α denotes the angular momentum quantum numbers allowed among the ungerade A_α states to be mixed into the d-d spectroscopic states. The large values of ℓ_α for $G_A \equiv O_h$ suggest why the first-order SC mechanism is quite often referred to as "unphysical" in its applications to complexes of erstwhile O_h symmetry. Sector rules pertaining to $R_{om}^{(1)}(SC)$, expressed in terms of the cartesian coordinates of the perturber groups, have been listed by Schellman[36] for a wide variety of achiral chromophoric point groups (G_A).

The dynamic-coupling contribution to the first-order rotatory strength, $R_{om}^{(1)}(DC)$, arises from a dissymmetric coupling between electric dipole transition vectors located in the ligand environment with multipolar transition moments associated with the chromophoric d-d transitions. The physical basis of this mechanism can be viewed in terms of an induction of (virtual) electric dipole transitions in the ligand environment by (radiation) field-induced excitation localized in the chromophoric group (A). The entire first-order electric dipole transition moment, $P_{om}^{(1)}(DC)$, appearing in eq. (34) is located in the ligand environment. $P_{om}^{(1)}(DC)$ represents a vector sum of electric dipole transition moments arrayed in the ligand environment of the metal complex.

The q-th (spherical polar) component of $P_{om}^{(1)}(DC)$ may be expressed in terms of the perturber group polarizabilities according to:

$$P_{om;q}^{(1)}(DC) = -\sum_r \sum_{\ell_A} \sum_{m_A} \sum_{q'} T_{m_A,q'}^{(\ell_A,1)}(r)\, \alpha_r(q,q')(A_o|D_{m_A}^{(\ell_A)}(A)|A_m), \tag{35}$$

where $\alpha_r(q,q')$ is the (q,q') component of the polarizability tensor for the r-th perturber group at the frequency $\nu_m = E_m/h$.

This quantity is defined by

$$\alpha_r(q,q') = -\sum_{i\neq o} (B_o|D_q^{(1)}(r)|B_{ri})(B_{ri}|D_{q'}^{(1)}(r)|B_o)[2E_{ri}/(E_m^2 - E_{ri}^2)]. \tag{36}$$

The first-order dynamic-coupling rotatory strength may now be written as,

$$R_{om}^{(1)}(DC) = \text{Im} \sum_q P_{om;q}^{(1)} M_{mo;q}^{(0)} \quad , \tag{37}$$

where $M_{mo;q}^{(0)}$ is the q-th component of the magnetic dipole transition moment, $M_{mo}^{(0)}$, and $P_{om;q}^{(1)}$ is given by eq. (35).

It is clear from eq. (28) that only the magnetic dipole allowed d-d transitions gain net rotatory strength to first-order on the ISP model presented here. Furthermore, eq. (32) expresses the fact that to first-order the static-coupling and dynamic-coupling mechanisms are complementary with respect to their contributions to the net rotatory strength. For a side-by-side comparison of the SC versus DC mechanisms as applied to a series of chiral Co(III) diamine complexes, one is referred to an excellent paper recently published by Mason and Seal.[55]

3.5. Second-Order Rotatory Strengths

The second-order contribution to the rotatory strength of a magnetic dipole forbidden d-d transition is given by,

$$R_{on}^{(2)} = \text{Im}\ (P_{on}^{(1)} \cdot M_{no}^{(1)}) \ . \tag{38}$$

Proceeding as before, we may partition both $P_{on}^{(1)}$ and $M_{no}^{(1)}$ into their respective SC and DC components so that:

$$P_{on}^{(1)} = P_{on}^{(1)}(SC) + P_{on}^{(1)}(DC) \quad , \tag{39}$$

$$M_{no}^{(1)} = M_{no}^{(1)}(SC) + M_{no}^{(1)}(DC) \quad , \tag{40}$$

and finally,

$$R_{on}^{(2)} = \mathrm{Im}[P_{on}^{(1)}(SC)\cdot M_{no}^{(1)}(SC)] + \mathrm{Im}[P_{on}^{(1)}(DC)\cdot M_{no}^{(1)}(DC)]$$

$$+ \mathrm{Im}[P_{on}^{(1)}(SC)\cdot M_{no}^{(1)}(DC) + P_{on}^{(1)}(DC)\cdot M_{no}^{(1)}(SC)] \quad , \tag{41}$$

or,

$$R_{on}^{(2)} = R_{on}^{(2)}(SC) + R_{on}^{(2)}(DC) + R_{on}^{(2)}(SC,DC) \quad . \tag{42}$$

Unlike the first-order case where the SC and DC contributions are strictly additive, the second-order rotatory strength $R_{on}^{(2)}$ includes contributions from mechanisms involving combinations of (simultaneous) static-coupling and dynamic-coupling between the metal ion chromophore and the ligand environment.

The first-order electric dipole transition moments, $P_{on}^{(1)}(SC)$ and $P_{on}^{(1)}(DC)$, appearing in eq. (41) have been written down previously (see Section 3.4.). The first-order magnetic dipole transition moment, $M_{no}^{(1)}(SC)$, may be expressed as,

$$M_{no}^{(1)}(SC) = -\sum_{\gamma\neq o} M_{n\gamma}^{(0)}\; V(o,\gamma)E_{\gamma}^{-1} - \sum_{\gamma\neq n} M_{\gamma o}^{(0)}\; V(n,\gamma)\Delta E_{\gamma n}^{-1} \quad , \tag{43}$$

where $\sum_{\gamma}$ runs over a set of <u>gerade</u> states localized on A. The dynamic-coupling contribution to the first-order magnetic dipole transition moment may be expressed as,

$$M_{no}^{(1)}(DC) = \sum_{r}\sum_{\ell_A}\sum_{m_A}\sum_{m_r}\sum_{i\neq o} T_{m_A,m_r}^{(\ell_A,1)}(r)(B_{ri}|\hat{m}'(r)|B_o)$$

$$\times\; (A_n|D_{m_A}^{(\ell_A)}(A)|A_o)(B_o|D_{m_r}^{(1)}(r)|B_{ri})[2E_n/(E_n^2 - E_{ri}^2)], \tag{44}$$

where,

$$\hat{m}'(r) = \hat{m}(r) + (e/2m_e c)\vec{R}_r \times \hat{p}(r) \ . \tag{45}$$

Here, m_e denotes eléctron mass, $\hat{m}(r)$ is a magnetic dipole operator located on the r-th perturber group, $\hat{p}(r)$ is a linear momentum operator located on the r-th group, and the vector $\vec{R}_r$ defines the position of perturber site r with respect to an origin within the A chromophore.

The magnetic dipole matrix elements in eq. (44) may be rewritten as follows:

$$(B_{ri}|\hat{m}'(r)|B_o) = (B_{ri}|\hat{m}(r)|B_o) + (e/2m_e c)\vec{R}_r \times (B_{ri}|\hat{p}(r)|B_o)$$

$$= m_{io}(r) + (i\pi E_{ri}/hc)\vec{R}_r \times \mu_{io}(r) \ , \tag{46}$$

where $\mu_{io}(r) = (B_{ri}|\hat{\mu}(r)|B_o)$ and $m_{io}(r) = (B_{ri}|\hat{m}(r)|B_o)$. In writing eq. (46) we have made use of the relation,

$$(\psi_i|\hat{p}|\psi_j) = -(\psi_j|\hat{p}|\psi_i) = (2\pi i \, m_e/e)\,(E_i - E_j)(\psi_i|\hat{\mu}|\psi_j) \ . \tag{47}$$

Since we are restricting the dynamic-coupling chromophore (A) - perturber (B) interaction potential to A (multipole) - B (dipole) terms, the only ligand (perturber) transitions contributing to eq. (46) are those which possess non-vanishing electric dipole character.

The second-order contribution to the rotatory strength of a <u>magnetic dipole allowed</u> d-d transition is given by,

$$R_{om}^{(2)} = \mathrm{Im}(P_{om}^{(1)} \cdot M_{mo}^{(1)} + P_{om}^{(2)} \cdot M_{mo}^{(0)}) \ . \tag{48}$$

This equation may be expanded to the following form:

$$R_{om}^{(2)} = \mathrm{Im}[P_{om}^{(1)}(SC)\cdot M_{mo}^{(1)}(SC) + P_{om}^{(2)}(SC)\cdot M_{mo}^{(0)}]$$

$$+ \mathrm{Im}[P_{om}^{(1)}(DC)\cdot M_{mo}^{(1)}(DC) + P_{om}^{(2)}(DC)\cdot M_{mo}^{(0)}]$$

$$+ \mathrm{Im}[P_{om}^{(1)}(SC)\cdot M_{mo}^{(1)}(DC) + P_{om}^{(1)}(DC)\cdot M_{mo}^{(1)}(SC) + P_{om}^{(2)}(SC,DC)\cdot M_{mo}^{(0)}] . \quad (49)$$

Expressions for the first-order transition moments have been given previously in this and the preceding section (3.4). The next step is to obtain general expressions for the second-order electric dipole transition moments, $P_{om}^{(2)}$. In doing this we shall use the "contracted" notation listed in Table 1.

The static-coupling contribution to $P_{om}^{(2)}$ is given by

$$P_{om}^{(2)}(SC) = \sum_{\alpha}\sum_{\alpha'} [P_{\alpha m}^{(0)} \; V(\alpha o,\alpha' o)V(\alpha' o,oo)E_{\alpha}^{-1}\,E_{\alpha'}^{-1}$$

$$+ P_{o\alpha}^{(0)} \; V(\alpha o,\alpha' o)\; V(\alpha' o,mo)\Delta E_{m\alpha}^{-1}\;\Delta E_{m\alpha'}^{-1}$$

$$- P_{\alpha\alpha'}^{(0)} \; V(\alpha o,oo)\; V(oo,\alpha' o)\; E_{\alpha}^{-1}\;\Delta E_{m\alpha'}^{-1}]$$

$$- \sum_{\alpha} \mu_{oo} \; V(oo,\alpha o)\; V(\alpha o,mo)\; E_{\alpha}^{-1}\;\Delta E_{m\alpha}^{-1} \; , \quad (50)$$

where it is understood that the summations exclude all terms in which the energy denominators are zero. The dynamic-coupling contribution to $P_{om}^{(2)}$ is given by

$$P_{om}^{(2)}(DC) = \sum_{\alpha\neq o} P_{\alpha m}^{(0)} \sum_{\beta\neq o} \sum_{\alpha'\neq o,\alpha} V(\alpha o,\alpha'\beta)V(\alpha'\beta,oo)E_{\alpha'\beta}^{-1} E_{\alpha}^{-1}$$

$$+ \sum_{\alpha\neq m} P_{o\alpha}^{(0)} \sum_{\beta\neq o} \sum_{\alpha'\neq m,\alpha} V(\alpha o,\alpha'\beta)V(\alpha'\beta,mo)\Delta E_{m\alpha}^{-1}(E_m-E_{\alpha'\beta})^{-1}$$

$$+ \sum_{\beta\neq o} \mu_{o\beta} \sum_{\beta'\neq o,\beta} \sum_{\alpha\neq o,m} [V(o\beta,\alpha\beta')V(\alpha\beta',mo)\Delta E_{m\beta}^{-1}(E_m-E_{\alpha\beta'})^{-1}$$

$$+ V(m\beta,\alpha\beta')V(\alpha\beta',oo)E_{m\beta}^{-1} E_{\alpha\beta'}^{-1}]$$

$$- \sum_{\alpha\neq o} \sum_{\alpha'\neq m,o} P_{\alpha\alpha'}^{(0)} \sum_{\beta\neq o} V(\alpha\beta,oo)V(\alpha'\beta,mo)E_{\alpha\beta}^{-1}(E_m-E_{\alpha'\beta})^{-1}$$

$$- \sum_{\beta\neq o} \sum_{\beta'\neq o} \mu_{\beta\beta'} \sum_{\alpha\neq o,m} V(\alpha\beta,oo)V(\alpha\beta',mo)E_{\alpha\beta}^{-1}(E_m-E_{\alpha\beta'})^{-1} . \quad (51)$$

The expression for the "mixed" static-coupling/dynamic-coupling contributions to $P_{om}^{(2)}$ is somewhat more complicated than eqs. (50) and (51) and will not be reproduced here.

The expressions (50) and (51) can be reduced to simpler form only upon consideration of specific systems with prescribed symmetry properties. We note that $P_{om}^{(2)}(DC)$ includes terms with metal-localized electric dipole transition moments as well as terms with ligand-localized electric dipole transition moments.

3.6. Comments on Second-Order Rotatory Strengths

The details of the physical mechanisms subsumed in the second-order contributions to the d-d rotatory strengths are indeed quite complex. Some discussion of these physical mechanisms and their implications with regard to structure elucidation have appeared in the literature.[24,26-29,31,37,54-56,59] Richardson and coworkers[26-29] have dealt primarily (but not entirely) with the $R^{(2)}$(SC) terms, whereas Mason and coworkers[55] have dealt primarily with the dynamic-coupling contributions. Carrying the dynamic-coupling model to second order, Strickland and Richardson[56] calculated the d-d optical activity for a series of Cu(II) complexes of amino acids, dipeptides, and tripeptides. More recently, Schipper[59] has presented a theoretical treatment of d-d optical activity based on the independent systems/perturbation model and focused his attention primarily on what we have denoted as $R^{(2)}$(DC) and $R^{(2)}$(SC,DC), that is, the second-order dynamic-coupling and "mixed" static-coupling/dynamic-coupling contributions. Of special interest in the Schipper study is the assertion that the rotatory strengths of magnetic dipole allowed d-d transitions arise predominantly from $R^{(2)}$(DC) contributions involving dipole (metal ion chromophore)-dipole (ligand) interactions.

Schipper[59] discounts the "pure" static-coupling and "mixed" static-coupling/dynamic-coupling mechanisms as dominant contributors to the d-d rotatory strengths in most metal complexes since the chiral components of the static part of the ligand field potential are generally expected to be weak. Exceptions to this would occur in those cases where charged ligand groups are dissymmetrically disposed about the metal ion chromophoric unit. Schipper also discounts the $R^{(1)}$(DC) contributions as being dominant due to the necessity in this case of including multipolar ($\ell_A > 1$)-dipole terms in the metal ion-ligand interaction potential. The dominance of $R^{(2)}$(DC) over $R^{(1)}$(DC), according to Schipper, may be attributed to the dominance of dipole-dipole over multipole ($\ell_A > 1$)-dipole interaction terms in the metal ion chromophore-ligand dynamic-coupling mechanism.

Although Schipper's arguments for the dominance of $R^{(2)}$(DC) via dipole-dipole coupling would appear to be plausible from qualitative physical considerations, no numerical or quantitative calculations have yet been reported to support these arguments. To date, all calculations of d-d rotatory strengths based on the dynamic-coupling model have relied on chiral multipole ($\ell_A > 1$)-dipole metal-ligand interactions to induce optical activity. Furthermore, all of the sector rules proposed for d-d CD spectra have been based on the expressions for $R^{(1)}$(SC), $R^{(2)}$(SC), $R^{(1)}$(DC), and the multipole ($\ell_A > 1$)-dipole parts of $R^{(2)}$(DC). These theoretical analyses (in which the dipole-dipole coupling terms in $R^{(2)}$(DC) have been ignored) have enjoyed some moderate success in accounting for the empirically observed CD data, both qualitatively and semi-quantitatively. However, given the extensive parameterization of the assumed models this success cannot be taken as conclusive evidence for the correctness of the models. Quantitative analyses based on the dipole-

dipole coupling terms in $R^{(2)}(DC)$ are required before further conclusions can be reached regarding the dominant mechanism in d-d optical activity.

An example of a dipole-dipole coupling term in $R^{(2)}(DC)$ is given by (for a magnetic dipole allowed transition o→m),

$$-i\,V_{dd}(o\beta,\alpha\beta')V_{dd}(\alpha\beta',mo)\Delta E_{m\beta}^{-1}(E_m-E_{\alpha\beta'})^{-1}[\mu_{o\beta}\cdot M_{mo}^{(0)}],$$

where $\beta \neq o$, $\beta' \neq o,\beta$, and α is coupled to both the ground state (o) and the excited state (m) via an electric dipole operator $\hat{\mu}(A)$ located on the chromophore A. Here, V_{dd} denotes the dipole-dipole coupling term in the interaction potential. In this case, the matrix element product on B (the ligand environment) has the form,

$$\mu_{o\beta}\;\mu_{\beta\beta'}\;\mu_{\beta'o} \quad ,$$

and the matrix element product on A has the form,

$$P_{o\alpha}^{(0)}\;P_{\alpha m}^{(0)}\;M_{mo}^{(0)} \quad .$$

This contribution to the rotatory strength of the o→m transition will be maximized when

$$\vec{\mu}_{o\beta} \times \vec{\mu}_{\beta\beta'} \cdot \vec{\mu}_{\beta'o}$$

is maximized; i.e., the electric dipole transition moments on B are all strongly allowed and mutually orthogonal. A detailed account of the electronic selection rules and stereochemical sector rules inherent to the dipole-dipole coupling terms of $R^{(2)}(DC)$ may be found in Schipper's article.[59]

Another example of a dipole-dipole coupling term in $R^{(2)}(DC)$ is given by,

$$-\,i\,V_{dd}(\alpha o,\alpha'\beta)V_{dd}(\alpha'\beta,mo)\Delta E_{m\alpha}^{-1}(E_m-E_{\alpha'\beta})^{-1}[P_{o\alpha}^{(0)}\cdot M_{mo}^{(0)}] \quad ,$$

where $\beta \neq o$, $\alpha' \neq \alpha$, and α' is connected to both α and m via an electric dipole operator. In this case, an electric dipole transition moment ($P^{(0)}_{o\alpha}$) is induced on the chromophore A via mediation of the ligand environment.

3.7. Applicability of the Independent Systems/Perturbation Model

Despite its obvious limitations and shortcomings in providing an accurate and complete representation of the electronic structure of transition metal coordination compounds, the ISP model would appear to give a useful and reasonably reliable representation of the d-d spectroscopic states of these systems. This is especially so if the chromophoric unit of the model (in zeroth-order) is expanded to include the metal ion and the directly coordinated donor atoms of the ligands. By defining the zeroth-order chromophoric unit to include the ligand donor atoms, the ISP model may, for many systems, also be applied to treating metal-ligand charge-transfer transitions localized within the metal-donor atom cluster.

The ISP model is perhaps even more appropriate for treating the f-f spectroscopic transitions of lanthanide ion complexes than for treating the d-d transitions of transition metal complexes. This is due to the presumably very weak nature of f-orbital/ligand interactions. However, the treatment of f-f optical activity in chiral lanthanide ion complexes remains in a very primitive state of development.[61]

4. VIBRONIC INTERACTIONS

Concern about the possible influence of vibronic interactions upon the CD spectra of metal complexes was first expressed by R.G. Denning.[62] Denning proposed that the $^1T_{1g}$ excited state of $Co(en)_3^{3+}$ undergoes a strong (tetragonal) Jahn-Teller distortion via coupling to an e_g vibrational mode of the CoN_6 cluster. This strong tetragonal Jahn-Teller (JT) distortion was then presumed to be effective in "quenching" the crystal-field-induced trigonal splitting of the $^1T_{1g}$ state (a manifestation of the so-called Ham effect[63]) in $Co(en)_3^{3+}$. Denning[62] further suggested that the two CD bands observed in the $^1A_{1g} \rightarrow {}^1T_{1g}$ region arose from two different JT vibronic states derived from $^1T_{1g}$-e_g coupling, rather than from the two trigonal components (^1E and $^1A_2$) of the $^1T_{1g}$ electronic state. The influence of Jahn-Teller (JT) and pseudo Jahn-Teller (PJT) interactions upon the CD spectra of d-d transitions has been studied in considerable detail (theoretically) by Richardson and coworkers.[26,32,64-67] These studies included consideration of metal complexes belonging to trigonally symmetric structural classes[26,65] as well as metal complexes of pseudo-tetragonal symmetry.[32,64,66,67] The main conclusion of these studies was that whereas vibronic interactions of the JT and PJT types (within the manifold of d-d excited states) will not, in general, alter the net d-d rotatory strength for a given system, they can play a dominant role in determining how CD intensity is distributed throughout the d-d transition region. These studies thus pointed to an additional reason why only net d-d CD intensity (or rotatory strength) can be safely used in making spectra-structure correlations. In the presence of strong JT or PJT interactions among the d-d states, it becomes impossible (or meaningless) to assign specific features in the CD spectra to specific d-d electronic transitions. The individual CD bands, in such cases, will generally reflect "mixed" electronic parentage.

Vibronically induced coupling of the d-d spectroscopic states to odd-parity (ungerade) electronic states of metal complexes plays a significant (and sometimes dominant) role in determining the observed dipole strengths and absorption intensities of d-d transitions. The possible influence of these vibronic interactions upon d-d rotatory strengths has been considered qualitatively by M.J. Harding[68] using the vibronic coupling formalism of Weigang and coworkers.[69-72]

Hilmes, Caliga, and Richardson[67] investigated the influence of (simultaneous) spin-orbit and vibronic interactions upon the chrioptical properties of nearly degenerate d-d transitions in metal complexes of pseudo-tetragonal symmetry. A model system was considered in which three nearly degenerate d-d excited states are coupled via both spin-orbit and vibronic interactions. Vibronic interactions among the three nearly degenerate electronic states were assumed to arise from a PJT coupling mechamism involving three different vibrational modes of the undistorted metal complex system. The model adopted was conservative with respect to the total (or net) rotatory strength associated with transitions to the vibronic sub-levels of the three perturbed d-d excited states. The vibronic rotatory strengths and simulated CD spectra calculated in this study again demonstrated the extreme sensitivity of d-d CD intensity distributions to vibronic and spin-orbit coupling within the manifold of d-d excited states.

5. SPIN-FORBIDDEN TRANSITIONS

The (essential) role of spin-orbit coupling in generating optical activity in the spin-forbidden d-d transitions of chiral transition metal complexes has been studied by a number of workers.[73-78] Most of the experimental data reported to date on CD within spin-forbidden transitions have been acquired on Co(III) and Cr(III) complexes. As might be expected for metal ions belonging to the first row of the transition metal series, the $\Delta S = 0$ spin selection rule remains rather strong (due to the relatively small spin-orbit coupling constants) and the CD intensities observed for the spin-forbidden transitions are generally found to be several orders of magnitude weaker than those observed for the corresponding spin-allowed transitions. Stronger spin-forbidden optical activity is expected to occur in complexes involving metal ions belonging to the second and third rows of the transition metal series (where the spin-orbit coupling constants are larger). For lanthanide ions, the f-electron states must be described in terms of a jj-coupling scheme and S is no longer a "good" (well-defined) quantum number. In this case, selection rules involving ΔJ values are dominant.[61]

6. SINGLE CRYSTAL SPECTRA

Let us consider a sample of fixed orientation in space having uniaxial (cylindrical) macroscopic symmetry about an axis γ (also fixed in space). Let us further assume that this (macroscopic) sample is comprised of an assembly of non-interacting chromophoric units each of which is orientationally fixed. If we perform a CD experiment with radiation propagating along the γ-axis of the sample, then to lowest order in the expansion of the vector potential of the radiation field the rotatory strength of an electronic transition $i \rightarrow j$ (localized on a chromophoric unit) is given by

$$R_{ij} = (3/2)\mathrm{Im}[\langle\psi_i|\hat{\mu}_\alpha|\psi_j\rangle\langle\psi_j|\hat{m}_\alpha|\psi_i\rangle + \langle\psi_i|\hat{\mu}_\beta|\psi_j\rangle\langle\psi_j|\hat{m}_\beta|\psi_i\rangle]$$

$$- (3/4)(2\pi\nu_{ji}/c)\mathrm{Re}[\langle\psi_i|\hat{\mu}_\alpha|\psi_j\rangle\langle\psi_j|\hat{q}_{\beta\alpha}|\psi_i\rangle - \langle\psi_i|\hat{\mu}_\beta|\psi_j\rangle\langle\psi_j|\hat{q}_{\gamma\alpha}|\psi_i\rangle], \tag{52}$$

where $\nu_{ji} = (E_j - E_i)/h$, the indices (α, β, γ) refer to a space-fixed orthogonal coordinate system, and $\hat{q}_{\beta\gamma}$ and $\hat{q}_{\gamma\alpha}$ are, respectively, the $\beta\gamma$- and $\gamma\alpha$-components of the electric quadrupole tensor operator. If the sample is a crystal of uniaxial macroscopic symmetry, then γ must correspond to the principal (or optic) axis of the crystal. If the sample is a crystal of cubic symmetry, then the choice of γ (with respect to the crystallographic axes) is entirely arbitrary. The components of the electronic operators in eq. (52) are referred to the (α, β, γ) coordinate system. Furthermore, the operators may be defined within the respective chromophoric units since the chromophoric units have been assumed to be non-interacting. The wave functions, ψ_i and ψ_j, correspond to localized chromophoric state functions and reflect the site symmetry (and dissymmetry) of the chromophoric units.

If the chromophoric units of the sample are interacting (coupled), then eq. (52) must be modified to reflect excitation delocalization (exciton motion) or electron delocalization among the coupled chromophoric units. In this case the wave functions, ψ_i and ψ_j, reflect crystal (or unit cell) symmetry and correspond to crystal state functions. Furthermore, the operators in eq. (52) can no longer be defined simply with respect to origins located within individual chromophoric units. The general problems attendant to treating the optical activity of crystalline samples comprised of interacting or non-interacting chromophoric sub-systems have been addressed elsewhere[30,38,79,80] and will not be dealt with further here.

Two principal questions have arisen in the theoretical analyses of d-d optical activity exhibited by crystalline samples of transition metal complexes. One question regards the relative contributions of the electric dipole-magnetic dipole versus the electric dipole-electric quadrupole terms in eq. (52), and the other question regards the relative contributions of site (local) dissymmetry versus unit cell dissymmetry to the overall, observed d-d optical activity. In their detailed theoretical treatment (formal and computational) of the d-d optical activity exhibited by crystalline $NiSO_4 \cdot 6H_2O$[30] and crystalline Cu^{2+}: $ZnSeO_4 \cdot 6H_2O$,[32] Richardson and co-workers concluded that the electric dipole-electric quadrupole terms contributed only minimally (less than 5%) to the rotatory strengths, even for transitions which are formally electric-quadrupole-allowed according to cubic (O_h) selection rules. They further concluded that the d-d states in these systems could be considered localized on the MO_6 clusters and that site dissymmetry is entirely responsible for the observed optical activity. For $NiSO_4 \cdot 6H_2O$, the d-d optical activity was attributed entirely to inherent dissymmetry within the $Ni(H_2O)_6^{2+}$ unit and to the dissymmetric field created at each Ni^{2+} site by the four nearest-neighbor SO_4^{2-} anions. Similarly for Cu^{2+}: $ZnSeO_4 \cdot 6H_2O$, all of the optical activity of the Cu^{2+} d-d transitions was attributed to $Cu(H_2O)_6^{2+}$ dissymmetry (inherent) and to the four nearest-neighbor SeO_4^{2-} anions. The calculations carried out by Richardson and co-workers[30,32] were based on the static-coupling variant of the independent systems/

perturbation model carried to second-order, and considerable attention was given to the sensitivity of the results to parameter variations within the constraints of this model. The calculated results were in substantial agreement with experimental observation.

Kato[38] has carried out a theoretical analysis of the d-d optical activity of a whole series of M^{2+}: $ZnSeO_4 \cdot 6H_2O$ systems (where $M^{2+} = Cu^{2+}, Ni^{2+}, Co^{2+}, Fe^{2+}$, and Mn^{2+}). This analysis was also based on the static-coupling terms of the general independent systems/perturbation model carried to second-order. Local (site) dissymmetry was found to be the dominant source of optical activity in these systems, and contributions from the electric dipole-electric quadrupole terms of eq. (52) were found to be negligable.

The optical activity caused by exciton dispersion in chiral crystals has been treated formally by Kato, Tsujikawa, and Murao,[81] and crystalline $CsCuCl_3$ belonging to the optically active space group $D_6^2(P6_122)$ or $D_6^3(P6_522)$ was considered as a case where such optical activity might be observed. No calculations were reported in this study, however.

Barron[79] has proposed that the electric dipole-electric quadrupole terms of eq. (52) may contribute significantly to (or dominate) the optical activity of d-d transitions which are magnetic-dipole-forbidden (or only weakly allowed) but which are electric-quadrupole-allowed. He suggested that the $^1A_1(^1A_{1g}) \rightarrow {}^1E(^1T_{2g})$ transition of Co(III) in crystalline $2[(+)\text{-}Co(en)_3Cl_3] \cdot NaCl \cdot 6H_2O$ might be a case where the electric dipole-electric quadrupole contribution would dominate. Order-of-magnitude calculations and a comparison of solution versus crystal CD intensity in the region of this transition suggest that Barron's proposal is certainly plausible. Barron further suggested that the $^3A_{2g} \rightarrow {}^3T_{1g}$ transitions in crystalline $NiSO_4 \cdot 6H_2O$ may acquire most of their optical activity via the electric dipole-electric quadrupole term in eq. (52). This latter suggestion is not supported by the calculations performed by Strickland and Richardson.[30] At this point it is safe to say that the relative importance of the electric dipole-magnetic dipole versus the electric dipole-electric quadrupole terms of eq. (52) in determining the optical activity of electric-quadrupole-allowed transitions has not yet been decided for any particular case.

7. SPECTRA-STRUCTURE RELATIONSHIPS

The sources of dissymmetry in an optically active metal complex may generally be classified as follows: (1) <u>inherent dissymmetry</u> within the metal ion-donor atom coordination cluster;

(2) configurational dissymmetry due to a chiral arrangement of chelate systems about the metal ion; (3) conformational dissymmetry due to chiral conformations within individual chelate rings; and, (4) vicinal dissymmetry due to asymmetric sites located within the coordinated ligands. Identification and separation of the contributions made by these types of dissymmetry to the observed d-d chiroptical spectra can provide at least "rough-grained" stereochemical information about a metal complex.

One class of metal complexes which are inherently dissymmetric due to the distribution of achiral monodentate ligands about the metal ion is represented by all-cis-$[M(A)_2(B)_2(C)_2]$, where M denotes the metal ion and A,B, and C denote chemically dissimilar monodentate ligands (see Figure 1). Two examples of metal complexes of this type are: R-(+)-all-cis-$[Co(NH_3)_2(H_2O)_2(CN)_2]^+$ and S-(+)-all-cis-$[Co(NH_3)_2(H_2O)_2(NO_2)_2]^+$. These two complexes have recently been synthesized and resolved,[82] and their electronic absorption and circular dichroism spectra have been reported.[82] Mason[83] has calculated the optical activity

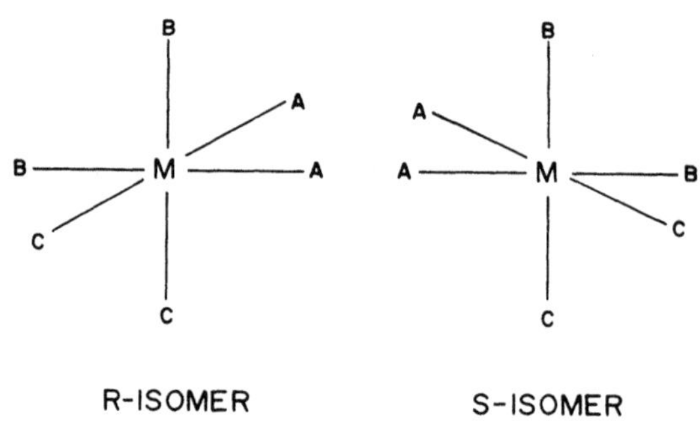

Figure 1: Enantiomeric forms of an all-cis-$[M(A)_2(B)_2(C)_2]$ complex in which A,B, and C denote chemically dissimilar, achiral, monodentate ligands.

of the octahedral $^1A_{1g} \rightarrow ^1T_{1g}$ Co(III) d-electron transition for these systems using the independent systems/perturbation model carried to third- and fourth-order in the dynamic-coupling terms. (Mason refers to the dynamical coupling aspects of the independent systems/perturbation model as the ligand-polarization model.) Mason found that the third-order rotatory strengths vanish if the coordination octahedron of the complex is geometrically regular, but not those of the fourth-order model, which is based upon the pairwise mixing of the three components of the octahedral $^1A_{1g} \rightarrow ^1T_{1g}$ d-d transition, mediated by the Coulombic potential between the individually correlated induced dipoles in different ligands. The sum of the third- and fourth-order contributions were found to reproduce the signs and a significant fraction of the magnitude of the observed rotatory strengths exhibited by the two complexes for which data are available. In performing these calculations, Mason neglected all static-coupling and "mixed" static-coupling/dynamic-coupling contributions. He further neglected all mixing between the $^1T_{1g}$ and $^1T_{2g}$ octahedral d-electron states of Co(III).

Examples of complexes which possess inherent dissymmetry within the metal ion-donor atom cluster as well as configurational dissymmetry due to chiral distributions of chelate rings are $Co(ox)_3^{3-}$ and $Cr(ox)_3^{3-}$, where ox denotes an oxalato dianion. The MO_6 clusters in these complexes are trigonally distorted octahedra possessing exact trigonal dihedral (D_3) symmetry, and the planar ox ligands are arrayed about the metal ion with exact D_3 symmetry. The relative contributions of the donor atoms versus the chelate bridging atoms (or groups) to the optical activity of these systems have never been sorted out or determined conclusively. Trigonal distortion operations on the ML_6 cluster of a tris(bidentate ligand) metal complex with D(Λ) absolute configuration are depicted in Figure 2. These distortion operations are classified as azimuthal twists about the C_3 axis of the system or as polar elongations (or compressions) along the C_3 axis. As depicted in Figure 2, the azimuthal twists exert an ungerade perturbation upon the (erstwhile) octahedral system, whereas the polar distortions exert a gerade perturbation. An aximuthal twist within the ML_6 cluster is essential to producing optical activity within the d-d transitions of such complexes so long as all the donor atoms are identical and the chelate bridging atoms (or groups) are ignored. The polar distortions cannot produce d-d optical activity in the absence of an azimuthal twist.[26,33] The tris (oxalato) and tris(ethylenediamine) complexes of Co(III) are examples of systems in which the ML_6 cluster suffers polar compression/azimuthal contraction (at least in crystalline media). The tris(malonato) complex of Cr(III) in crystalline form reveals a polar elongation/azimuthal expansion of the CrO_6 cluster. The tris(trimethylenediamine) complex of Co(III) also exhibits polar elongation/azimuthal expansion in its crystalline form.

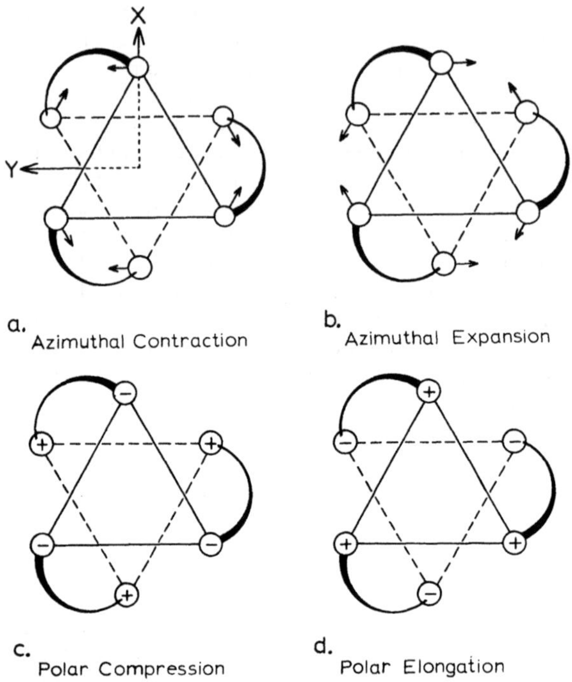

Figure 2: Trigonal distortion operations on the ML_6 cluster of a tris (bidentate ligand) metal complex with D(Λ) absolute configuration.

All of the structures shown in Figure 2 have a D(Λ) absolute configuration defined with respect to the distribution of chelate rings about the metal ion. If the chelate bridging atoms determine (or make the dominant contributions to) the d-d rotatory strengths, then one may readily correlate the d-d optical activity observables to the absolute configuration of tris(symmetric bidentate ligands) complexes. On the other hand, if the signs and magnitudes of the d-d rotatory strengths are primarily determined by distortions within the ML_6 cluster then great care must be exercised in deducing absolute configurations from the optical activity observables. Uncertainties regarding the relative importance of inherent dissymmetry (within the ML_6 cluster) versus configurational dissymmetry (due to the distribution of chelate rings) to producing optical activity in the d-d transitions of tris (bidentate ligand) complexes have retarded progress in developing reliable and generally applicable spectra-structure relationships for such systems.

The independent systems/perturbation model neglects metal ion-donor atom bonding interactions and ignores the detailed orbital nature of the ligand electronic distributions. For these reasons this model is not very suitable for treating effects due to distortions within the ML_6 cluster.

An example of a complex in which there are three sources of dissymmetry (inherent, configurational, and conformational) is $Co(en)_3^{3+}$, where en ≡ ethylenediamine ligand. In this complex the CoN_6 cluster is trigonally distorted with a polar compression/azimuthal contraction distortion (see Figure 2), the distribution of chelate rings reflects one of two enantiomeric configurations (Λ or Δ), and each five-membered chelate ring may exist in one of two enantiomeric conformational types, k(λ) or k'(δ). The possible configurational-conformational isomers for $Co(en)_3^{3+}$ are: Λ(δδδ), Λ(δδλ), Λ(δλλ), Λ(λλλ), Δ(δδδ), Δ(δδλ), Δ(δλλ), and Δ(λλλ). The conformational isomers of coordinated ethylenediamine are depicted in Figure 3. In crystalline media, $Co(en)_3^{3+}$ has been found to exist in the Λ(δδδ) or Δ(λλλ) enantiomeric forms. Conformational analysis calculations indicate that these should be the most stable isomers of free $Co(en)_3^{3+}$.[84] In the Λ(δδδ) isomer, the axes of the C-C bonds of the en ligands are parallel (lel) to the C_3 axis of the complex. Likewise, in the Δ(λλλ) isomer the C-C bonds of the en ligands are parallel to the C_3 symmetry axis. In the Λ(λλλ) and Δ(δδδ) isomers, the C-C bonds of the en ligands are oblique (ob) to the C_3 symmetry axis (see Figure 4). Conformational analysis indicates that for free $Co(en)_3^{3+}$ the order of stability for the various isomers is Λ(δδδ) = Δ(λλλ)>Λ(δδλ) = Δ(λλδ)>Λ(δλλ) = Δ(λδδ)>Λ(λλλ) = Δ(δδδ), with Λ(δδδ) being about 1.8 kcal/mole more stable than Λ(λλλ). There remains considerable uncertainty regarding the relative stabilities of these isomers in solution media.

Examples of metal complexes in which there are four sources of dissymmetry (inherent, configurational, conformational, and asymmetric centers within the ligands) are $Co(S\text{-}pn)_3^{3+}$ and $Co(R\text{-}pn)_3^{3+}$, where pn ≡ propylenediamine ligand. These systems can assume any of the eight configurational-conformational isomeric forms discussed above for $Co(en)_3^{3+}$ and each ligand in these complexes has one asymmetric carbon atom (which is a bridging atom in the five-membered chelate rings). The methyl substituent in R-pn is equatorial to the chelate ring in the λ conformation and is axial to the chelate ring in the δ conformation. The methyl substituent in S-pn is equatorial to the chelate ring in the δ conformation and is axial to the chelate ring in the λ conformation. The most stable isomer of $Co(R\text{-}pn)_3^{3+}$ is Δ(λλλ), whereas the most stable isomer of $Co(S\text{-}pn)_3^{3+}$ is Λ(δδδ).[84] The conformational isomers of coordinated S-pn are depicted in Figure 5.

Examples of complexes possessing dissymmetry only from chiral chelate ring conformations and asymmetric centers within the ligands are trans-$[Co(S\text{-}pn)_2(NH_3)_2]^{3+}$ and trans-$[Co(S\text{-}ala)_2(H_2O)_2]^+$, where ala ≡ alaninato ligand. An example of a metal

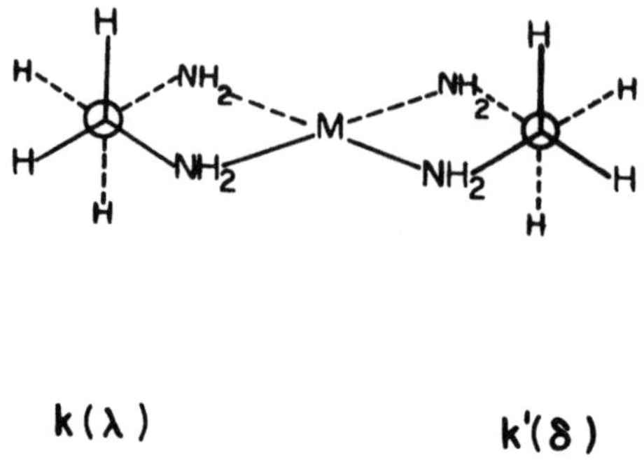

Figure 3: Conformational isomers of coordinated ethylenediamine. The view is along a direction parallel to the C-C bonds in the en ligands.

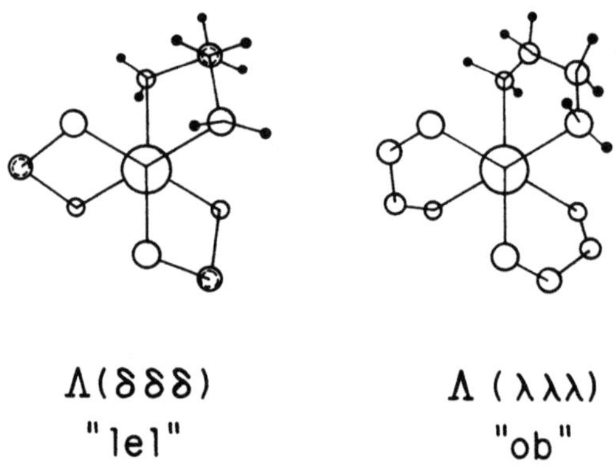

Figure 4: The "lel" and "ob" forms of a Λ-$[M(en)_3]$ complex.

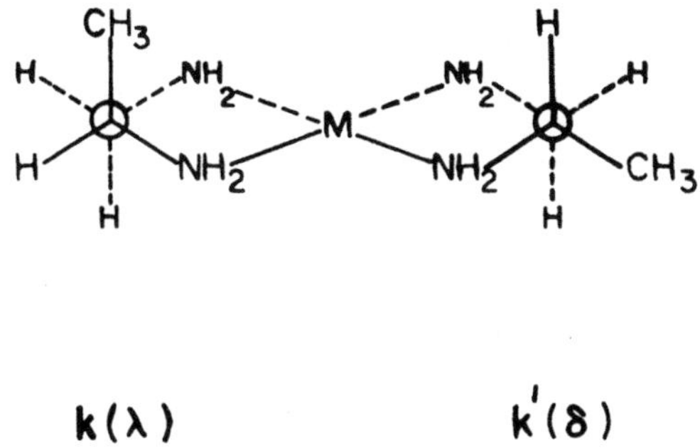

Figure 5: Conformational isomers of coordinated S-propylenediamine. The view is along a direction parallel to the C-C bonds in the S-pn ligands.

complex which has only one source of dissymmetry, an asymmetric carbon atom in one of its ligands, is $[Co(NH_3)_5(S\text{-}amH)]^{2+}$, where S-amH is an amino acid ligand coordinated only through its carboxylate group (i.e., it is bound unidentate).

One of the principal objectives of theoretical research dealing with the d-d optical activity of chiral transition metal complexes is to show how each of the sources of dissymmetry discussed above is manifested in the observed CD spectra of these systems. More specifically, one would like to relate the band splittings, sign patterns, relative band intensities, and total (net) CD intensity to the structural characteristics of the metal complex. These structural characteristics include distortions within the metal ion-donor atom cluster, the distribution of donor atoms about the metal ion, absolute configuration, conformational features of chelate rings, and the spatial disposition of ligand substituent groups. Developing such spectra-structure relationships is an ambitious task and the current theories of d-d optical activity are not yet entirely adequate for this purpose. The theories have been reasonably successful in providing a'posteriori interpretations and rationalizations of the experimental data, but their predictive value has been somewhat more limited. However, there now exist a number of reasonably reliable spectra-structure relationships for several structural classes of chiral metal complexes. Most of these relationships may be considered semiempirical in the sense that they were derived from systematic correlations of empirical data and were made to conform to (or be consistent with) general theoretical principles.

The most comprehensive investigations of spectra-structure relationships within the context of the independent systems/perturbation model with static-coupling (ISP-SC) were carried out by Richardson.[26-29,33] These studies included consideration of all sources of dissymmetry in four-coordinate and six-coordinate complexes belonging to a variety of structural and symmetry classes. Given the simplicity of the model employed (ISP-SC), these studies led to rather remarkable qualitative correlations between the CD observables and various structural features of metal complexes. Applications of the sector rules derived for six-coordinate dissymmetric complexes were given special attention in ref. 29, and applications of the sector rules derived for four-coordinate dissymmetric complexes were considered in ref. 28. Complexes of trigonal dihedral (D_3) symmetry were treated in ref. 26, and complexes of pseudo-tetragonal symmetry were the subject of ref. 27. Mason[31,37] has also examined the sector rules and other spectra-structure relationships inherent in the static-coupling parts of the general independent systems/perturbation model of d-d optical activity.

The most comprehensive investigations of spectra-structure relationships within the context of the independent systems/perturbation model with dynamic-coupling (ISP-DC) have been carried out by Mason and co-workers.[31,37,53-55,83,85] Comparisons of the sector rules derived from the DC and SC parts of the general independent systems/perturbation model were given explicit consideration, and a number of quantitative calculations were carried out within the polarizability approximation for representing ligand contributions to the electric dipole transition moments of the perturbed d-d transitions. Richardson[27,56] has also employed the ISP-DC model to examine the spectra-structure relationships appropriate to a series of pseudo-tetragonal complexes. Detailed calculations based on the ISP-DC model carried to second-order were reported for the d-d optical activity in Cu(II) complexes of amino acids, dipeptides, and tripeptides.[56]

The most complete set of symmetry rules applicable to the d-d optical activity of chiral transition metal complexes have been set forth in the work of Schipper.[59,86,87] These symmetry (selection) rules apply to the spectroscopic moments appearing in the d-d rotatory strength expressions and may be related in a straightforward way to sector rules pertaining to ligand structural (stereochemical) features. Schipper's work is based on the general independent systems/perturbation model (which he refers to as the "separable chromophore model"), and includes consideration of both static-coupling and dynamic-coupling.

The only source of dissymmetry in metal complexes which poses a serious problem for the general ISP model is that of inherent dissymmetry within the metal ion-donor atom cluster. The neglect of covalent bonding interactions between the metal ion and the ligand donor atoms (or groups) may be expected to be a poor approximation when dealing with ML_n dissymmetry (where L denotes donor atoms). However, Mason[83] has achieved some success in treating the d-d optical activity of <u>all-cis</u>-$[M(A)_2(B)_2(C)_2]$ complexes using the ISP-DC model, and Richardson[26-29] has simply incorporated the ligand donor atoms into the ligand perturber set in his applications of the ISP-SC model. The extent to which the general ISP model can be successfully and usefully applied to sorting out the configurational, ligand conformational, and ligand vicinal contributions to d-d optical activity is determined in large part by how one chooses to partition the ligand environment into perturber fragments. The greater the number of perturber fragments (atoms, groups, <u>or</u> bonds) represented in the model, the better the chances are of accurately reflecting all of the stereochemical subtleties of the ligand environment. Of course, the greater the number of ligand perturber fragments included, the greater the scope of the calculations and the greater the demand for spectroscopic information about the ligands (and their component parts).

The molecular orbital and related methods for calculating d-d optical activity[41-43,45,46,48-50] have not proved to be particularly useful in sorting out spectra-structure relationships. The models and calculations dealing only with the metal ion-donor atoms cluster (ML_n) have, perhaps, yielded some insights regarding the sensitivity of d-d rotatory strengths to metal-donor atom orbital interactions (especially with regard to orbital overlaps and orbital hybridization - on the metal ion <u>and</u> on the ligand donor atoms).[41-43,45,50] These models are possibly the most appropriate for treating inherent dissymmetry within the ML_n cluster. The models and calculations which have included <u>all atoms</u> of the complexes[46,48,49] have yielded results which are useful primarily as checks or tests of results obtained empirically or from alternative calculational methods. These calculations have produced few, it any, new spectroscopic or structural insights.

8. OTHER TOPICS

There are several aspects of d-d optical activity theory which will not be discussed here. These include (a) solvent induced optical activity (achiral complex in a chiral solvent); (b) the Pfeiffer effect; and, (c) optical activity induced in an achiral complex by outer-sphere association of chiral species. Theoretical studies related to these phenomena have been reported by a number of workers[88-101] over the past five years, and significant progress has been made in understanding the underlying interaction mechanism and spectroscopic processes. The reader is especially referred to the recent series of papers authored by P. Schipper.[88-93,99-101] This series of papers encompasses all of the above-mentioned phenomena, and represents the most comprehensive and detailed work on the related theory.

As was mentioned in the Introduction, the optical activity associated with metal↔ligand charge-transfer transitions and with ligand-localized transitions lies outside the scope of this lecture. We have dealt only with the ligand-field (d-d) transitions localized on the metal ion chromophore. However, ligand-localized transitions entered into our independent systems/ perturbation (ISP) model by providing electric dipole character to the d-d transitions via the dynamic-coupling interaction mechanism. Similarly, the metal↔ligand charge-transfer transitions entered into our ISP model by providing electric dipole character to the d-d transitions via the static-coupling interaction mechanism. By the ISP model, therefore, the spectroscopic properties of the d-d, metal↔ligand charge-transfer, and ligand-localized transitions are related (interdependent), and there is some artificiality involved in discussing their chiroptical properties separately. However, the d-d contributions to the CD spectra of metal complexes can generally be separately identified and assigned, and it remains useful to treat these contributions with only indirect reference to other types of transitions which may occur in the complexes.

ACKNOWLEDGMENTS

This work was supported by the National Science Foundation and by a Teacher-Scholar Award from the Camille and Henry Dreyfus Foundation.

REFERENCES

1. A. Cotton, C.R.H. Acad. Sci., 120, 989, 1044 (1895).

2. Alfred Werner 1866-1919, Helv. Chim. Acta, Commemoration Volume IX ICCC, Zurich, 1966.

3. F.M. Jaeger, Spatial Arrangements of Atomic Systems and Optical Activity, George Fisher Baker Lectures, Vol. 7, Cornell University, McGraw-Hill, New York, 1930.

4. J.P. Mathieu, Les Theories Moleculaires du Pouvoir Rotatoire Naturel, Gauthier-Villars, Paris, 1946.

5. W. Kuhn and K. Bein, Z. Phys. Chem., Abt. B, 24, 335 (1934).

6. W. Kuhn and K. Bein, Z. Anorg. Allg. Chem., 216, 321 (1934).

7. W. Kuhn, Naturwissenschaften, 19, 289 (1938).

8. W. Kuhn, Z. Phys. Chem., Abt. B, 4, 14 (1929).

9. W. Kuhn, Trans. Faraday Soc., 26, 293 (1930).

10. W. Kuhn, Z. Phys. Chem., Abt. B, 20, 325 (1933).

11. Y. Saito, K. Nakatsu, M. Shiro, and H. Kuroya, Acta Crystallogr., 8, 729 (1955).

12. For the accepted nomenclature and notation regarding absolute configuration designations for coordination compounds, see:
(a) I.U.P.A.C. Information Bulletin No. 33, 68 (1968);
(b) Inorganic Chem., 9, 1 (1970).

13. W. Moffitt, J. Chem. Phys., 25, 1189 (1956).

14. E.U. Condon, W. Altar, and H. Eyring, J. Chem. Phys., 5, 753 (1937).

15. S. Sugano, J. Chem. Phys., 33, 1883 (1960).

16. See, for example: Y. Saito, Coord. Chem. Rev., 13, 305 (1974).

17. F.S. Richardson and J.P. Riehl, Chem. Rev., 77, 773 (1977).

18. (a) C.K. Luk and F.S. Richardson, J. Am. Chem. Soc., 97, 6666 (1975);

(b) H.G. Brittain and F.S. Richardson, Inorg. Chem., 15, 1507 (1976);
(c) H.G. Brittain and F.S. Richardson, J. Am. Chem. Soc., 98, 5858 (1976);
(d) H.G. Brittain and F.S. Richardson, J. Am. Chem. Soc., 99, 65 (1977);
(e) H.G. Brittain and F.S. Richardson, Bioinorg. Chem., 7, 233 (1977).

19. L. Rosenfeld, Z. Phys., 52, 161 (1928).

20. (a) H. Bethe, Ann. Phys., 5, 133 (1929);
(b) J.H. Van Vleck, Phys. Rev., 41, 208 (1932);
(c) W.G. Penney and R. Schlapp, Phys. Rev., 41, 194 (1932).

21. N.K. Hamer, Mol. Phys., 5, 339 (1962).

22. H. Poulet, J. Chim. Physique, 59, 584 (1962).

23. T.S. Piper and A. Karipedes, Mol. Phys., 5, 475 (1962).

24. M. Shinada, J. Phys. Soc. Jap., 19, 1607 (1964).

25. D.J. Caldwell, J. Phys. Chem., 71, 1907 (1967).

26. F.S. Richardson, J. Phys. Chem., 75, 692 (1971).

27. F.S. Richardson, J. Chem. Phys., 54, 2453 (1971).

28. F.S. Richardson, Inorg. Chem., 10, 2121 (1971).

29. F.S. Richardson, Inorg. Chem., 11, 2366 (1972).

30. R.W. Strickland and F.S. Richardson, J. Chem. Phys., 57, 589 (1972).

31. S.F. Mason, J. Chem. Soc. A, 667 (1971).

32. F.S. Richardson and G. Hilmes, Mol. Phys., 30, 237 (1975).

33. G. Hilmes and F.S. Richardson, Inorg. Chem., 15, 2582 (1976).

34. S. Sugano and M. Shinada, Abstracts International Symposium on Molecular Structure and Spectroscopy, Tokyo, 1962.

35. W. Moffitt, R.B. Woodward, A. Moscowitz, W. Klyne, and C. Djerassi, J. Am. Chem. Soc., 83, 4013 (1961).

36. J.A. Schellman, J. Chem. Phys., 44, 55 (1966).

37. S.F. Mason, in Fundamental Aspects and Recent Developments in Optical Rotatory Dispersion and Circular Dichroism, F. Ciardelli and P. Salvadori, Ed., Heyden and Son Ltd., New York, N.Y., 1973, Chapter 3.6.

38. T. Kato, J. Phys. Soc. Japan, 32, 192 (1972).

39. A.J. McCaffery and S.F. Mason, Mol. Phys., 6, 359 (1963).

40. Th. Bürer, Mol. Phys., 6, 541 (1963).

41. A. Liehr, J. Phys. Chem., 68, 665 (1964).

42. A. Liehr, J. Phys. Chem., 68, 3629 (1964).

43. A. Karipedes and T.S. Piper, J. Chem. Phys., 40, 674 (1964).

44. T.S. Piper and A.G. Karipedes, Inorg. Chem., 4, 923 (1965).

45. R.W. Strickland and F.S. Richardson, Inorg. Chem., 12, 1025 (1973).

46. R.S. Evans, A.F. Schreiner, and P.J. Hauser, Inorg. Chem., 13, 2185 (1974).

47. K. Nakatsu, Bull. Chem. Soc. Jap., 35, 832 (1962).

48. C.Y. Yeh and F.S. Richardson, Inorg. Chem., 15, 682 (1976).

49. G. Hilmes, C.Y. Yeh, and F.S. Richardson, J. Phys. Chem., 80, 1798 (1976).

50. C.E. Schäffer, Proc. Roy. Soc. London, Ser. A, 297, 96 (1968).

51. C.E. Schäffer, Struct. Bonding (Berlin), 5, 68 (1968).

52. C.E. Schäffer and C.K. Jorgensen, Mol. Phys., 9, 401 (1965).

53. R.H. Seal, Ph.D. Thesis, London University, 1974.

54. S.F. Mason and R.H. Seal, J.C.S. Chem. Comm., 331 (1975).

55. S.F. Mason and R.H. Seal, Mol. Phys., 31, 755 (1976).

56. R.W. Strickland and F.S. Richardson, J. Phys. Chem., 80, 164 (1976).

57. I. Tinoco, Advan. Chem. Phys., 4, 113 (1962).

58. E.G. Höhn and O.E. Weigang, Jr., J. Chem. Phys., 48, 1127 (1968).

59. P.E. Schipper, J. Am. Chem. Soc., 100, 1433 (1978).

60. What is referred to as the independent systems/perturbation model in the present article is designated the "separable chromophore model" by Schipper.[59]

61. T.R. Faulkner and F.S. Richardson, work in progress.

62. R.G. Denning, Chem. Comm., 120 (1967).

63. F.S. Ham, Phys. Rev., 138, A1727 (1965).

64. D. Caliga and F.S. Richardson, Mol. Phys., 28, 1145 (1974).

65. F.S. Richardson, D. Caliga, G. Hilmes, and J.J. Jenkins, Mol. Phys., 30, 257 (1975).

66. F.S. Richardson, G. Hilmes, and J.J. Jenkins, Theoret. Chim. Acta (Berl.), 39, 75 (1975).

67. G. Hilmes, D. Caliga, and F.S. Richardson, Chem. Phys., 13, 203 (1976).

68. M.J. Harding, J.C.S. Faraday II, 68, 234 (1972).

69. O.E. Weigang, J. Chem. Phys., 42, 2244 (1965).

70. O.E. Weigang, J. Chem. Phys., 43, 3609 (1965).

71. S.E. Harnung, E.C. Ong, and O.E. Weigang, J. Chem. Phys., 55, 5711 (1971).

72. O.E. Weigang and E.C. Ong, Tetrahedron, 30, 1783 (1974).

73. S. Kaizaki, J. Hidaka, and V. Shimura, Inorg. Chem., 12, 142 (1973).

74. G.L. Hilmes, H.G. Brittain, and F.S. Richardson, Inorg. Chem., 16, 528 (1977).

75. S. Kaizaki and Y. Shimura, Bull. Chem. Soc. Japan, 48, 3611 (1975).

76. S.F. Mason and B.J. Peart, J.C.S. Dalton, 937 (1977).

77. C.J. Ballhausen, Mol. Phys., 5, 461 (1963).

78. S. Kaizaki, J. Hidaka, and Y. Shimura, Bull. Chem. Soc. Japan, 43, 1100 (1970).

79. L.D. Barron, Mol. Phys., 21, 241 (1971).

80. C.W. Deutsche, J. Chem. Phys., 53, 1134 (1970).

81. T. Kato, I. Tsujikawa, and T. Murao, J. Phys. Soc. Japan, 34, 763 (1973).

82. T. Ito and M. Shibato, Inorg. Chem., 16, 108 (1977).

83. S.F. Mason, Mol. Phys., in press.

84. C.J. Hawkins, Absolute Configuration of Metal Complexes, Wiley-Interscience, New York, 1971, Chapter 3.

85. J.A. Hearson, S.F. Mason, and R.H. Seal, J.C.S. Dalton, 1026 (1977).

86. P.E. Schipper, Chem. Phys., 23, 159 (1977).

87. P.E. Schipper, J. Am. Chem. Soc., 100, 3658 (1978).

88. P.E. Schipper, J. Am. Chem. Soc., 98, 7938 (1976).

89. P.E. Schipper, Mol. Phys., 29, 1705 (1975).

90. P.E. Schipper, Inorg. Chim. Acta, 14, 161 (1975).

91. P.E. Schipper, Chem. Phys., 12, 15 (1976).

92. P.E. Schipper, Chem. Phys. Lett., 30, 323 (1975).

93. P.E. Schipper, Inorg. Chim. Acta, 12, 199 (1975).

94. S.F. Mason, Chem. Phys. Lett., 32, 201 (1975).

95. D.P. Craig, E.A. Power, and T. Thirunamachandran, Chem. Phys. Lett., 27, 149 (1974).

96. D.P. Craig and P.J. Stiles, Chem. Phys. Lett., 41, 225 (1976).

97. D.P. Craig, E.A. Power, and T. Thirunamachandran, Proc. Roy. Soc. London A, 348, 19 (1976).

98. B. Norden, Chem. Scripta, 7, 14 (1975).

99. P.E. Schipper, Inorg. Chim. Acta, 12, 199 (1975).

100. P.E. Schipper, Chem. Phys., 23, 159 (1977).

101. P.E. Schipper, J. Am. Chem. Soc., 100, 1079 (1978).

THE LIGAND POLARIZATION MODEL FOR THE SPECTRA OF METAL COMPLEXES

S. F. Mason

Chemistry Department, King's College,
London WC2R 2LS, England.

1. INTRODUCTION

Investigations of the optical activity of chiral metal complexes, discussed in the preceding Chapter, have the notable feature that the one-electron static-field model virtually monopolised the independent-systems approach to the problem over the two decades following the pioneer work of Moffitt [1] in 1956. In contrast, the dynamic-coupling and the static-field methods received comparable attention in discussions of chiral organic compounds over the same period, and any bias there was lay more towards the dynamic-coupling treatment, although the two models were generally taken to be complementary.

The one-electron static-coupling model for optical activity derived its inspiration, in the hands of Condon and coworkers [2,3], from the earlier development of crystal-field theory [4], and the general success of that theory over a range of problems connected with the transition energies of metal complexes ensured the preferred choice of the static-coupling model in the treatment of chiral coordination compounds. Crystal-field theory was developed by physicists primarily concerned with the line-spectra of the atomic ions and with the Stark-splitting of the individual lines in a static electric field, notably the lanthanide(III) spectra, which often remain line-like in the condensed phase [4]. The spectroscopy of organic π-systems, on the other hand, was based upon the earlier work of the organic-dye theorists with their concept of a basic light-absorbing group, the chromophore, modified by auxochromic substituents, which produced hyperchromic intensity-increases and bathochromic red-shifts in proportion to their

Stephen F. Mason (ed.), Optical Activity and Chiral Discrimination. 161–187.

polarizability. Organic molecular spectroscopy retained to a relatively late date [5] classical coupled-oscillator theory, from which the dynamic-coupling substituent-polarization model for optical activity derived [6,7].

In crystal field theory, the main spectroscopic role played by the large, soft, and highly-polarizable ligands, such as the halide ions, is that of the electrodes in a molecular Stark apparatus. The principal perturbation envisaged is that of the small, hard, metal cation by the ligands which, even in the case of neutral species, are much the more polarizable. The general tenor of crystal-field theory runs counter to the earlier rules of Fajans [8] who specified that the polarization of a cation by the anion in an ionic substance is negligible compared with the converse effect. Indeed Fajans argued that, since anhydrous $CuSO_4$ is colourless, the blue colour of the hydrate is due largely to the polarization of the water molecules by the metal cation, as is further demonstrated by the substantial intensity-increase and wavelength-shift of the light absorption resulting from the replacement of the water molecule by the more polarizable ammonia ligand [8].

2. TRANSITION-PROBABILITY ANOMALIES OF CRYSTAL FIELD THEORY.

Extensions of crystal-field theory from transition-energies, where the theory enjoyed considerable success, to transition-probabilities retained the basic model of a perturbation of the metal-ion electronic states by the static ligand field. The d-d or the f-f excitations of a transition-metal or lanthanide complex, electric-dipole forbidden by the rule of Laporte [9], acquire a first-order moment by mixing with the electric-dipole allowed transitions of the metal ion under a non-centrosymmetric ligand field, either static, from the stereochemistry of the complex, or oscillatory, due to the non-totally symmetric vibrational modes in a complex with a centre of inversion [10,11,12].

When the forced electric dipole mechanism was placed upon a quantitative basis for both transition-metal [13,14] and lanthanide complexes [15,16] a class of anomalous cases became evident in each of the series. While the oscillator strengths of the d-d excitations in octahedral transition metal complexes are accommodated by the crystal-field model [13], those of the corresponding transitions in tetrahedral analogues have calculated values which are as much as two orders of magnitude too small [14]. Similarly a class of hypersensitive f-f transitions in lanthanide complexes has been identified [17] with oscillator strengths ranging up to two orders of magnitude larger than the values calculated from the crystal-field model.

These f-f transitions are electric-quadrupole allowed, like the $^4I_{9/2} \rightarrow {}^4G_{5/2}$ transition of Nd(III) near 600 nm, which has an intensity compatible with the crystal-field treatment for the aquo-ion, but not for a range of other complexes, particularly the trigonal-planar halides NdX_3 in the vapour phase [18].

Further anomalies became evident from studies of the temperature dependence of the d-d absorption intensities in the tetrahedral series of transition metal complexes where a marked fall in the oscillator strength with increasing temperature is general [19]. As follows from the forced electric dipole mechanism in the crystal field treatment, the d-d oscillator strengths of octahedral transition metal complexes are found to increase as the temperature is raised, due to the increasing population of the higher vibrational levels of the non-totally-symmetric modes in the electronic ground state of the complex [20]. The similar trends expected for tetrahedral complexes are not observed, and the visible band of $[CoCl_4]^{2-}$, for example, has an oscillator strength at 1000°C little more than one-half of the corresponding ambient temperature value [19].

An additional transition-probability anomaly in the crystal field theory of tetrahedral complexes emerged from studies of the magnetically-induced circular dichroism studies of the cobalt(II) tetrahalides [21,22,23]. Measurements of the B and the C Faraday-effect terms from the MCD spectra of the halide complexes $[CoX_4]^{2-}$ afford a signed ratio, q, between the electric dipole moment of a transition connecting two different d-orbitals of t_2 symmetry in T_d to the corresponding moment of an excitation connecting an e with a t_2 d-orbital [21]. The Faraday effect involves a determinate phase relationship between these two types of d-electron promotion, the phase being expressed in the sign of the ratio, q. Experimentally the ratio, q, is found to be negative [23], whereas a positive value for the ratio is expected from the crystal-field model or from simple MO theory [21].

As is discussed in detail in the preceding Chapter, the problems of the crystal field model are profound in the treatment of the optical activity of chiral transition metal complexes. The possibility of a finite first-order crystal-field rotational strength was eliminated by Sugano [24] who, in his critique of the pioneer work of Moffitt [1], showed that the simplest pseudoscalar ligand field potential, with A_{1u} symmetry in O_h, has a nineth-order dependency on the electronic coordinates. The exploration of the corresponding second-order models, employing simultaneous T_{2u} and T_{2g} crystal field potentials in O_h symmetry, culminated in the demonstration that, for a wide range of parameter sets, the computed rotational strength ratios conflict qualitatively with the corresponding experimental values, even with the calculations extended to all orders of perturbation theory [25].

3. TRANSITION PROBABILITIES IN THE LIGAND-POLARIZATION MODEL.

The anomalies of the crystal-field treatment of d-d and f-f transition probabilities are resolved, within the independent-systems scheme, by the complementary dynamic-coupling model which allows for the perturbation of the ligand atoms by a potential originating from the central metal ion of a coordination compound [26-32]. Essentially the ligand polarization model for metal complexes is a generalisation of the $m_1 - \mu_2$ mechanism for optical activity [7], discussed in Chapter I, covering the dipole strengths, and the A, B and C terms of the Faraday effect, as well as the rotational strengths of chiral coordination compounds.

According to the ligand polarization model a d-d or f-f transition of a metal complex has its first-order electric dipole moment located wholly in the ligand atoms. The transient electric dipoles induced by the radiation field in the individual ligand atoms or groups are Coulombically correlated under the potential of the leading electric multipole, which is necessarily even, of the metal-ion transition, giving a non-vanishing resultant electric dipole in the first-order for determinate coordination symmetries. The selection rules of the ligand polarization model are given by the generalisation of equation (15) in Chapter I to,

$$(M_o M_m | V | L_o L_\ell) = \sum_L \mu^\alpha_{o\ell} \, M^{\beta\cdots\omega}_{om} G^L_{\alpha,(\beta\cdots\omega)} \qquad (1)$$

where $M^{\beta\cdots\omega}_{om}$ is one of the (2n + 1) components of the leading 2^n-pole electric moment of the metal-ion transition, $M_o \rightarrow M_m$, dependent upon the particular pair of d-orbitals, or of f-orbitals, which are connected, and $\mu^\alpha_{o\ell}$ is a component of the contribution of the ligand transition, $L_o \rightarrow L_\ell$, to the electric dipole induced by the radiation field. The geometric tensor, $G^L_{\alpha,(\beta\cdots\omega)}$, of equation (1) represents the radial and angular factors governing the Coulombic potential between the multipole component $M^{\beta\cdots\omega}_{om}$ centred on the metal ion and the dipole component $\mu^\alpha_{o\ell}$ located in a particular ligand atom or group, the sum being taken over all ligand groups, L. The Greek suffixes in equation (1) denote Cartesian components, $\theta_{\alpha\beta}$ representing the xy, or yz, or xz, etc., component of an electric quadrupole, and $H_{\alpha\beta\gamma\delta}$ a given component of a hexadecapole. Only these two types of charge-distribution, the electric 2^2-pole and the 2^4-pole, are allowed for d-d transitions, while a 2^6-pole is additionally permitted for f-f excitations.

The ligand-polarization matrix element of equation (1) is non-vanishing if the multipole component, $M_{om}^{\beta\cdots\omega}$, of the metal ion excitation and the dipole component, $\mu_{o\ell}^{\alpha}$, of the ligand groups are spanned by the same representation, or the same row of a degenerate representation, in the point group to which the coordination compound belongs [26]. If the condition is satisfied, the potential of the multipole component of the metal ion produces a constructive correlation of the dipoles induced in the several ligand groups, giving a non-vanishing resultant electric-dipole transition moment in the first-order.

The general selection rule indicates that the ligand polarization mechanism is forbidden for the equilibrium nuclear configuration of centrosymmetric complexes, since all even multipole components are *gerade* while the dipole components are all *ungerade*. In these cases the ligand polarization mechanism is dependent upon the loss of inversion symmetry through the non-totally symmetric vibrational modes of the complex, as in crystal field theory. Where the metal-ion transition is electric-quadrupole allowed, the ligand-polarization mechanism is effective at the equilibrium nuclear configuration in complexes belonging to the dihedral groups, D_p, and the C_{pv} groups, with p unrestricted, and to the groups T_d, D_{3h}, C_{3h}, and their subgroups. If the leading multipole moment of the metal-ion transition is a hexadecapole component, the mechanism is allowed over a similar range of point group symmetries, with the notable exception of T_d, although hexadecapole-dipole coupling is allowed in the isomorphous group, O.

The further development of the ligand polarization model follows that of the $m_1 - \mu_2$ mechanism discussed in Chapter I. For an electric-quadrupole allowed d-d or f-f transition of a metal complex, containing ligand groups with the mean polarizability $\overline{\alpha}(L)$ at the frequency ν_{om} of the metal-ion transition, the α-component of the first-order electric dipole moment is given by the expression,

$$\mu_{om}^{\alpha} = -\sum_{L}\sum_{\beta\gamma} \theta_{om}^{\beta\gamma}\,\overline{\alpha}(L)\,G_{\alpha\beta\gamma}^{L} \qquad (2)$$

The square of the first order electric dipole moment (equation 2) represents the dipole strength of the metal ion transition in the complex, D_{om}, and the product ($D_{om}\,\nu_{om}$) is related through universal constants to the corresponding oscillator strength, f_{om}.

Equation (2) accounts satisfactorily for the observed oscillator-strength of the hypersensitive f-f transitions of lanthanide complexes, which are generally electric-quadrupole allowed, and of the quadrupolar d-electron excitations in tetrahedral transition metal complexes [26,27]. The environmentally-sensitive $^4I_{9/2} \rightarrow {}^4G_{5/2}$ transition of Nd(III) near 600 nm, for example, has the oscillator strength (10^6f) of 5.6 for NdF_3 in the LaF_3 lattice, where the metal ion has C_2 site symmetry, compared with the corresponding value of 530 for NdI_3 in the vapour phase, where the molecule has D_{3h} symmetry. These oscillator strengths and those of other hypersensitive f-electron transitions are satisfactorily reproduced by the ligand-polarization mechanism [26,32], while the larger values remain anomalous in the crystal-field treatment.

The ligand-polarization mechanism accounts not only for the observed d-d oscillator strengths of tetrahedral transition metal complexes but also for the unusual decrease in those strengths with an increase in temperature [27,28]. In the cobalt(II) tetrahalides quadrupolar d-electron transitions occur from the $^4A_2$ ground state to each of the excited states, $^4T_1(F)$, observed in the near infrared, and $^4T_1(P)$, found in the visible region. The $d_{z^2} \rightarrow d_{xy}$ excitation, with the quadrupole component θ_{xy} as its leading moment, contributes to one of the three components of each of these two transitions. In a tetrahedral complex, the potential of the electric quadrupole transition moment, θ_{xy}, of the metal ion produces a constructive correlation of the z-component of the electric dipole induced in each of the four ligands (Figure 1). The other two ligand dipole components are analogously aligned by the field of the corresponding excitation quadrupole components, θ_{yz} and θ_{zx}. The resultant first-order electric-dipole moments of the three components of a given $[CoX_4]^{2-}$ d-electron transition differ only in polarization direction and have the same magnitude, being dependent upon a common geometric tensor of radial and angular factors,

$$\sum_L G^L_{\alpha\beta\gamma} = -15 \sum_L (XYZ)_L \, R^{-7}$$

$$= -(15/2) \sum_L (\cos\theta \sin^2\theta \sin 2\phi)_L \, R^{-4} \qquad (3)$$

where R is the metal-ligand bond length and $(XYZ)_L$ expresses the product of the Cartesian coordinates of the ligand atom L in the tetrahedral coordinate frame (Figure 1), with R, θ_L, ϕ_L, representing the equivalent spherical polar coordinates.

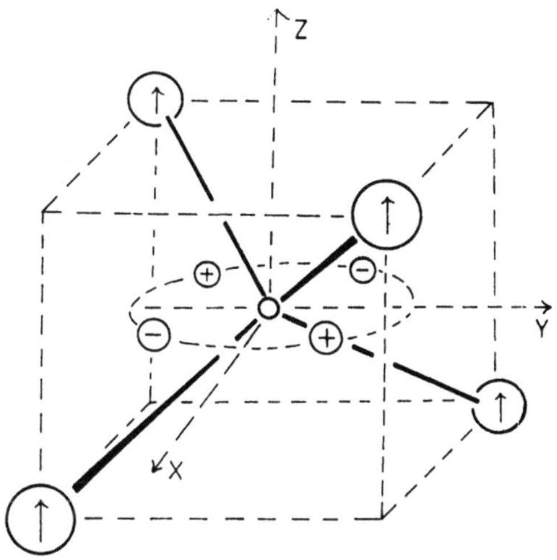

Figure 1. The Coulombic correlation of the z-component of the transient dipole induced in each of the ligands produced by the xy-component of the quadrupole moment of a metal-ion d-d transition in a tetrahedral metal complex.

For a given metal-ligand bond distance R, the geometric tensor of equation (3) has an optimum value in four-coordinate complexes at tetrahedral symmetry. All departures of the angles θ_L and ϕ_L from the tetrahedral values reduce the magnitude of the sum in equation (3). The sum goes to zero, for example, in the limit of square-planar four-coordination. The unusual decrease in the intensity of quadrupolar d-d transitions in tetrahedral complexes with an increase in temperature is thus a consequence of the progressive population of the higher levels of all of the tetrahedral vibrational modes as the temperature is raised [28]. The anharmonicity of the stretching modes, $\nu_1(a_1)$ and $\nu_3(t_2)$, results in a progressively longer mean metal bond-length with increasing temperature, which entails additionally progressively larger angular excursions of the ligands from their tetrahedral equilibrium nuclear configuration through the bending modes, $\nu_2(e)$ and $\nu_4(t_2)$. The loss of intensity is quantitatively accommodated by a calculation of the mean-square amplitudes of vibration in the electronic ground state as a function of temperature for each of the four tetrahedral vibrational modes. The relation considered is the temperature dependence of the dipole strength sum of the transitions, $^4A_2 \rightarrow {}^4T_1(F)$, $^4T_1(P)$ in the cobalt(II) tetrahalides where the configurational excitation probability,

Table 1. The ligand-polarization oscillator strengths ($f \times 10^4$) from equation (2), and the temperature-coefficient of the dipole strength ($C \times 10^4$) from equation (4), and the corresponding experimental values, for the d-d transitions to the $^4T_1(F)$ and $^4T_1(P)$ states from the $^4A_2$ ground state in the tetrachloride and the tetrabromide of cobalt(II), together with the observed transition moment ratio, q (equation 5), obtained from the B-term and the C-term of the corresponding MCD spectra.

		$[CoCl_4]^{2-}$		$[CoBr_4]^{2-}$	
Upper state		$^4T_1(F)$	$^4T_1(P)$	$^4T_1(F)$	$^4T_1(P)$
$\nu/10^3\ cm^{-1}$		5.5	14.7	5.0	14.0
$10^4 f$	calc	5.15	29.3	4.90	30.5
	obs	7.21	50.9	7.10	58.9
q	B-term	-0.6	∿0	-1.2	-0.6
	C-term	-1.4	-1.1	-1.1	–
$10^4 C$	calc	3.5		4.6	
	obs	5.6		6.2	

$^4A_2(e^4t_2^3) \rightarrow {}^4T_1(e^3t_2^4)$, is shared by the two state-transitions. At temperatures above ∿ 100°K the relation is linear, having the form,

$$D(T) = D_e(1 - CT) \quad (4)$$

where D(T) and D_e refer to the dipole-strength at the temperature T(K) and at the equilibrium tetrahedral nuclear configuration, respectively. The calculated value of the factor C in equation (4) compares satisfactorily with the observed value (Table 1).

The form of equation (2) indicates that a phase relationship observed between the first-order electric dipole moments of two d-d transitions in a tetrahedral metal complex is a reflection of a corresponding phase relationship between the zero-order quadrupolar moments of those transitions. The MCD spectra of the d-d transitions from the $^4A_2$ to the $^4T_1(F)$ and

$^4T_1(P)$ excited state of the cobalt(II) tetrahalides afford the reduced dipole moment ratio, q, given by,

$$\langle t_2||\mu||t_2\rangle/\langle e||\mu||t_2\rangle = q \tag{5}$$

The crystal field model requires that q = +2, and simple MO theory indicates that the ratio, q, has a positive value [21]. Measurements of the B and the C Faraday effect terms, from the temperature-variation of the MCD spectra of the cobalt(II) tetrahalides [22,23,29], show that the ratio, q, is generally negative, with a value (Table 1) close to that expected, according to the ligand polarization model, from the corresponding ratio of the reduced quadrupole moments [29],

$$\langle t_2||\theta||t_2\rangle/\langle e||\theta||t_2\rangle = -[3/2]^{\frac{1}{2}} \tag{6}$$

4. LIGAND POLARIZATION OPTICAL ACTIVITY

The most widely investigated chiral metal complexes are the diamine chelates of cobalt(III), containing the octahedral $[Co^{(III)}N_6]$ chromophore. The principal transition of interest for the d-electron optical activity of these chiral complexes is the $^1A_1 \rightarrow {}^1T_1$ octahedral excitation near 465 nm, made up of the three single-orbital promotions, $d_{xy} \rightarrow d_{x^2-y^2}$, and the analogues obtained by the cyclic permutation of the electronic coordinates.

The leading moments of the transition, $d_{xy} \rightarrow d_{x^2-y^2}$, are the z-component of a magnetic dipole and the $[xy(x^2-y^2)]$-component of an electric hexadecapole. The potential of the electric hexadecapole component, $H_{xy(x^2-y^2)}$, produces a constructive correlation of the z-component of the induced electric dipole moment in each ligand group which does not lie in an octahedral symmetry plane of the $[Co^{(III)}N_6]$ cluster (Figure 2). The resultant first-order electric dipole transition moment is collinear with the zero-order magnetic dipole moment of the $d_{xy} \rightarrow d_{x^2-y^2}$ transition, and the scalar product of these two moments affords the z-component of the rotational strength, R_{om}^{z}, of the $^1A_1 \rightarrow {}^1T_1$ octahedral excitation,

$$R_{om}^{z} = i\, m_{mo}^{z}\, H_{om}^{xy(x^2-y^2)} \sum_{L} \overline{\alpha}(L)\, G^{L}_{z,xy(x^2-y^2)} \tag{7}$$

If it is assumed that the d-d transitional charge distribution is a point-hexadecapole and that the moment induced in the ligand group is a point-dipole at the position, X,Y,Z in the octahedral Cartesian frame (Figure 2) at a distance R from the

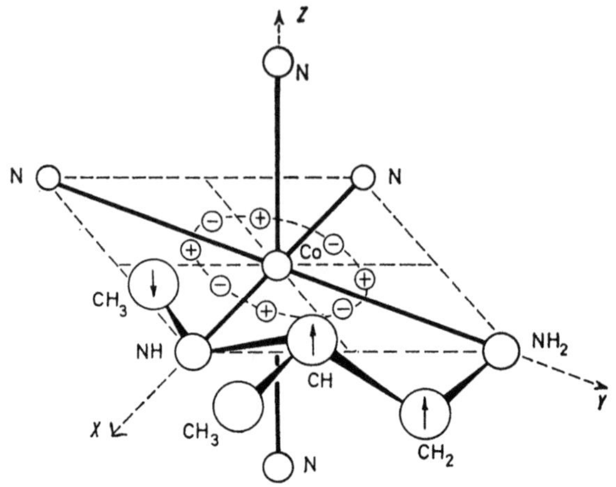

Figure 2. The correlation produced by the electric hexadecapole moment of the $d_{xy} \rightarrow d_{x^2-y^2}$ transition of the $[Co^{(III)}N_6]$ chromophore of the transient electric dipole induced in each ligand group of a 1,2-diamine chelate ring in the δ-conformation with a N-methyl and a C-methyl substituent (N-methyl-(S)-(+)-propylene-diamine).

metal ion, the geometric tensor in equation (7) has the form,

$$G_{z,xy(x^2-y^2)} = 315\ XYZ(Y^2-X^2)/2R^{11} \qquad (8)$$

The other two components of the octahedral $^1A_1 \rightarrow {}^1T_1$ transition, due to the single-orbital promotions, $d_{yz} \rightarrow d_{y^2-z^2}$, and $d_{zx} \rightarrow d_{z^2-x^2}$, give rise analagously to the rotational strength components, R^x_{om} and R^y_{om}, respectively. The forms of the latter two components are obtained in the point-multipole approximation by cyclic permutation of the coordinates in equations (7) and (8).

For a chiral complex containing the $[Co^{(III)}N_6]$ chromophore and a chiral chelate ring spanning the X and the Y axes (Figure 2), the sum of R^x_{om} and R^y_{om} is equal in magnitude and opposite in sign to R^z_{om}, and the net first-order ligand-polarization rotational strength, $R(T_1)$, is zero in the point-multipole approximation for a complex of O symmetry. If the symmetry of the chromophore is reduced, so that the components of the $^1T_1$ octahedral state are separated in energy, the oppositely-signed component rotational strengths no longer mutually cancel, appearing at different frequencies in the CD spectrum. Thus equation (8) provides the dynamic-coupling basis for the hexadecant regional rule relating the d-electron optical activity of chiral *trans*-dihalo-bis(diamine)cobalt(III) complexes to the location of the ligand groups in the tetragonal frame [33]. The marked contribution of N-alkyl groups, opposed in sign to that of the chelate-ring alkyl groups in a given chiral tetragonal complex [34], is rationalised by the ligand-polarization model (Figure 2), as is their minor contribution to $R(T_1)$ in complexes containing the $[Co^{(III)}N_6]$ octahedral chromophore [35].

The point-multipole approximation is satisfactory only when the distance R separating the charge-distributions considered is large compared with their individual dimensions. The radial factor for the hexadecapole moment of a 3d-electron transition is $\langle 3d|r^4|3d\rangle$, and the double-exponent 3d-functions of Richardson and coworkers [36] give radial extensions in the range from 0.956 to 0.684 Å for neutral and tripositively-charged cobalt, respectively. These extensions provide the upper and the lower bound of the hexadecapole moment radial maximum, and neither value is negligible relative to the Co-N bond length (2.0 Å) or the metal-carbon distance (3.0 Å) in the tris-ethylenediamine complex, $[Co(en)_3]^{3+}$.

An allowance for the finite radial extension is made by summing the potential between each pole of the hexadecapole moment and a particular ligand-group point dipole, and a subsequent sum is taken over all ligand groups [30,35]. By this procedure the net rotational strength, $R(T_1)$, of a chiral complex containing an octahedral chromophore is found to be non-vanishing. The dominant component of the octahedral $^1A_1 \rightarrow {}^1T_1$ cobalt(III) transition has its charge distribution in the mean plane of the chelate ring considered, e.g. the $d_{xy} \rightarrow d_{x^2-y^2}$ component for a ring spanning the X and the Y axis of the chromophore (Figure 2). The sign of the net rotational strength, $R(T_1)$, due to a 1,2-diamine chelate ring is governed by the ring chirality, being positive and negative for the λ- and the δ-conformation, respectively, and $R(T_1)$ is additive over the number of rings with the same chirality, as is observed experimentally. A hypothetical complex with all octahedral edges of the $[Co^{(III)}N_6]$ chromophore spanned by equivalent rings with a common chirality, having O symmetry, is expected to display a non-vanishing first-order net rotational strength, $R(T_1)$, according to the ligand polarization model [30,35] in contrast to the corresponding crystal field treatments [24,25].

In the dihedral tris-diamine complexes of cobalt(III) the upper state of the octahedral $^1A_1 \rightarrow {}^1T_1$ d-electron transition is broken down into components with $^1A_2$ and ^1E symmetry in the D_3 group. The sum of the rotational strengths of the two

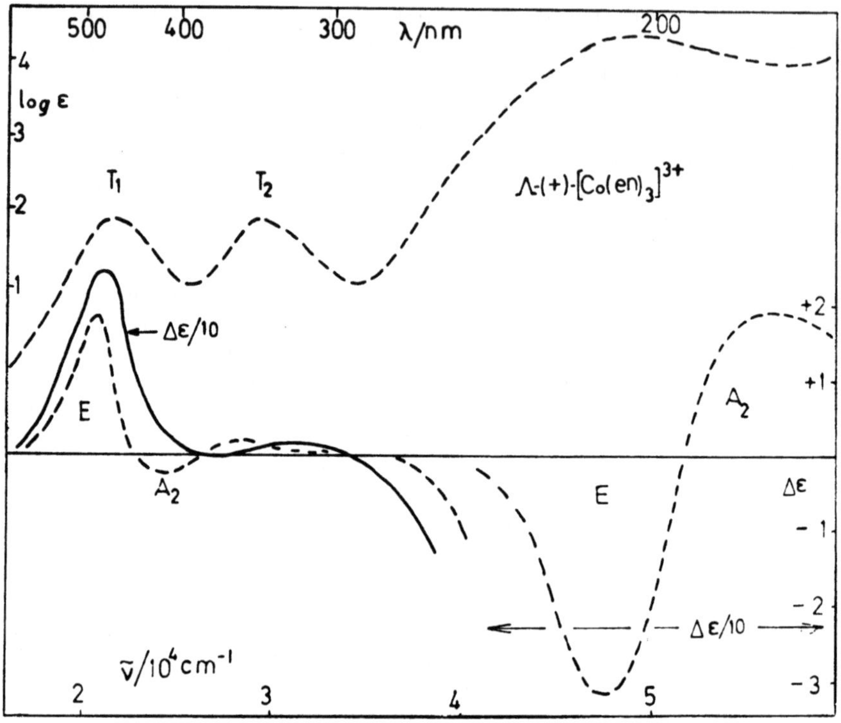

Figure 3. The axial single-crystal CD spectrum (solid curve, $\Delta\varepsilon/10$) of $\{\Lambda\text{-}(+)\text{-}[Co(en)_3]Cl_3\}_2.NaCl.6H_2O$ and the absorption and CD spectrum (upper and lower broken curves, respectively) of $\Lambda\text{-}(+)\text{-}[Co(en)_3](ClO_4)_3$ in water (D_2O below 200 nm).

components, the net resultant, $R(T_1)$, is afforded by the CD spectrum of the complex ion randomly-orientated in solution. With the complex ion oriented in a uniaxial single crystal, the rotational strength of the individual component, R(E), is obtained directly by CD measurements with the radiation propagated along the optic axis (Figure 3), and that of the other component $R(A_2)$ is given indirectly by reference to the sum, $R(T_1)$ [37,38]. Recently the rotational strengths of both of the individual components, $R(A_2)$ and R(E), have been determined directly for the prototype case of $\Lambda\text{-}(+)\text{-}[Co(en)_3]^{3+}$ from ortho-axial single-crystal CD measurements [38].

The rotational strengths $R(A_2)$ and R(E) of $\Lambda\text{-}(+)\text{-}[Co(en)_3]^{3+}$, and the more precisely-measured sum, $R(T_1)$, are reproduced quantitatively by extending the ligand polarization treatment to second order [30]. While the crystal field and the ligand polarization methods are mutually exclusive and complementary to first order, hybrid treatments become feasible to higher orders. The particular second order treatment adopted considers the mixing of different electronic transitions of the $[Co^{(III)}N_6]$ chromophore through the Coulombic field of the induced electric dipoles in the individual ligand groups [30,35].

The vacuum UV and CD spectrum of Λ-(+)-$[Co(en)_3]^{3+}$, including the high wavenumber tail of the single-crystal CD spectrum, characterise two strong electric-dipole transitions in the UV region, the first to a ^1E state and the second, at higher energy, to a $^1A_2$ state (Figure 3). These two UV transitions provide the main source from which electric dipole strength is borrowed within the $[Co^{(III)}N_6]$ chromophore by the d-electron excitations in the visible region. In the second order model, the induced electric dipoles in the individual ligand groups, themselves aligned by the electric hexadecapole moment of a given D_3 component of the $^1A_1 \rightarrow {}^1T_1$ d-electron transition in the visible region, produce in turn a correlation of the electric dipole moment of the corresponding D_3 component in the high-intensity UV transition of the $[Co^{(III)}N_6]$ chromophore. Computations based on the second-order model account satisfactorily for the d-d rotational strengths, and the dipole strength enhancement relative to $[Co(NH_3)_6]^{3+}$, of the tris-diamine cobalt(III) complexes for which an X-ray crystal structure is available [39], covering some ten cases in all [30].

In all of these cases, the alkyl ligand groups are necessarily located close to the equatorial XY plane and are relatively distant from the polar $C_3(Z)$ axis of the D_3 complex. The addition of polarizable groups to the polar region of Λ-(+)-$[Co(en)_3]^{3+}$ (I) and its analogues changes the relative contribution of the two dihedral components of the octahedral $^1A_1 \rightarrow {}^1T_1$ cobalt(III) transition to the net rotational strength, $R(T_1)$, the component $R(A_2)$ being enhanced at the expense of the component $R(E)$. The change is apparent in the effects of N-alkyl substitution in Λ-(+)-$[Co(en)_3]^{3+}$ [35], ion-pair formation with multiply-charged oxyanions [40] e.g. phosphate, which has been shown to have the polar disposition (II) in an X-ray crystal structure [41], the alkyl-bridge capping of the complex in Λ-(+)-$[Co(sen)]^{3+}$ (III) [42], and the bicapping of Λ-(+)-$[Co(en)_3]^{3+}$, by treatment with formaldehyde and ammonia, to form Λ-(-)-$[Co(sep)]^{3+}$ (IV) [43] (Figure 4).

In all of these cases (II) - (IV) the alkyl groups of the ethylenediamine chelate rings, with a location in the equatorial region of the Λ-(+)-$[Co(en)_3]^{3+}$ moiety, make a contribution to the d-electron optical activity. Such contributions are absent in the complex $[Co(R\text{-}MeTACN)_2]^{3+}$ (V), containing the ligand

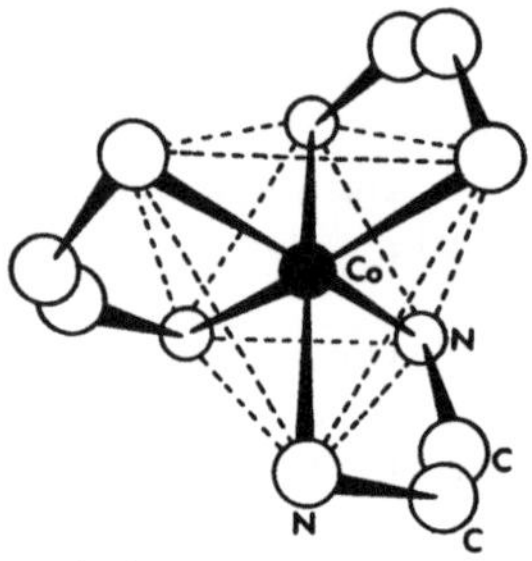

(I)

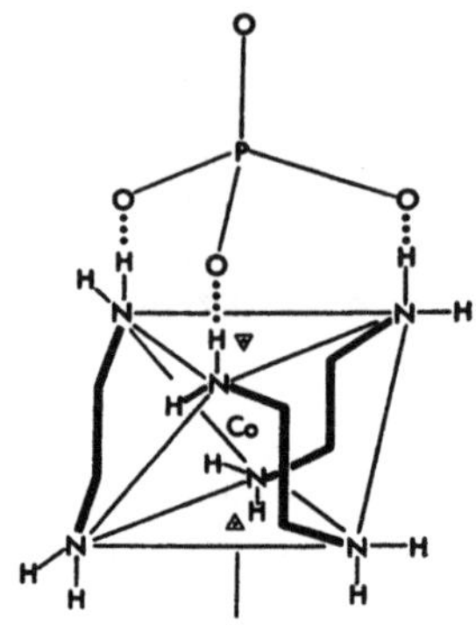

(II)

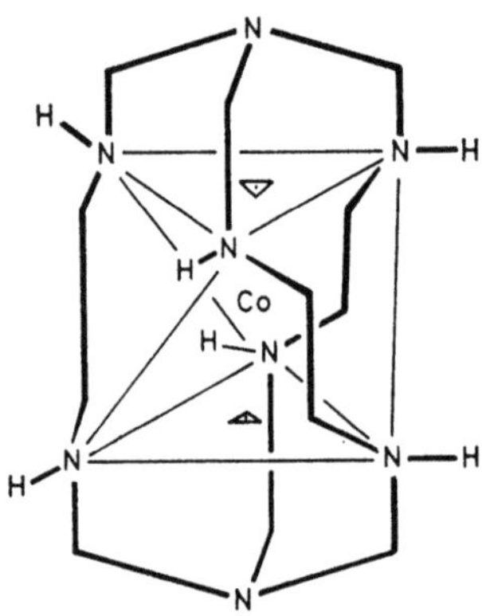

(IV)

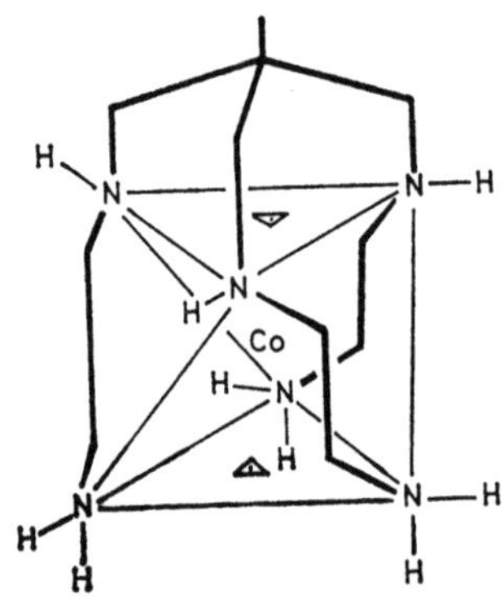

(III)

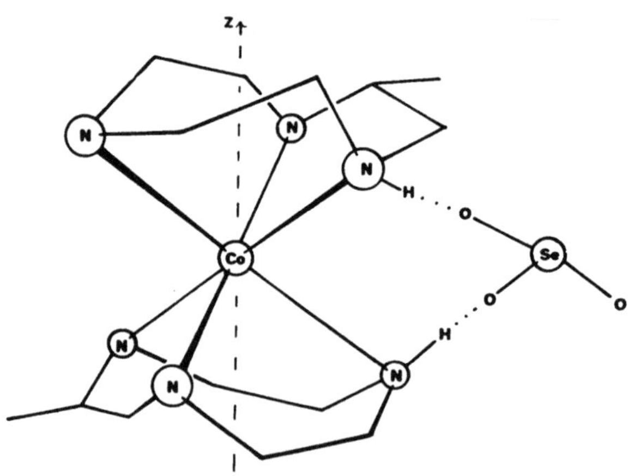

(VI)

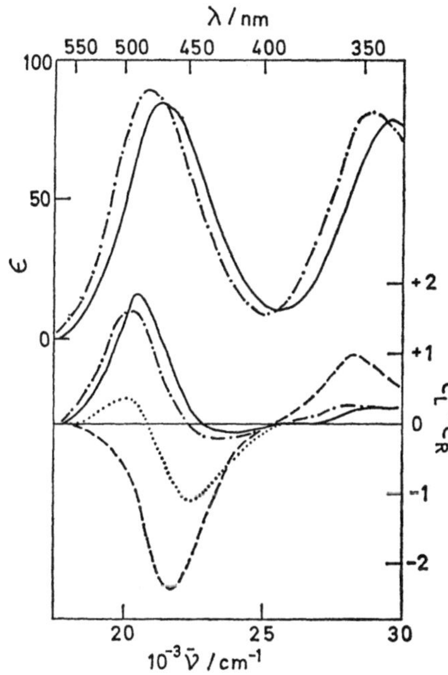

Figure 4. Absorption spectra (upper curves) and CD spectra (lower curves) of Λ-(+)-$[Co(en)_3]^{3+}$ (I) (———) and its NN'-dimethyl derivative (-•-•-), and the CD spectra of Λ-(+)-$[Co(sen)]^{3+}$ (III) (•••••), and of Λ-(-)-$[Co(sep)]^{3+}$ (IV) (-----).

R-(-)-2-methyl-1,4,7-triazacyclononane, derived from R-(-)-1,2-propylenediamine [44]. The individual chelate rings of each ligand in $[Co(R\text{-}MeTACN)_2]^{3+}$ have a common preferred λ-conformation, and all of the alkyl groups have a more polar location, nearer to the $C_3(Z)$ axis, as is shown by the X-ray crystal structure [45] of $[Co(R\text{-}MeTACN)_2]I_3.5H_2O$, which is optically-uniaxial, belonging to the space group R32.

The axial CD spectrum of a single crystal of $[Co(R\text{-}MeTACN)_2]I_3.5H_2O$, compared with the corresponding solution CD spectrum of the complex ion (V) (Figure 5), shows that, of the two D_3 components of the octahedral $^1A_1 \rightarrow {}^1T_1$ cobalt(III) d-electron transition near 470 nm, the major contribution to the optical activity arises from the $^1A_1 \rightarrow {}^1A_2$ component, with R(A) = +0.32 Debye-Bohr magneton, compared to R(E) = -0.16 $D\beta_M$ [46,47]. Multiply-charged oxyanions, in the case of $[Co(R\text{-}MeTACN)_2]^{3+}$ (V), enhance R(E) at the expense of R(A_2) (Figure 5), suggesting that the anion has a preferred equatorial location in the ion-pair (VI).

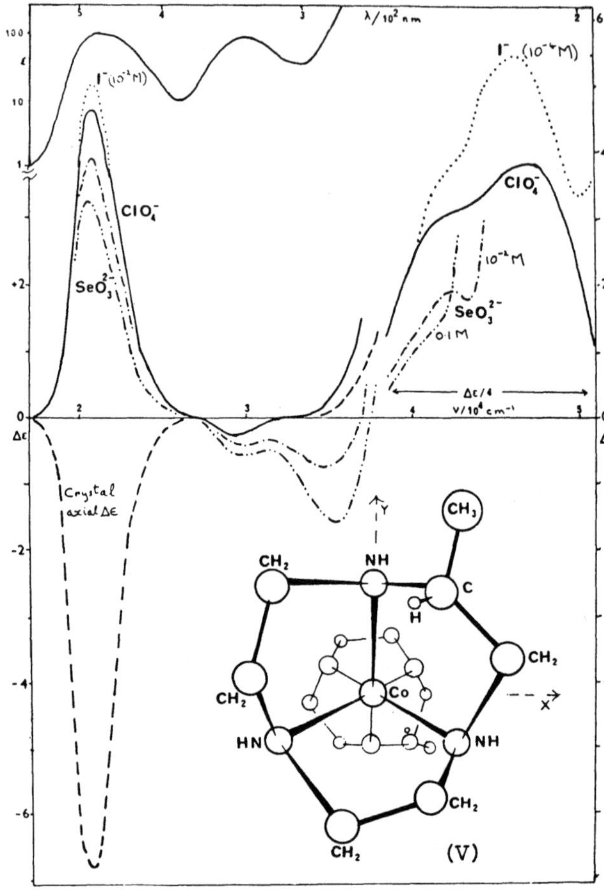

Figure 5. The axial single-crystal CD spectrum of $[Co(R\text{-}MeTACN)_2]I_3.5H_2O$ (broken curve) and the absorption (upper) and CD spectrum (lower solid curve) of $[Co(R\text{-}MeTACN)_2](ClO_4)_3$ in water and in aqueous solutions of the ions indicated.

First-order ligand polarization calculations of the rotational strengths of $[Co(R\text{-}MeTACN)_2]^{3+}$, based upon the atomic coordinates determined in the X-ray crystal structure analysis of the tri-iodide pentahydrate [45], give the values, +0.66, -0.64, and +0.02 $D\beta_M$ for $R(A_2)$, $R(E)$, and $R(T_1)$, respectively. While the calculated values have the correct signs and relative magnitudes, they are not quantitatively satisfactory [46,47]. The same is true for the corresponding calculations carried out on the bicapped complex

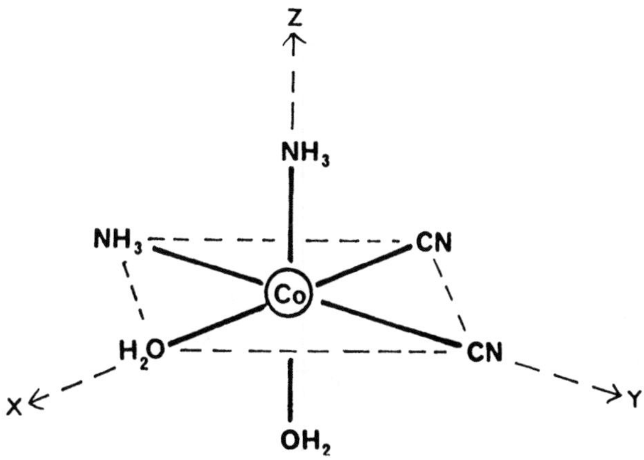

(VII) R-(+)-*all-cis*-$[Co(NH_3)_2(H_2O)_2(CN)_2]^+$

Λ-(-)-$[Co(sep)]^{3+}$ (IV), for which the theoretical and experimental values of $R(T_1)$ are -0.010 and -0.068 $D\beta_M$, respectively [47].

Further ligand polarization calculations have been carried out on the rotational strengths of the components of the octahedral $^1A_1 \rightarrow {}^1T_1$ d-electron transition in chiral cobalt(III) complexes containing solely monodentate ligands, of the *all-cis*-$[Co^{(III)}(A)_2(B)_2(C)_2]$ type, exemplified by the recent synthesis of R-(+)-*all-cis*-$[Co(NH_3)_2(H_2O)_2(CN)_2]^+$ (VII) and S-(+)-*all-cis*-$[Co(NH_3)_2(H_2O)_2(NO_2)_2]^+$ [48]. All three components of the octahedral $^1A_1 \rightarrow {}^1T_1$ transition are resolved in the CD spectrum of (VII) (Figure 6) [48].

The general problem of the origin of the d-electron optical activity in chiral complexes of the *all-cis*-$[M(A)_2(B)_2(C)_2]$ type, where M is a transition metal ion, A,B, and C, are monodentate, isotropic, and cylindrically-symmetric ligands of different chemical species, and the coordination is octahedral, requires at the minimum three pairwise interactions in the dynamic-coupling approach, necessitating recourse to third-order perturbation theory. In the third-order model adopted [49], the potential of the electric hexadecapole moment of each component of the octahedral $^1A_1 \rightarrow {}^1T_1$ cobalt(III) transition produces a primary alignment of the electric dipole induced in each ligand sharing a common σ_h plane of the coordination octahedron with the transitional charge-distribution of the metal ion. Concomitant ligand-ligand interactions give rise to a secondary and a tertiary correlation of the ligand dipoles, the latter affording an electric dipole alignment collinear

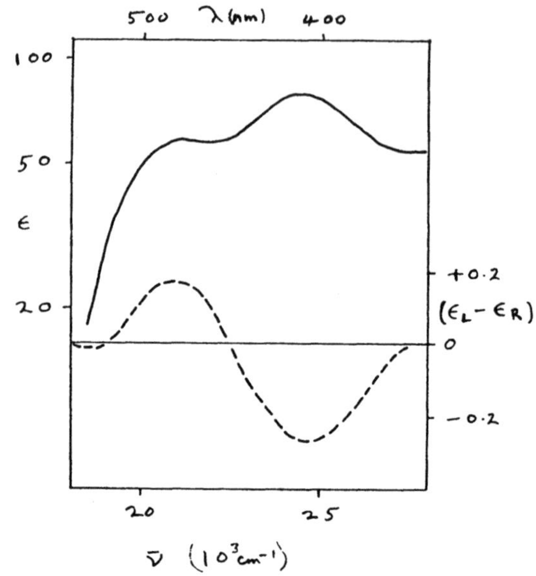

Figure 6. The absorption spectrum (upper curve) and the circular dichroism (lower curve) of R-(+)-*all-cis*-$[Co(NH_3)_2(H_2O)_2(CN)_2]^+$ (VII) in the region of the $^1A_1 \rightarrow {}^1T_1$ octahedral cobalt(III) d-electron transition (adapted from reference [48]).

with the zero-order magnetic dipole of the metal-ion transition.

Thus, the $d_{xy} \rightarrow d_{x^2-y^2}$ transition, with the z-component of a magnetic dipole and the electric hexadecapole, $H_{xy(x^2-y^2)}$, as its leading moments, entails through the potential of the latter an alignment of the x-component of the electric dipoles induced in the ligands located on the Y-molecular axis, and of the y-component for the ligands situated on the X-axis of the complex (Figure 7). The induced dipoles of the ligands on the Z-axis remain uncorrelated to the first order. The primary correlation of, e.g., the y-component of the dipole induced in the ligand B_1, namely, $\mu_y^{(1)}(B_1)$, results in the secondary orientations $\mu_y^{(2)}(A_1)$, $\mu_y^{(2)}(A_2)$, $\mu_y^{(2)}(C_1)$, and $\mu_y^{(2)}(B_2)$, Each of the latter in turn give rise to orthogonal tertiary correlations, e.g. $\mu_y^{(2)}(A_1)$ produces the alignments, $\mu_z^{(3)}(A_2)$ and $\mu_z^{(3)}(C_1)$, which are collinear with the zero-order magnetic moment, m_{mo}^z (Figure 7).

If all the metal-ligand bonds have equal lengths, the third-order treatment gives a vanishing d-electron optical activity for chiral complexes of the type, *all-cis*-

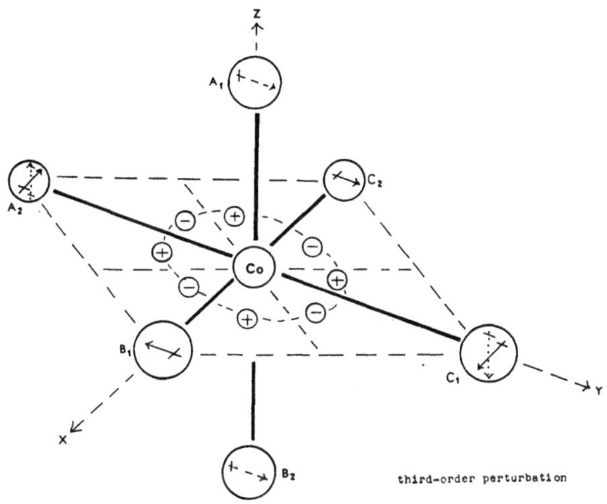

Figure 7. The primary correlation produced by the hexadecapole charge distribution of the $d_{xy} \rightarrow d_{x^2-y^2}$ component of the $^1A_1 \rightarrow {}^1T_1$ cobalt(III) transition of the induced dipoles in the ligands B_1, A_2, C_1, and C_2, (full arrows). Representative secondary alignments of induced dipoles with a common orientation, arising from the particular primary correlation of $\mu_y^{(1)}(B_1)$ are $\mu_y^{(2)}(A_1)$, and $\mu_y^{(2)}(B_2)$ (dashed arrows). The orthogonal tertiary correlations due to the particular secondary orientation $\mu_y^{(2)}(A_1)$ are $\mu_z^{(3)}(A_1)$ and $\mu_z^{(3)}(C_1)$ (dotted arrows).

$[Co(A)_2(B)_2(C)_2]$, even though the three types of ligand have different mean polarizabilities. In a complementary fourth-order treatment the d-electron rotational strengths are non-vanishing even if the coordination octahedron is regular. The fourth-order model considers the pairwise mixing of the three components of the octahedral $^1A_1 \rightarrow {}^1T_1$ cobalt(III) d-electron transition mediated by the potential between the induced ligand dipoles, which are individually aligned by the hexadecapole moment of a component of the d-d transition. Thus the potential of the hexadecapole moment of the $d_{xy} \rightarrow d_{x^2-y^2}$ transition produces the primary alignment of the ligand dipole components, μ_y in B_1 and C_2, and μ_x in A_2 and C_1 (Figure 8). However, the x- and the y-component of the induced dipoles in a different ligand set, those situated on the Z-axis, are analogously orientated by the field of the hexadecapole moment of the $d_{zx} \rightarrow d_{z^2-x^2}$ and the $d_{yz} \rightarrow d_{y^2-z^2}$ transition, respectively (Figure 8). The field of both of the latter hexadecapole moments also produce a correlation of the z-component of the

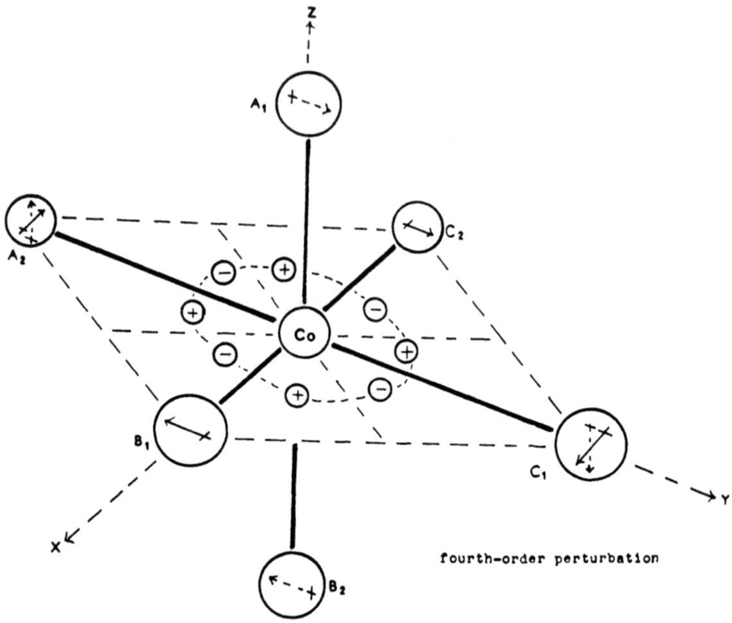

Figure 8. The correlation produced by the hexadecapole charge distribution of the $d_{xy} \rightarrow d_{x^2-y^2}$ component of the $^1A_1 \rightarrow {}^1T_1$ cobalt(III) transition of the induced dipoles in the ligands B_1, A_2, C_1, and C_2 (full arrows), and the corresponding dipole orientation in the ligands B_1, A_1, A_2, and C_1 (broken arrows) due to the field of the hexadecapole moment of the $d_{yz} \rightarrow d_{y^2-z^2}$ component.

ligand dipoles, e.g. $\mu_z(A_2)$ and $\mu_z(C_1)$ by the field of $H_{yz(y^2-z^2)}$ (Figure 8), so that the $d_{xy} \rightarrow d_{x^2-y^2}$ excitation attains a fourth-order electric dipole moment collinear with its zero-order magnetic moment, m^z_{mo}, by mixing with the other two components of the octahedral $^1A_1 \rightarrow {}^1T_1$ cobalt(III) transition through the potential between the individually-correlated ligand dipoles with a common orientation.

Following the crystal-field or the MO energy-perturbation treatment of Yamatera [50] and subsequent workers, or the angular-overlap model [51], the three CD bands exhibited over the octahedral $^1A_1 \rightarrow {}^1T_1$ region (Figure 6) by R-(+)-*all-cis*-$[Co(NH_3)_2(H_2O)_2(CN)_2]^+$ (VII) are assigned [48] to the following components, in serial order of increasing frequency, $d_{zx} \rightarrow d_{z^2-x^2}$ (ν_1), $d_{yz} \rightarrow d_{y^2-z^2}$ (ν_2), and $d_{xy} \rightarrow d_{x^2-y^2}$ (ν_3). The observed wavenumber and rotational strength of each of the components, and the corresponding sum of the calculated third and fourth order rotational strengths, are as listed (Table 2).

Table 2. The frequency and the calculated and the observed rotational strength of each component of the $^1A_1 \rightarrow {}^1T_1$ d-electron transition of (VII).

Component		ν_1	ν_2	ν_3
$\nu/10^3$ cm^{-1}		18.6	21.1	24.6
$R/10^{-4} D\beta_M$	obs	-1.3	+44	-75
	calc	-2.3	+28	-16

The dynamic coupling between the electric hexadecapole moment of the metal ion d-electron transition and the induced electric dipole in the ligand groups remains the dominant ligand polarization mechanism for optical activity in chiral four-coordinate complexes containing a square-planar chromophore, such as the bis(diamine) complexes of palladium(II) and platinum(II). This particular mechanism is forbidden in four-coordinate complexes containing a tetrahedral chromophore, although it is formally allowed on reduction to D_{2d} chromophoric symmetry. The dynamic coupling between the d-electron quadrupole transition moment and the dipoles of the ligand groups becomes the principal ligand polarization mechanism for d-electron optical activity in the tetrahedral coordination case.

The main series of chiral complexes with tetrahedral coordination investigated as yet are the $[M^{(II)}(\text{diamine})X_2]$, where M is cobalt, nickel, or copper, X is a halide or pseudo-halide ion, and the diamine is chiral and di-tertiary, notably, R-(-)-N,N,N',N'-tetramethyl-1,2-propylenediamine and its analogues, or (-)-spartein [52] and its epimers [53,54]. The X-ray crystal structures of [Cu{(-)-β-isospartein}Cl_2] [55], [Co{(-)-spartein}Cl_2] [56], and [Co{(-)-α-isospartein}Cl_2] [57], show that the coordination is tetrahedral in the series although the chromophoric symmetry at the highest is C_2 (Figure 9), with an average dihedral angle between the MCl_2 and the MN_2 plane of 3°, 8°, and 19°, in the three complexes, respectively.

The quadrupole-allowed d-electron transitions of tetrahedral cobalt(II) from the $^4A_2$ ground state and to the $^4T_1(F)$ and $^4T_1(P)$ excited state, lying in the near infrared and the

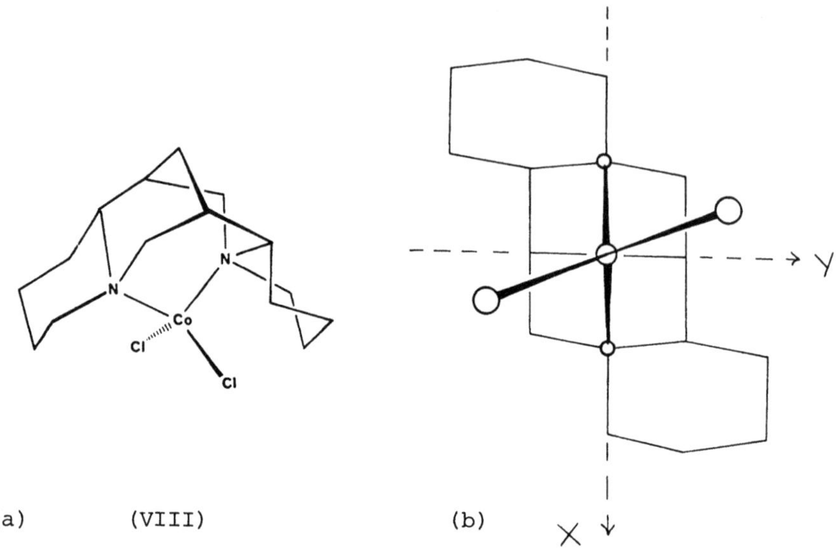

(a) (VIII) (b)

Figure 9. Perspective view (a) and projection on the plane perpendicular to the $C_2(Z)$ axis (b) of the molecular structure of [Co{(-)-α-isospartein}Cl_2] (VIII).

visible region, respectively. All three components of the transition to the tetrahedral $^4T_1(F)$ state are evident in the CD spectrum of [Co{(-)-α-isospartein}Cl_2] (VIII) over the 5000 to 12,000 cm^{-1} range, and two of the three $^4T_1(P)$ components are prominent in the range 15,000 to 22,000 cm^{-1} (Figure 10). The components of the hexadecapole-allowed transition to the tetrahedral $^4T_2$ state near 3000 cm^{-1} are overlaid by the stronger absorption due to the ligand C-H vibrational fundamentals.

A number of single d-orbital promotions are both magnetic-dipole and electric-quadrupole allowed, e.g. $d_{z^2} \rightarrow d_{yz}$ has the x-component of a magnetic dipole and the yz-component of a quadrupole as its leading moments, but none of the component transitions of (VIII) correspond largely to a particular single-orbital excitation. The d-orbital composition of each component of the tetrahedral $^4T_1(F)$ and $^4T_1(P)$ excited state of (VIII) is afforded by the angular overlap model [51], employing a comparison of the theoretical and the observed transition frequencies as a criterion in the choice of parameters. With the d-orbital composition of each component excited state of (VIII), a ligand polarization calculation gives, from the correlation of the induced ligand-group dipoles produced by the quadrupolar transition moment of the metal ion, the first-order electric dipole moment (equation 2) collinear with the

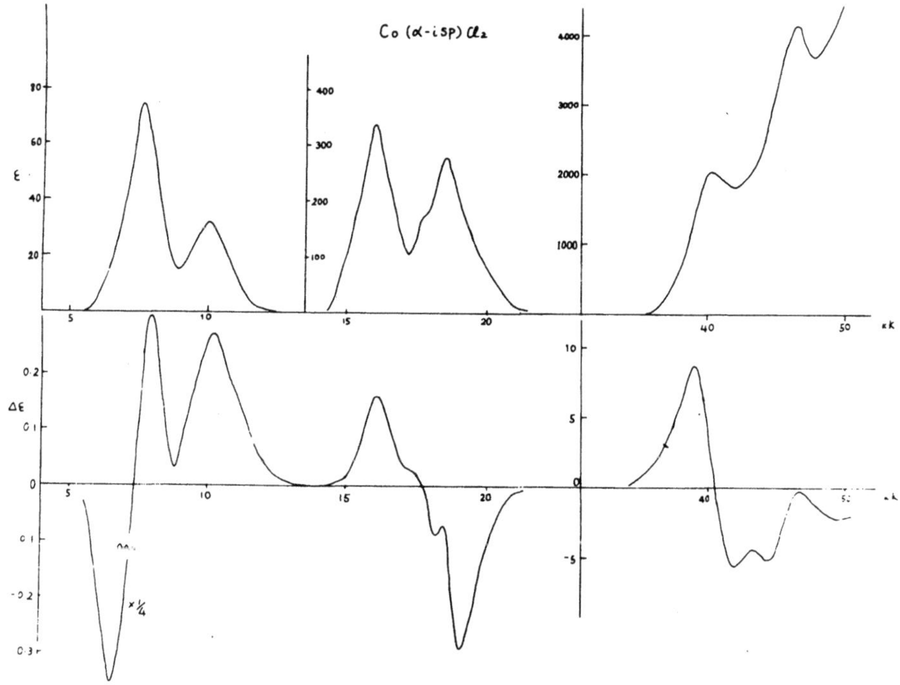

Figure 10. The absorption spectrum (upper curve) and CD spectrum (lower curve) of [Co{(-)-α-isospartein}Cl_2] (VIII) in chloroform solution (5000 to 30,000 cm^{-1}) and acetonitrile solution at higher frequencies.

zero-order magnetic moment of each component d-electron transition. The d-electron rotational strengths of (VIII) and of the isomeric complex [Co{(-)-spartein}Cl_2] are compared with the corresponding ligand polarization values based, firstly, upon all the ligand atoms located in the crystal structure analysis of the latter complex [56] and, secondly, on the [$Co^{(II)}N_2Cl_2$] chromophore alone with a dihedral angle of 15° between the $CoCl_2$ and the CoN_2 planes in the same sense as in (VIII) (Figure 9). The comparison shows (Table 3) that the dihedral angular distortion of the chromophore is the major stereochemical source of d-electron optical activity in (VIII) and its analogues.

Table 3. The frequencies ($10^3 cm^{-1}$) and the calculated and observed d-electron rotational strengths (10^{-3} Debye-Bohr magneton) of {Co{(-)-α-isospartein}Cl_2] (VIII) and [Co{(-)-spartein}Cl_2].

Td state		$^4T_1(F)$			$^4T_1(P)$		
C_2 state		B	A	B	B	A	B
$\nu/10^3 cm^{-1}$		6.5	7.9	10.2	16.1	-	19.1
R_{obs}	[Co(isp)Cl_2]	-50	+7.0	+10.0	+4.1	-	-4.9
	[Co(sp)Cl_2]	-36	+2.7	+5.0	+1.9	-	-3.5
R_{calc}	All atoms	-37	+11.5	+3.3	+0.8	+5.4	-15.6
	15° twist [CoN_2Cl_2]	-30	+1.6	+3.9	+4.5	+0.8	-4.2

REFERENCES:

[1] W. Moffitt, J. Chem. Phys., 25, 1189 (1956).

[2] E.U. Condon, Rev. Mod. Phys., 9, 432 (1937).

[3] E.U. Condon, W. Altar and H. Eyring, J. Chem. Phys., 5, 753 (1937).

[4] H. Bethe, Ann. Phys., [5] 3, 133 (1929).

[5] G.N. Lewis and M. Calvin, Chem. Rev., 25, 273 (1939).

[6] J.G. Kirkwood, J. Chem. Phys., 5, 479 (1937).

[7] E.G. Höhn and O.E. Weigang, Jr., J. Chem. Phys., 48, 1127 (1968).

[8] K. Fajans, Naturwiss, 11, 165 (1923).

[9] O. Laporte, Z. Phys., 51, 512 (1924).

[10] J.H. Van Vleck, J. Phys. Chem., 41, 67 (1937).

[11] L.J.F. Broer, C.J. Gorter and J. Hoogschafen, Physica, 11, 231 (1945).

[12] Y. Tanabe and S. Sugano, J. Phys. Soc. Japan, 9, 753 and 766 (1954).

[13] A.D. Liehr and C.J. Ballhausen, Phys. Rev., 106, 1161 (1957).

[14] C.J. Ballhausen and A.D. Liehr, J. Mol. Spectros., 2, 342 (1958); 4, 190 (1960).

[15] B.R. Judd, Phys. Rev., 127, 750 (1962).

[16] G.S. Ofelt, J. Chem. Phys., 37, 511 (1962).

[17] C.K. Jorgensen and B.R. Judd, Molec. Phys., 8, 281 (1964).

[18] D.M. Gruen and C.W. De Kock, J. Chem. Phys., 45, 455 (1966).

[19] D.M. Gruen and R.L. McBeth , Pure Appl. Chem., 6, 23 (1963).

[20] O. Holmes and D.S. McClure, J. Chem. Phys., 26, 1686 (1957).

[21] P.J. Stephens, J. Chem. Phys., 43, 4444 (1965).

[22] R.G. Denning, J. Chem. Phys., 45, 1307 (1966).

[23] R.G. Denning and J.A. Spencer, Sym. Faraday Soc., 3, 84 (1969).

[24] S. Sugano, J. Chem. Phys., 33, 1883 (1960).

[25] G. Hilmes and F.S. Richardson, Inorg. Chem., 15, 2582 (1976).

[26] S.F. Mason, R.D. Peacock, and B. Stewart, Molec. Phys., 30, 1829 (1975).

[27] R. Gale, R.E. Godfrey, S.F. Mason, R.D. Peacock, and B. Stewart, J. Chem. Soc. Chem. Comm., 329 (1975).

[28] R. Gale, R.E. Godfrey, and S.F. Mason, Chem. Phys. Letters, 38, 441 (1976).

[29] R. Gale, R.E. Godfrey, and S.F. Mason, Chem. Phys. Letters, 38, 446 (1976)

[30] S.F. Mason and R.H. Seal, Molec. Phys., 31, 755 (1976).

[31] R.D. Peacock, Molec. Phys., 33, 1239 (1977).

[32] R.D. Peacock, J. Mol. Struc., 46, 203 (1978).

[33] S.F. Mason, J. Chem. Soc. (A), 667 (1971).

[34] J.A. Tiethof and D.W. Cooke, Inorg. Chem., 11, 315 (1972).

[35] J.A. Hearson, S.F. Mason and R.H. Seal, J. Chem. Soc. Dalton, 1026 (1977).

[36] J.W. Richardson, W.C. Nieuport, R.R. Powell, and W.F. Edgell, J. Chem. Phys., 36, 1057 (1962).

[37] S.F. Mason and A.J. McCaffery, Molec. Phys., 6, 359 (1963).

[38] R. Kuroda and Y. Saito, Bull. Chem. Soc. Japan, 49, 433 (1976).

[38] H.P. Jensen and F. Galsbøl, Inorg. Chem., 16, 1294 (1977).

[39] Y. Saito, Coordination Chem. Rev., 13, 305 (1974).

[40] S.F. Mason and B.J. Norman, J. Chem. Soc. (A) 307 (1966).

[41] E.N. Duesler and K.N. Raymond, Inorg. Chem., 10, 1486 (1971).

[42] J.E. Sarneski and F.L. Urbach, J. Amer. Chem. Soc., 93, 884 (1971).

[43] I.I. Creaser, J.M. Harrowfield, A.J. Herlt, A.M. Sargeson, J. Springborg, R.J. Gene and M.R. Snow, J. Amer. Chem. Soc., 99, 3181 (1977).

[44] S.F. Mason and R.D. Peacock, Inorg. Chim. Acta, 19, 75 (1976).

[45] M. Mikami, R. Kuroda, M. Konno and Y. Saito, Acta Cryst., B33, 1485 (1977).

[46] R. Kuroda and S.F. Mason, Chem. Letters, 1045 (1978).

[47] A.F. Drake, R. Kuroda and S.F. Mason, J. Chem. Soc. Dalton, (1979) in press.

[48] T. Ito and M. Shibata, Inorg. Chem., 16, 108 (1977).

[49] S.F. Mason, Molec. Phys., 37, in press (1979)

[50] H. Yamatera, Bull. Chem. Soc. Japan, 31, 95 (1958).

[51] C.E. Schäffer and C.K. Jorgensen, Molec. Phys., 9, 401 (1965).

[52] S.F. Mason and R.D. Peacock, J. Chem. Soc. Dalton, 226 (1973).

[53] B. Stewart, Ph.D. Thesis, University of London, (1976).

[54] S.J. Hirst, Ph.D. Thesis, University of London (1979).

[55] L.S. Childers, K. Folting, L.L. Merritt, Jr., and W.E. Streib, Acta Cryst., B31, 924 (1975).

[56] R. Kuroda and S.F. Mason, J. Chem. Soc. Dalton, 371 (1977).

[57] R. Kuroda and S.F. Mason, J. Chem. Soc. Dalton, in press (1979).

CIRCULAR POLARIZATION DIFFERENTIALS IN THE LUMINESCENCE OF CHIRAL SYSTEMS

F.S. Richardson

Department of Chemistry, University of Virginia
Charlottesville, Virginia, USA

1. INTRODUCTION

The optical activity of chiral molecular systems arises from differential interactions with left and right circularly polarized radiation. The various manifestations of optical activity may be classified according to whether they involve the absorption, scattering, or emission of radiation. Circular dichroism (CD) is an absorptive phenomenon; Rayleigh and Raman circular intensity differentials (CID's)[1] may be classified as scattering phenomena; and circular polarization of luminescence, or circularly polarized luminescence (CPL), refers to the differential (spontaneous) emission of left and right circularly polarized radiation. The optical activity manifested in fluorescence detected circular dichroism (FDCD) measurements arises from the differential absorption of left and right circularly polarized radiation, although the experimental observables are emission intensities.[2] The fundamental feature common to all molecular chiroptical phenomena is the ability of the (chiral) molecular systems to sense the helicity of incident radiation (in absorption and scattering) or to emit radiation with a net helicity (ellipticity or partial circular polarization). Here we shall consider circularly polarized luminescence spectroscopy as a probe of molecular stereochemistry and electronic structure.

All molecular systems which are chiral in their ground states exhibit CD inside absorptive regions. Similarly, all molecular systems which are chiral in their emitting (luminescent) states exhibit CPL. In considering luminescence we are dealing only with spontaneous emission (as opposed to induced emission), so that the CPL associated with a particular emissive transition can

Stephen F. Mason (ed.), Optical Activity and Chiral Discrimination. 189–217.

be related to the difference in magnitudes of the Einstein A coefficients for left and right circularly polarized (emitted) radiation. Analogously, the CD associated with a particular absorptive transition can be related to the difference in magnitudes of the Einstein B coefficients for left and right circularly polarized (absorbed) radiation. The well-known relationship between the Einstein A and B coefficients suggests that the CD and CPL associated with a particular molecular transition will depend upon the same set of molecular structural parameters and will, therefore, provide similar (if not identical) structural information. For molecular electronic transitions, however, differences between the CD and CPL observables will occur when the structural features of the ground electronic state of the molecular system differ from those of the emitting (luminescent) electronic state. In these cases, the vertical (Franck-Condon) nature of the electronic transitions requires that the CD observables reflect the ground state molecular structural details, whereas the CPL observables will be determined by the structural characteristics of the molecular emitting state. Thus, CD may be considered to be a ground state structure probe and CPL an excited (emitting) state structure probe. Under certain sample conditions, photoselection and reorientational relaxation processes occurring prior to (or simultaneous with) emission may also introduce disparities between the CD and CPL observables for a particular molecular electronic transition. Under these conditions, CPL may be used to probe excited state molecular dynamics.

The use of CPL as a structure probe is a relatively recent development. The first significant exploitation of CPL in this regard can be found in the pioneering studies of Oosterhoff and coworkers[3-6] at the University of Leiden (The Netherlands). The University of Leiden remains an active center for CPL studies under the current direction of H.P.J.M. Dekkers.[7,8] Most of the CPL studies carried out at Leiden have been on optically active ketones. The second major program in CPL spectroscopy was established by Steinberg and coworkers at the Weizmann Institute of Science (Israel). Steinberg and coworkers have used CPL to study a wide variety of molecular systems - including small organic molecules, biopolymer systems, biopolymer-dye complexes, and metal-protein complexes.[9-11] The third major program in CPL spectroscopy was established at the University of Virginia (USA) by Richardson and coworkers. In the latter program, principal emphasis has been given to metal complexes involving both small and large (biopolymer) ligands and to the use of lanthanide ions as luminescent probes.

The formal theoretical aspects of molecular CPL have been presented by Snir and Schellmen,[12] Steinberg and Ehrenberg,[13] and Riehl and Richardson.[14] Richardson and Riehl[15] have recently presented a comprehensive review article dealing with the theory, measurement, and applications of CPL spectroscopy.

As mentioned above, all chiral luminescent systems will exhibit a CPL spectrum. Additionally, all achiral luminescent systems when placed in a static magnetic field will exhibit CPL along a direction parallel to the direction of the applied magnetic field.[16] This latter phenomenon is referred to as magnetically-induced circularly polarized luminescence (MCPL). MCPL is just the emission analogue of magnetic circular dichroism (MCD). A discussion of MCPL spectroscopy lies outside the scope of this presentation and the reader is referred to the review article of Richardson and Riehl[15] for an outline of MCPL theory and applications.

2. THEORY

2.1. General Aspects

The principal observable in CPL spectroscopy is the difference in intensities between the left and right circularly polarized components of the emitted radiation. This difference in intensities is usually denoted by $\Delta I = I_L - I_R$, where the subscripts L and R refer to left circularly polarized and right circularly polarized, respectively. The differential intensity variable may be related to the differential transition rate for the spontaneous emission of left and right circularly polarized radiation according to

$$\Delta I(\omega) = \hbar\omega_\ell N_n \Delta W_{ng} f_{CPL}(\omega) , \tag{1}$$

where N_n is the number of molecules in the emitting state n, ω_ℓ is the resonance frequency of the molecular transition $n \to g$ (where g denotes molecular ground state), $f_{CPL}(\omega)$ is a normalized line-shape function, and ΔW_{ng} is the differential transition rate (for the $n \to g$ transition) defined by

$$\Delta W_{ng} = W_{ng}^{L} - W_{ng}^{R} . \tag{2}$$

Retaining only the electric dipole, magnetic dipole, and electric quadrupole terms in the radiation field-molecule interaction Hamiltonian, the differential transition rate may be expressed as[15]

$$\Delta W_{ng} = K(\omega_\ell^3)\,[\mathrm{Im}(\mu_1^{gn} m_1^{ng} + \mu_2^{gn} m_2^{ng}) + (\omega_\ell/c)\mathrm{Re}(\mu_2^{gn} Q_{13}^{gn} - \mu_1^{gn} Q_{23}^{gn})], \tag{3}$$

where

$$K(\omega_\ell^3) = (\omega_\ell^3/\pi c^3 \hbar) , \tag{4}$$

$$\mu_i^{gn} = \langle g|\hat{\mu}_i|n\rangle = \mu_i^{ng} \quad , \tag{5}$$

$$m_i^{ng} = \langle n|\hat{m}_i|g\rangle = -m_i^{gn} \quad , \tag{6}$$

$$Q_{ij}^{gn} = \langle g|\hat{Q}_{ij}|n\rangle = Q_{ij}^{ng} \quad , \tag{7}$$

and the electric dipole ($\hat{\mu}_i$), magnetic dipole ($\hat{m}_i$), and electric quadrupole ($\hat{Q}_{ij}$) operator components are defined with respect to the laboratory coordinate axes ($i,j,k \equiv 1,2,3$) shown in Figure 1. Equation (3) is appropriate for the experimental configuration shown in Figure 1 where emission detection is along the laboratory 3-direction.

To evaluate eq. (3) one must know the orientational distribution of the molecular transition moments in the sample at the time of emission. This is determined by photoselection processes in excitation, molecular orientational and electronic relaxation processes occurring between excitation and emission, and the inherent polarization properties of the molecular emissive transitions. Photoselection depends upon the excitation-emission detection geometry employed, the polarization of the exciting radiation, the orientational distribution of the molecules in

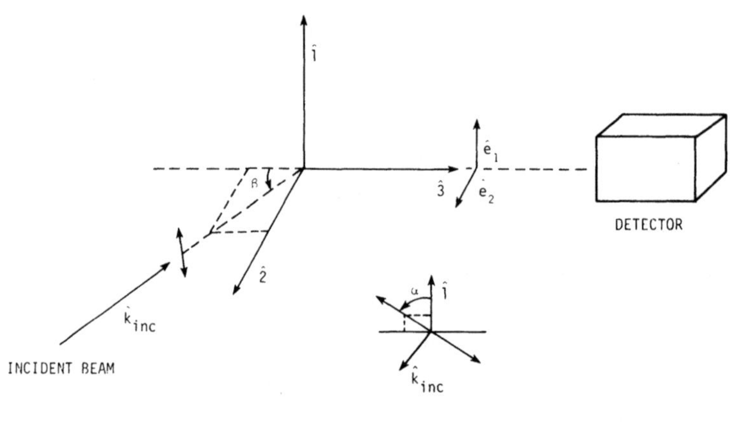

Figure 1: Schematic of a generalized CPL experiment depicting the incident (excitation) radiation, scattering center, emitted radiation, and detector. The laboratory coordinates are denoted by $\hat{1}, \hat{2}$, and $\hat{3}$; the polarization vectors for the emitted radiation are denoted by $\hat{e}_1$ and $\hat{e}_2$; and the angles α and β defined and discussed in the text.

their ground state (i.e., prior to excitation), and the polarization properties of the molecular absorptive transitions involved in the excitation event. Orientational relaxation will depend upon the physical state of the bulk sample (e.g., gas phase, fluid medium, glass, or crystalline state), and electronic relaxation refers to the internal conversion and intersystem crossing pathways leading from the initial excited state of the molecules to the luminescent state. The polarization properties of the emissive transitions are, of course, determined by the electronic and stereochemical structural details of the molecules as they exist in their luminescent state(s).

2.2. Photoselection and Orientational Relaxation

In order to relate the differential transition rate defined in eq. (3) to observed CPL intensities one must take into account the following features which are common to all emission theories for molecules.

(a) The spatial distribution of molecular orientations in the emitting sample will, in general, be anisotropic even if the sample is isotropic in the ground state (as, for example, in a fluid or glassy medium). This anisotropy arises from excitation photoselection. The precise initial distribution created by the exciting radiation may be calculated from the polarization and direction of the exciting radiation <u>and</u> the polarization properties of the molecular absorption dipoles (or multipoles).

(b) In considering a fluid sample, the reorientational relaxation of the molecules due to rotary Brownian motion in the time interval between excitation and emission must be suitably described.

(c) The depletion of the emitting state population due to competing nonradiative relaxation processes must be considered.

We begin our treatment of photoselection and orientational relaxation by defining a function $\Omega(\theta,\phi,t)$ as the probability that the unit vector $\vec{\gamma}$ (which is defined to point along the molecular z axis) is in the direction (θ,ϕ) at time t. The angles (θ,ϕ) are defined with respect to the laboratory 1,2,3-coordinate axes as shown in Figure 2. From these definitions, it follows that

$$\Omega(\theta,\phi,t) = \int_0^{2\pi} d\phi_o \int_0^{\pi} \sin\theta_o \, d\theta_o \; \Omega(\theta_o,\phi_o)\, G(\theta_o,\phi_o|\theta,\phi,t), \qquad (8)$$

where $G(\theta_o,\phi_o|\theta,\phi,t)$ is a function that describes the time evolution of the molecular orientation; $\Omega(\theta_o,\phi_o) = \Omega(\theta,\phi,t{=}o)$, and absorption (excitation) takes place at $t = o$. We further specify that the excitation involves a simple (electric) dipole transition,

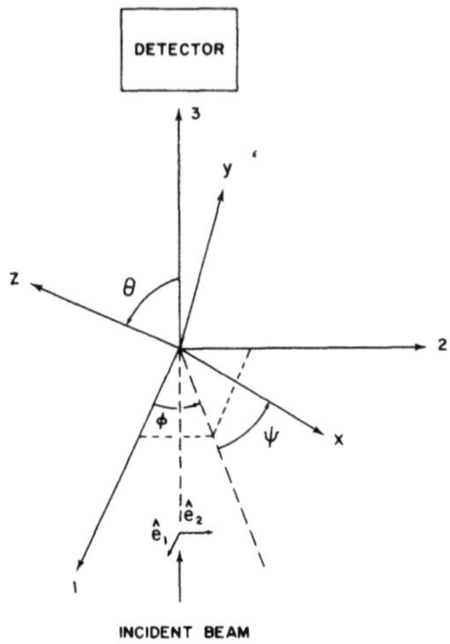

Figure 2: Relationship between the laboratory axes (1,2,3) and the molecular axes (x,y,z) as defined by the angles of transformation (θ,φ,ψ).

and that the associated transition dipole vector defines the molecular z axis (i.e., it points in the $\vec{\gamma}$ direction). The initial (t = o) distribution of absorption dipoles (excited molecules) in the sample is calculated from the polarization and direction of the incident (exciting) light beam. Any arbitrarily polarized light beam may be decomposed into components polarized along the laboratory axes. For a linearly polarized beam with angles (α, β) (see Figure 1) measured from the laboratory axis system, the decomposition may be represented as follows:

$$\vec{e}_1 \cos\alpha + \vec{e}_2 \sin\alpha\cos\beta + \vec{e}_3 \sin\alpha\sin\beta$$

where $\vec{e}_1$, $\vec{e}_2$, and $\vec{e}_3$ are unit polarization vectors along the laboratory 1,2, and 3 directions, respectively. The probability that a molecule will be excited at t = o may now be expressed as

$$P_{abs} = \kappa^2\Omega(\theta_o,\phi_o) = (3/4\pi)\kappa^2(|\vec{e}_1\cdot\vec{\gamma}|^2\cos^2\alpha + |\vec{e}_2\cdot\vec{\gamma}|^2\sin^2\alpha\cos^2\beta + |\vec{e}_3\cdot\vec{\gamma}|^2\sin^2\alpha\sin^2\beta) \quad , \qquad (9)$$

where κ^2 contains the geometry independent parts of the absorption probability and is, of course, a function of exciting light intensity and the dipole strength of the absorbing transition. Evaluating the scalar products appearing in eq. (9) leads to

$$P_{abs} = \kappa^2(3/8\pi)(\cos^2\alpha\sin^2\theta_o + \sin^2\alpha\cos^2\beta\sin^2\theta_o + 2\sin^2\alpha\sin^2\beta\cos^2\theta_o), \quad (10)$$

or

$$P_{abs} = \kappa^2(1/4\pi)[1-F(\alpha,\beta)P_2(\cos\theta_o)] , \quad (11)$$

where F is a geometrical factor defined by

$$F = (1 - 3\sin^2\alpha\sin^2\beta) , \quad (12)$$

and $P_2(\cos\theta_o)$ is the Legendre polynomial of order 2.

Equation (8) may be solved by expanding $G(\theta_o,\phi_o|\theta,\phi,t)$ in terms of spherical harmonics, $Y_{\ell,m}(\theta,\phi)$, with time-dependent coefficients, $C_{\ell,m}(t)$, and by using the appropriate normalization and boundary conditions. The result is easily shown to be:

$$\Omega(\theta,\phi,t) = (1/4\pi)[1-FC_{2,o}(t)P_2(\cos\theta)] . \quad (13)$$

The time-dependent coefficient $C_{2,o}(t)$ [which we shall henceforth abbreviate to C(t)] may be viewed as describing the reorientational relaxation in the time-correlation function formalism. In time-correlation notation,

$$C(t) \equiv C_{2,o}(t) = \langle P_2[\vec{\mu}(o)\cdot\vec{\mu}(t)]\rangle_{ens} , \quad (14)$$

where $\vec{\mu}$ is a unit vector aligned along the transition polarization direction ($\vec{\gamma}$) and the brackets denote an ensemble average.

The time dependence of any quantity, R, that is a function of orientation and which is moving in space is given by

$$R(t,\eta) = R(\eta)\,\Omega(\eta,t) \quad (15)$$

where η denotes the collection of spatial (orientational) coordinates and $\Omega(\eta,t)$ is given by eq. (13). In the present context, $R(\eta)$ will be comprised of molecular matrix elements defined with respect to the laboratory coordinate system. In the special case where we are dealing with a sample of randomly oriented molecular systems, it is appropriate to find the spatial average of the property R by integrating eq. (15) over all orientation space. In this case,

$$<R(t)> = \int R(\eta)\Omega(\eta,t)d\eta$$

$$= (1/8\pi^2) \int_0^{2\pi}\int_0^{2\pi}\int_0^{\pi} d\psi d\phi \sin d\theta \, R(\theta,\phi,\psi)[1-FC(t)P_2(\cos\theta)] \,, \tag{16}$$

where (θ,ϕ,ψ) are the set of Euler angles which take the laboratory coordinate system (1,2,3) into the molecular coordinate system (x,y,z) [see Figure 2].

In what follows, we shall consider an isotropic sample (random orientation of molecules in their ground state) excited by a light beam characterized by the angles (α,β) [see Figure 1]. The differential emission intensity, ΔI, may be expressed as

$$\Delta I = h\omega_\ell \, K(\omega_\ell^3) \, N_n \, f_{CPL}(\omega) < R_{ng} > \,, \tag{17}$$

where R_{ng} is defined as the <u>rotatory strength</u> of the $n \rightarrow g$ transition, and is given by

$$R_{ng} = \mathrm{Im}(\mu_1^{gn} m_1^{ng} + \mu_2^{gn} m_2^{ng}) + (\omega_\ell/c)\mathrm{Re}(\mu_2^{gn} Q_{13}^{gn} - \mu_1^{gn} Q_{23}^{gn}) \,. \tag{18}$$

The brackets <> in eq. (17) denote a spatial average appropriate for the orientational distribution of emitting molecules in the sample. Assuming random orientations of ground state molecules and molecular absorption dipoles polarized along the molecular z coordinate axis, the following expression may be obtained for the time-dependent rotatory strength:

$$<R_{ng}(t)> = i\{[(2/3)-(1/15)FC(t)](\mu_x^{gn} m_x^{gn} + \mu_y^{gn} m_y^{gn})$$

$$+[(2/3)+(2/15)FC(t)] \, \mu_z^{gn} m_z^{gn}\} + (1/15)FC(t)k_\ell(\mu_x^{gn} Q_{yz}^{gn} - \mu_y^{gn} Q_{xz}^{gn}) \,, \tag{19}$$

where $k_\ell = \omega_\ell/c$, and the spectroscopic transition moments have been expressed in terms of components referred to the molecular x,y,z-coordinate system. Equation (19) was obtained by substituting eq. (18) into eq. (16) and then performing the necessary integrations. For a rigid sample in which there is no molecular reorientational motion subsequent to excitation, $C(t) = 1$ and R_{ng} has no time dependence.

Time dependence of the differential emission intensity, ΔI, may arise not only from the time dependence of R_{ng} (due to orientational relaxation processes), but also from the time-dependent nature of N_n. Assuming that excitation (at $t = o$) is directly into the emitting state n of the molecule and that the decay of this state follows first-order kinetic processes, then the population of state n at time t is given by

$$N_n(t) = N_n^o \exp(-t/\tau_n) \quad , \tag{20}$$

where N_n^o is the state population at $t = o$ (excitation) and τ_n is the total "lifetime" of the state. The lifetime parameter τ_n may be expressed as

$$\frac{1}{\tau_n} = \frac{1}{\tau_{rad}} + \frac{1}{\tau_q} \quad , \tag{21}$$

where $(1/\tau_{rad})$ is the rate constant for radiative decay and $(1/\tau_q)$ is the rate constant for radiationless decay processes. When excitation is into a molecular excited state higher in energy than the emitting state n, eq. (20) remains valid <u>only</u> if relaxation processes leading to population of n are much faster than $(1/\tau_n)$.

Substitution of eqs. (19) and (20) into eq. (17) yields the following expression for the time-dependent differential CPL intensity:

$$\Delta I(t) = h\omega_\ell K(\omega_\ell^3) N_n^o \exp(-t/\tau_n) f_{CPL}(\omega)\{i[(2/3)-(1/15)FC(t)](\mu_x^{gn} m_x^{gn} + \mu_y^{gn} m_y^{gn}) + i[(2/3)+(2/15)FC(t)]\mu_z^{gn} m_z^{gn} + (1/15)FC(t)k_\ell(\mu_x^{gn} Q_{yz}^{gn} - \mu_y^{gn} Q_{xz}^{gn})\} \quad . \tag{22}$$

Recall that in utilizing eq. (19) we have assumed the sample to be isotropic in the ground state and we have taken the absorption transition dipoles to be oriented along the molecular z-axis. In utilizing eq. (20) we have assumed that the excited state manifold of the luminescent molecular system has reached a thermal equilibrium prior to emission and that N_n^o may be treated as a constant (independent of time). A more detailed description of the time dependence of ΔI (as given in eq. (22)) requires an explicit specification of the time-dependent coefficient C(t). In the present context C(t) describes the average random thermal motion (rotary Brownian motion) of the molecules in the sample (assumed to be in a fluid state). For molecules of a completely asymmetric ellipsoidal shape, C(t) may be expressed in terms of five exponential decay functions.[17] If the molecules have a symmetric ellipsoidal shape, C(t) may be expressed in terms of three exponentials, and for spherically shaped molecules C(t) is given by a single exponential function.[17] For rotating spheres (so-called Einstein spheres) of volume V in a fluid medium of viscosity η, C(t) has the simple form,

$$C(t) = \exp(-6Dt) \; , \tag{23}$$

where D, the diffusion constant, is given by

$$D = kT/6V_\eta \; . \tag{24}$$

The time-dependent differential CPL intensity given by eq. (22) may be determined using time-resolved emission measurement techniques. However, all CPL measurements reported to date have been conducted using steady-state excitation/detection techniques. The appropriate expression for the steady-state CPL experiment is obtained by averaging eq. (22) over long times:

$$\begin{aligned}\Delta I &= \int_o^\infty \Delta I(t)dt \\ &= \hbar\omega_\ell K(\omega_\ell^3) N_n^o \, f_{CPL}(\omega) \, \{ i\tau_n (2/3)(\mu_x^{gn} m_x^{gn} + \mu_y^{gn} m_y^{gn} + \mu_z^{gn} m_z^{gn}) \\ &+ (F/15)\int_o^\infty dt \, \exp(-t/\tau_n) C(t) \, [\, 2i \, \mu_z^{gn} m_z^{gn} - i(\mu_x^{gn} m_x^{gn} + \mu_y^{gn} m_y^{gn}) \\ &+ k_\ell (\mu_x^{gn} Q_{yz}^{gn} - \mu_y^{gn} Q_{xz}^{gn})]\} \; ,\end{aligned} \tag{25}$$

or,

$$\Delta I = \hbar\omega_\ell K(\omega_\ell^3)N_n^o \; f_{CPL}(\omega)\tau_n(2/3)i\,\vec{\mu}^{gn}\cdot\vec{m}^{gn}$$

$$+ \hbar\omega_\ell K(\omega_\ell^3)N_n^o \; f_{CPL}(\omega)(F/15)\int_0^\infty dt \; \exp(-t/\tau_n)C(t)\,[i(3\mu_z^{gn}m_z^{gn} - \vec{\mu}^{gn}\cdot\vec{m}^{gn})$$

$$+ k_\ell(\mu_x^{gn}\,Q_{yz}^{gn} - \mu_y^{gn}\,Q_{xz}^{gn})] \quad , \tag{26}$$

where $\vec{\mu}^{gn}$ and $\vec{m}^{gn}$ denote electric dipole and magnetic dipole transition vectors.

For a rigid sample, C(t) = 1, and eq. (26) reduces to

$$\Delta I = \hbar\omega_\ell K(\omega_\ell^3)N_n^o \; \tau_n \; f_{CPL}(\omega) \; [(2/3)i \; \vec{\mu}^{gn}\cdot\vec{m}^{gn}$$

$$+ (iF/15)(3\mu_z^{gn}\,m_z^{gn} - \vec{\mu}^{gn}\cdot\vec{m}^{gn})$$

$$+ (k_\ell F/15)(\mu_x^{gn}\,Q_{yz}^{gn} - \mu_y^{gn}\,Q_{xz}^{gn})] \; . \tag{27}$$

For a fluid sample comprised of spherically shaped luminescent molecules, C(t) = exp(-6 D t), and eq. (26) reduces to

$$\Delta I = \hbar\omega_\ell \; K(\omega_\ell^3)N_n^o \; \tau_n \; f_{CPL}(\omega) \; \{(2/3)i\,\vec{\mu}^{gn}\cdot\vec{m}^{gn}$$

$$+ [iF/(90\,D\,\tau_n + 15)](3\mu_z^{gn}\,m_z^{gn} - \vec{\mu}^{gn}\cdot\vec{m}^{gn})$$

$$+ [k_\ell F/(30\,D\tau + 5)](\mu_x^{gn}\,Q_{yz}^{gn} - \mu_y^{gn}\,Q_{xz}^{gn})\} \; . \tag{28}$$

In the case where the emission transition dipole $\vec{\mu}^{gn}$ is polarized along the molecular z-axis, $|\vec{\mu}^{gn}| = \mu_z^{gn}$, and eq. (27) reduces to the form,

$$\Delta I = \hbar\omega_\ell K(\omega_\ell^3) N_n^o \ \tau_n (2/3) i \ \mu_z^{gn} m_z^{gn} f_{CPL}(\omega) [1+(F/5)]. \qquad (29)$$

(Recall that the molecular z-axis was defined to be along the direction of the dipole transition vector associated with the excitation process.)

In the limit that $D\tau_n >> 1$, eq. (28) reduces to

$$\Delta I = \hbar\omega_\ell K(\omega_\ell^3) N_n^o \ \tau_n (2/3) i (\vec{\mu}^{gn} \cdot \vec{m}^{gn}) \ f_{CPL}(\omega) \quad . \qquad (30)$$

This corresponds to the case where the initial, photoselected distribution of excited molecules has completely "relaxed" (become random) by the time of emission. This will occur when the emitting molecules can reorient rapidly (large rotary diffusion coefficient) and/or have a long lifetime τ_n. Note that in this "relaxed" limit, the terms involving the electric quadrupole moments vanish, and that the differential CPL intensity is independent of the geometrical factor $F = 1-3\sin^2\alpha \sin^2\beta$. The sample is isotropic at the time of emission and there is no "memory" of the excitation direction (specified by angle β) and excitation polarization (specified by angle α).

In the limit that $D\tau_n << 1$, eq. (28) reduces to eq. (27). This corresponds to the case where the excited molecules are "frozen" or "locked-in" to their initial ($t = o$), photoselected distribution. This will occur when the luminescent molecules are in a highly viscous or rigid medium and/or when the emitting state lifetime is very short relative to the reorientational relaxation time.

Up to this point in the presentation we have assumed that the incident beam is linearly polarized with the plane of polarization tilted by an angle α from the laboratory 1-axis. The factor F (see eq. (12)) may also be used to describe an unpolarized excitation beam by summing two perpendicular (linear) polarizations. For example, an unpolarized beam whose direction of propagation is along the laboratory 2-axis ($\beta = 90°$) may be decomposed into components linearly polarized in the 1- and 3-directions [i.e., $(\frac{1}{2})^{\frac{1}{2}}(\vec{e}_1 + \vec{e}_3)$]. F, in this case, is determined by summing the contributions for $\alpha = 0°$ and $\alpha = 90°$ and then dividing the sum by two:

$$F = (\tfrac{1}{2})[(1-3\sin^2 0° \sin^2 90°) + (1-3\sin^2 90° \sin^2 90°) = -(\tfrac{1}{2}) \ . \qquad (31)$$

To eliminate the effects of photoselection from the CPL intensity variables, it is necessary to choose α and β such that $F = o$. For $F = o$, the differential CPL intensity is given by

$$\Delta I = \hbar\omega_\ell K(\omega_\ell^3) N_n^o \tau_n (2/3)\, i\, (\vec{\mu}^{\,gn} \cdot \vec{m}^{\,gn})\, f_{CPL}(\omega) \quad . \tag{32}$$

The conditions for F to vanish are apparent from its definition (see eq. (12)). For linearly polarized light, these conditions are satisfied by

$$\sin^2\alpha = 1/3\sin^2\beta \quad . \tag{33}$$

Equation (33) can always be satisfied if $\sin^2\beta \geq (1/3)$, i.e., $35.26^o \leq \beta \leq 144.74^o$. For unpolarized excitation, the condition for $F = o$ is

$$\sin^2\beta = (2/3) \; ; \quad \beta = 54.74^o \text{ or } 125.26^o \; .$$

2.3. Luminescence Dissymmetry Factor

The measurement of absolute emission intensities is very difficult experimentally. Consequently, the usual procedure in CPL studies is to measure ΔI and $I (= I_L + I_R)$ in arbitrary (or relative) intensity units, and to determine the ratio,

$$g_{lum}(\omega) = 2\Delta I(\omega)/I(\omega) \quad , \tag{34}$$

in absolute units. We shall refer to I as the total luminescence (TL) intensity and to g_{lum} as the luminescence dissymmetry factor. At a given emission frequency, ω, the percent of circular polarization in the luminescence is given by $(100/2)g_{lum}(\omega) = 50\, g_{lum}(\omega)$. The luminescence dissymmetry factor provides a measure of the chirality present in the molecular emitting state(s).

For an emitting sample in which the excited (emitting) state molecular distribution is isotropic, the luminescence dissymmetry factor is given by

$$g_{lum}(\omega) = 4i\,[f_{CPL}(\omega)/f_{TL}(\omega)]\,[\vec{\mu}^{\,gn} \cdot \vec{m}^{\,gn} / |\vec{\mu}^{\,gn}|^2] \quad , \tag{35}$$

where $f_{TL}(\omega)$ is a normalized line-shape function appropriate for the total luminescence and we have assumed that the total luminescence intensity arises entirely from electric dipole mechanisms. Equation (35) also holds for photoselected samples in which the excited molecules are not fully relaxed (orientationally randomized) under the excitation conditions in which $F = o$. Additionally, eq. (35) holds for any case in which the absorption and emission electric dipole transition vectors are parallel. These conclusions are based on the results presented in Section 2.2 where it was assumed that the sample is completely isotropic in the ground state (prior to excitation).

From eq. (35) we note that $g_{lum}(\omega)$ will be constant across the emission band profile only if the line-shape functions, $f_{CPL}(\omega)$ and $f_{TL}(\omega)$, are identical. In general, $f_{CPL}(\omega)$ and f_{TL} will differ due to different vibronic mechanisms operative in the CPL and TL spectra. That is, certain vibronic components of a given electronic (emissive) transition may exhibit more or less percent circular polarization than others.

A more complete discussion of luminescence dissymmetry factors and their relationship to various molecular spectroscopic parameters may be found in reference 15.

2.4. Comparison of Electronic Rotatory Strengths Obtained from CPL and CD Band Spectra

In chiroptical spectroscopy, spectra-structure relationships are generally based on obtaining electronic rotatory strengths from the observed CD (or CPL) spectra and then relating the signs and magnitudes of these rotatory strengths to specific stereochemical and electronic structural features of the molecular system. Detailed models of molecular optical activity are designed to calculate electronic rotatory strengths in terms of parameters related directly to stereochemical and electronic structural details. Here we shall consider the rotatory strength associated with a particular molecular electronic transition, $g \to n$ in absorption (CD) and $g \leftarrow n$ in emission (CPL). Furthermore, we shall assume an isotropic absorbing sample and an isotropic emitting sample. Under these conditions, the absorption and emission rotatory strengths are defined (in the electric dipole-magnetic dipole approximation) by:

$$R^a(g \to n) = \mathrm{Im} \langle g|\hat{\vec{\mu}}|n\rangle \cdot \langle n|\hat{\vec{m}}|g\rangle \quad , \tag{36}$$

and,

$$R^e(g \leftarrow n) = \mathrm{Im}\langle g^*|\hat{\vec{\mu}}|n^*\rangle \cdot \langle n^*|\hat{\vec{m}}|g^*\rangle \quad , \tag{37}$$

where $\hat{\vec{\mu}}$ and $\hat{\vec{m}}$ are the electric and magnetic dipole operators, respectively. Assuming thermal equilibrium in both the ground state (prior to absorption) and the emitting state (prior to emission), $|g\rangle$ and $|n\rangle$ are taken as eigenstates of the ground state electronic Hamiltonian $H_e(r,Q)$, and $|g^*\rangle$ and $|n^*\rangle$ are taken as eigenstates of the excited state electronic Hamiltonian $H_e(r,Q^*)$. The ground state normal coordinates of the molecular system are denoted by {Q}, the excited state normal coordinates are denoted by {Q*}, and the electronic coordinates are denoted collectively by {r}. We assume that $H_e(r,Q)$ and $H_e(r,Q^*)$ have identical symmetry properties and are sufficiently similar to insure that the electronic orbital properties of $|g\rangle$ and $|g^*\rangle$ and of $|n\rangle$ and $|n^*\rangle$ are nearly identical. That is, the $g \rightarrow n$ and $g \leftarrow n$ transitions are assumed to connect the same set of electronic states.

The experimental observable in CD spectroscopy is $\Delta\varepsilon(\omega)$, and $R^a(g \rightarrow n)$ may be related to this quantity according to:

$$R^a(g \rightarrow n) = 23.0 \times 10^{-40} \int_{g \rightarrow n} \Delta\varepsilon(\omega)d\omega/\omega \ , \tag{38}$$

where the integration is over the frequency interval spanned by the $g \rightarrow n$ absorption band envelop and $R^a(g \rightarrow n)$ is expressed in cgs units (esu^2cm^2). The analogous experimental observable in CPL spectroscopy is $\Delta I(\omega)$, and $R^e(g \leftarrow n)$ may be related to this quantity according to:

$$R^e(g \leftarrow n) = C_n \int_{g \rightarrow n} \Delta I(\omega)d\omega/\omega^4 \ , \tag{39}$$

where the integration is over the frequency interval spanned by the $g \leftarrow n$ emission band envelop and (assuming steady-state emission conditions) C_n is a constant given by

$$C_n = \pi c^3/N_n \ , \tag{40}$$

where N_n is the "steady-state" population of the emitting state n. Equations (38) and (39) are the expressions required for relating the experimental observables, $\Delta\varepsilon(\omega)$ and $\Delta I(\omega)$, to the theoretical quantities, $R^a(g \rightarrow n)$ and $R^e(g \leftarrow n)$, under conditions of sample isotropy (in absorption and emission), complete thermal equilibrium (in the states $|g\rangle$ and $|n^*\rangle$), and "steady-state" emission observation.

Clearly, $R^a(g \to n) = R^e(g \leftarrow n)$ only when $H_e(r,Q) = H_e(r,Q^*)$. Insofar as $H_e(r,Q)$ and $H_e(r,Q^*)$ reflect molecular conformational (stereochemical) and other geometrical features in the ground state $|g\rangle$ and emitting state $|n^*\rangle$, respectively, observed differences between $R^a(g \to n)$ and $R^e(g \leftarrow n)$ may be traced to differences in molecular structural details in the ground and emitting states. Within the Born-Oppenheimer approximation and ignoring all "hot-band" vibronic contributions in both absorption and emission (T = OK), we may write:

$$R^a(g \to n) = \text{Im} \int \Phi_{go}(\vec{\mu}^{\,gn} \cdot \vec{m}^{\,ng})\Phi_{go}\, dQ \quad , \tag{41}$$

and,

$$R^e(g \leftarrow n) = \text{Im} \int \Phi_{n^*o}(\vec{\mu}^{\,g^*n^*} \cdot \vec{m}^{\,n^*g^*})\Phi_{n^*o}\, dQ^* \quad , \tag{42}$$

where Φ_{go} is the ground state vibrational wave function of the ground electronic state (g) and Φ_{n^*o} is the ground state vibrational wave function of the electronic emitting state (n*). Equation (41) shows $R^a(g \to n)$ to be the expectation value of $\text{Im}(\vec{\mu}^{\,gn} \cdot \vec{m}^{\,ng})$ in the ground vibrational level of the ground electronic state, and $R^e(g \leftarrow n)$ to be the expectation value of $\text{Im}(\vec{\mu}^{\,g^*n^*} \cdot \vec{m}^{\,n^*g^*})$ in the ground vibrational level of the electronic emitting state. Clearly, if the structure parameters of a molecule differ in the ground and emitting states, then $R^a(g \to n)$ will in general differ from $R^e(g \leftarrow n)$. Furthermore, if the nuclear coordinate dependence of $(\vec{\mu}^{\,gn} \cdot \vec{m}^{\,ng})$ differs from the nuclear coordinate dependence of $(\vec{\mu}^{\,g^*n^*} \cdot \vec{m}^{\,n^*g^*})$, then the vibronic intensity profiles of the CD and CPL spectral bands will differ.

Detailed spectra-structure correlations must be based on CPL/CD band-shape analyses. Differences observed in CPL and CD band-shapes or fine-structure features carry detailed information about relative ground state/excited state structural characteristics and about the nature of vibronic perturbations within the ground and excited electronic states. Exploitation of this aspect of comparative CPL/CD studies is nicely demonstrated in an early paper of Emeis and Oosterhoff[6] and in a recent study reported by Dekkers and Closs.[7]

3. CPL AS A STRUCTURE PROBE

As a structure probe CPL resembles CD in that it reflects the chirality or dissymmetry of the molecular system. Both CPL and CD are, therefore, related to similar kinds of molecular structural parameters (electronic <u>and</u> stereochemical). Like CD, the interpretation of CPL in terms of exact and detailed molecular structural features is, in most cases, an elusive goal. However, like CD, CPL can serve as an excellent diagnostic tool for conformational or structural problems where one is looking for similarities, differences, or changes in molecular conformational aspects on a qualitative or, in some cases, a semi-quantitative level.

3.1. Optical Activity of Chromophores in Excited (Emitting) States

As was pointed out in Section 2.4., CPL reflects the stereochemical and conformational features of the luminescent states of molecular systems. That is, it is a direct probe of molecular excited state structure. One example of the use of CPL to characterize the excited state structure of a molecular system can be found in the studies of Gafni and Steinberg[18] and of Schlessinger and Warshel[19] on 1,1'-bianthracene-2,2'-dicarboxylic acid in chloroform solution. This molecule does not possess any asymmetrically substituted atoms in its structure, but is optically active by virtue of inherent chirality about the carbon-carbon bond connecting the two anthracence moieties in the molecule. Using CPL data (g_{lum} versus ω)and the results obtained from a quantum mechanically based conformational analysis, it was possible to characterize the excited (emitting) state conformation of this system with respect to the dihedral angle between the planes of the two anthracene rings.

Another example of the application of CPL to the characterization of excited state structure is found in the work of Gafni, Hardt, Schlessinger, and Steinberg[20] on chlorophyll and chloroplasts. A pronounced difference was found between the CPL of chlorophyll *a* dimers in fluid organic solvents and that of sub-chloroplast particles and chloroplasts, indicating that the former have a different structure from, and cannot serve as a model for, associated chlorophyll in native structures in electronically excited states. Information was also obtained about the relation between the emitting molecules in chloroplasts when the active centers are open or closed.

3.2. Change in Conformation Upon Electronic Excitation

Changes in molecular conformation accompanying electronic excitation may be assayed by combination CPL/CD studies. In this case, the CD spectrum associated with an absorptive transition to the luminescent molecular electronic state is compared to the corresponding CPL spectrum. Differences in band shape, vibronic fine-structure, and the frequency variations of g_{lum} and g_{abs} (the absorption dissymmetry factor defined by $\Delta\varepsilon/\varepsilon$) may be related to ground state versus excited state structural differences. Examples of comparative CPL/CD studies for the purpose of obtaining relative ground state versus excited state structural information can be found in the work of Emeis and Oosterhoff[6] and of Dekkers and Closs[7] on ketones, the work of Luk and Richardson[21] on camphorquinone, and the work of Schlessinger, Gafni, and Steinberg[22] on a series of cyclic dipeptides containing tyrosine and tryptophan residues. Using comparative CPL/CD studies, Steinberg and coworkers[23,24] were also able to show that very pronounced changes take place in the mode of interaction of extrinsic chromophores bound to proteins and polypeptides upon electronic excitation of the chromophores.

3.3. Selectivity of CPL to Luminescent Chromophores

CPL studies are limited to chromophores which are luminescent. This obviously limits the applicability of CPL as a structure probe. However, this aspect of CPL may also be used to great advantage in certain situations. For example, in complex systems (such as proteins) one often finds several kinds of chromophores all of which exhibit absorptions in the same frequency range of the spectrum. In this situation, it is not always possible to analyze the measured CD spectra in terms of contributions from the various kinds of chromophores. However, if only one subset of these chromophores is luminescent, then CPL may be used to probe this subset without interference from the remaining (non-luminescent) chromophores in the system. This is illustrated by the CPL measurements of proteins, in which the optical activity of only the tryptophan residues, and sometimes also the tyrosine residues, contribute. The disulfide and peptide moieties in proteins are non-luminescent, and the phenylalanine sidechain chromophores are luminescent only in the absence of tryptophan and tyrosine residues.

The spectral overlap between different chromophores in absorption and emission is often different. It may thus happen that better resolution between the spectral contributions of different chromophores is gained in performing CPL measurements versus CD measurements. This is the case, for example, with the tryptophan and tyrosine aromatic sidechain chromophores in proteins.

3.4. CPL of Forbidden Transitions

Some electronic transitions, referred to here as forbidden transitions, are too weak to be observed or easily measured in absorption because of their extremely low extinction coefficients. Absorption/CD measurements on these transitions become possible only with the use of very concentrated samples and/or very long optical path lengths. However, these transitions may often be observed quite readily in emission, and CPL is ideally suited to study the optical activity associated with such transitions. The usual procedure in these cases is to excite the chromophore in a transition region exhibiting high absorptivity and then depend on internal conversion or intersystem crossing processes to populate the emitting level. The intensity of the long-lived emission from this level will depend upon the quantum yield of the transition as well as the amount of excitation light absorbed by the sample and subsequently transferred to the emitting level.

The f-f transitions of trivalent lanthanide ions are examples of transitions exhibiting very low extinction coefficients ($\varepsilon<1$) even when the lanthanide ion resides in a low-symmetry ligand environment. Absorption/CD measurements in the f-f transition region of optically active lanthanide ion complexes are very difficult to obtain except at relatively high concentrations of the lanthanide ion. However, several lanthanide ions (such as Eu^{3+} and Tb^{3+}) exhibit a strong luminescence even in solution media at room temperature. Complexes of Eu^{3+} and Tb^{3+} with a wide variety of optically active ligands have been studied by CPL/emission techniques,[25,26] and the CPL/emission spectra of Tb^{3+}-protein complexes have been used to probe the metal binding sites of a large number of protein molecules (many of which are known, or suspected, to be calcium binding systems).[27,28] CD/absorption studies are not feasible at the concentrations desired in the Tb^{3+}-protein experiments.

Most of the emissions observed for the lanthanide ion systems are assigned to transitions which terminate in energy levels other than the ground electronic state of the ion. In fact, the strongest luminescence observed for Tb^{3+} complexes in solution is of $^5D_4 \rightarrow {}^7F_5$ free-ion parentage, where the $^7F_5$ energy level lies about 2100 cm^{-1} above the $^7F_6$ ground state energy level of Tb^{3+}. CPL/emission spectra spectra have been reported for the $^5D_4 \rightarrow {}^7F_J$ (J = 6,5,4,3, and 2) Tb^{3+} transitions in a wide variety of Tb^{3+} complexes, and CPL/emission spectra have been reported for the $^5D_0 \rightarrow {}^7F_J$ (J = 0,1,2,3, and 4) Eu^{3+} transitions in a number of Eu^{3+} complexes. (See Figure 3 for an approximate energy level diagram for the Tb^{3+} and Eu^{3+} ions.) Among the Tb^{3+} and Eu^{3+} transitions listed above, only the $^5D_4 \leftarrow {}^7F_6$ transition (in Tb^{3+}) and the $^5D_0 \leftarrow {}^7F_0$, $^7F_1$ transitions (in Eu^{3+}) can be studied by CD/absorption techniques at room temperature.

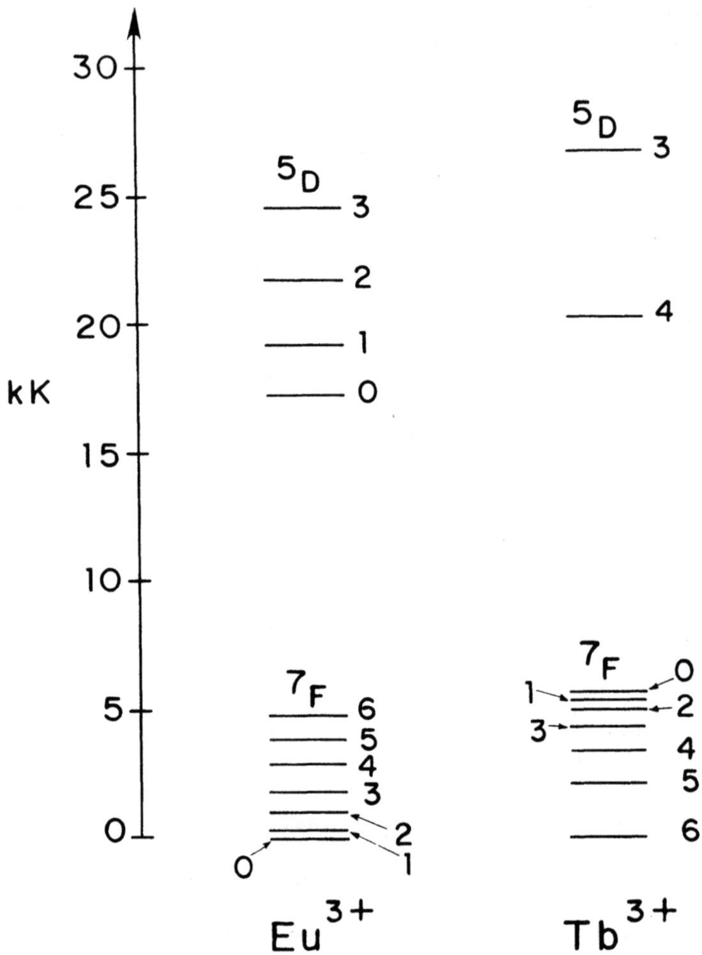

Figure 3: An approximate energy level diagram for Eu^{3+} and Tb^{3+} aquo ions.

3.5. Behavior of g_{lum} Across Emission Bands

For most molecular systems in condensed phases at room temperature, luminescence (which is generally a fluorescence under such conditions) originates with just one excited state. With the exception of the lanthanide ions mentioned above in Section 3.4., the luminescence can be identified with a single electronic transition involving the emitting level and the ground state of the molecular system. This is the case even when there exist several closely spaced excited states (so long as the energy spacings between the lowest lying excited state and the higher lying excited states are at least greater than kT). This may be contrasted to the case frequently encountered in CD/absorption spectra where several closely spaced electronic transitions may yield a spectrum comprised of several strongly overlapping CD and/or absorption bands which are resistant to resolution.

Since a given CPL band may generally be assigned to a single electronic transition, variations in g_{lum} across the band profile can usually be attributed to just two causes: (a) vibronic coupling mechanisms leading to variable intensity contributions (to CPL versus total luminescence) among the vibronic components of the electronic transition; and, (b) heterogeneity within the (emitting) chromophore population of the sample. The former cause (a) may arise when the intensity producing mechanisms for CPL and total luminescence depend upon different vibronic coupling mechanisms. The latter cause (b) may arise when the sample contains emitting molecules or chromophores which differ slightly in their CPL and/or total luminescence properties. For example, Steinberg, et al.,[11] found that g_{lum} varies significantly across the tryptophan fluorescence region for human serum albumin samples, although this protein molecule contains a single tryptophan residue per molecule. This behavior was attributed to the known heterogeneity of the albumin molecules in solution. Thus, different albumin molecules appear to have different g_{lum} values and slightly different emission spectra, leading to variations in g_{lum} across the composite emission band. This may be referred to as <u>intermolecular</u> heterogeneity. <u>Intramolecular</u> heterogeneity among tryptophan residues <u>within</u> a given protein molecule most likely can account for variations in g_{lum} across emission bands of proteins which contain several tryptophan residues per molecule. In this latter case the different tryptophanyl chromophores within the protein molecule "see" slightly different environments and exhibit, therefore, slightly different CPL/emission properties.[11,29]

In the study of the binding of a variety of dyes to polyglutamic acid and to poly (A), Steinberg, et al.,[30] also found variations in g_{lum} across the emission bands of the dyes. These variations were attributed to the presence of dimers and higher aggregates of the dye molecules on the polymers.

3.6. CPL of Photoselected Samples

It is important here to restate the somewhat obvious result given previously in Section 2.2.; namely, that in the limit that the distribution of excited molecules is isotropic about the emission detection direction, the measured differential and total luminescence intensities are entirely independent of the polarization of the excitation beam or the angle between the excitation and emission directions. As was illustrated in Section 2.2., an isotropic distribution of emitting molecules results from either selecting an experimental configuration such that $F = o$, or selecting a molecular system that completely relaxes (orientationally) in the time between emission and absorption. In this isotropic limit the measured quantities (ΔI and I) carry no information regarding the angle between the absorption and emission transition vectors.

For rigid ("frozen"), photoselected samples, the observed intensity quantities (ΔI and I) contain direct molecular information not only on the magnitude and direction of the emission transition vectors, but also on the magnitude and direction of the absorption transition vectors. In this case, the emitting molecules have complete memory [$C(t) = 1$] of the excitation event.

For an isotropic absorbing sample which is also isotropic during emission, the g_{lum} factor is given by

$$g_{lum} = 4i\, f_{CPL}(\omega)\vec{\mu}^{\,gn} \cdot \vec{m}^{\,gn} / f_{TL}(\omega)\,|\vec{\mu}^{\,gn}|^2 \quad , \tag{43}$$

for the $n \rightarrow g$ emissive transition. For an isotropic absorbing sample in which the excited molecules retain their photoselected orientational distribution during emission (i.e., a rigid sample), the g_{lum} factor has the form:

$$g_{lum} = \frac{4i\, f_{CPL}(\omega)}{f_{TL}(\omega)}$$

$$\times \frac{[(10-F)(\mu_x^{gn} m_x^{gn} + \mu_y^{gn} m_y^{gn}) + (10+2F)\mu_z^{gn} m_z^{gn} - 3i\, Fk_\ell\, (\mu_x^{gn} Q_{yz}^{gn} - \mu_y^{gn} Q_{xz}^{gn})]}{[(10-F)(|\mu_x^{gn}|^2 + |\mu_y^{gn}|^2) + (10+2F)|\mu_z^{gn}|^2]} \tag{44}$$

where $F = (1 - 3\sin^2\alpha \sin^2\beta)$ for linearly polarized excitation light (see Figure 1), and the molecular z-axis is defined by the absorption transition dipole vector. If $F = o$, or if $\mu_x^{gn} = \mu_y^{gn} = o$, then eq. (44) reduces to eq. (43). Equation (44) may be rewritten in the following form:

$$g_{lum} = \frac{4i\, f_{CPL}(\omega)\, [10(\vec{\mu}^{gn} \cdot \vec{m}^{gn}) + F(3\mu_z^{gn} m_z^{gn} - \vec{\mu}^{gn} \cdot \vec{m}^{gn})]}{f_{TL}(\omega)\, [10|\vec{\mu}^{gn}|^2 + F(3|\mu_z^{gn}|^2 - |\vec{\mu}^{gn}|^2)]}, \qquad (45)$$

where the electric dipole-electric quadrupole terms have been neglected.

Considering the special case in which the excitation light is unpolarized and is directed along the laboratory 3-axis ($\beta = 0°$), $F = 1$ and eq. (45) reduces to

$$g_{lum} = \frac{4i\, f_{CPL}(\omega)\, [3\,\vec{\mu}^{gn} \cdot \vec{m}^{gn} + \mu_z^{gn} m_z^{gn}]}{f_{TL}(\omega)\, [3\,|\vec{\mu}^{gn}|^2 + |\mu_z^{gn}|^2]}. \qquad (46)$$

Equation (46) may be re-expressed as,

$$g_{lum} = \frac{4i\, f_{CPL}(\omega)\, [3\,\vec{\mu}^{gn} \cdot \vec{m}^{gn} + \mu_z^{gn} m_z^{gn} \cos\delta]}{f_{TL}(\omega)|\vec{\mu}^{gn}|^2 (3 + \cos^2\delta)}, \qquad (47)$$

where δ is the angle between the absorption and emission electric dipole transition vectors. Assuming that the magnitude of $\cos\delta$ may be obtained from measurements of linear polarization,[13] and that $\vec{\mu}^{gn} \cdot \vec{m}^{gn}$ may be determined from CPL measurements on an isotropic (emitting) sample, then $\mu_z^{gn} m_z^{gn}$ may be deduced from eq. (47) to within a factor of $|\vec{\mu}^{gn}|^2$, the dipole strength of the transition.

3.7. Summary Comments

It is apparent that CPL both supplements and complements CD as a technique for investigating molecular optical activity. In some cases it may be used to obtain precisely the same kind of structural information as that provided by CD; in other cases it can be used to obtain unique structural information (especially about molecular excited states). The choice of method, CPL or CD, to be used in a particular instance must be determined on the basis of the kind of structural information desired, the types of systems to be studied, and the relative sensitivity levels offered by the respective measurement techniques.

4. MEASUREMENT TECHNIQUES

4.1. General Aspects

The measurement procedures and techniques employed in CPL spectroscopy are very similar to those used in any other kind of polarized emission experiment. The only difference between linearly polarized luminescence measurements and CPL measurements is that, in the latter case, the emitted light is analyzed in terms of its left and right circularly polarized components, whereas, in the former case, the emitted light is analyzed in terms of linear polarization in one or two (orthogonal) directions. The unique feature, then, of any CPL instrumentation is the presence of a circular analyzer between the emitting sample and the detector unit. This circular analyzer must be capable of discriminating between left and right circularly polarized radiation.

The simplest design for a circular analyzer involves the use of just two optical components: a quarter-wave retardation element followed by a linear polarizer (actually an analyzer in this case). The quarter-wave retardation element serves to convert circularly polarized light into linearly polarized light. The left and right circularly polarized components of the incident light will be linearly polarized in two mutually orthogonal directions. The linear analyzer then functions to select one or the other of these two linearly polarized components for detection and measurement. To measure $\Delta I (= I_L - I_R)$, one must take the difference in intensities measured when the linear analyzer is set in the two appropriate mutually orthogonal orientations, or one may keep the orientation of the linear analyzer fixed and measure the difference in intensities produced by (+) or (-) quarter-wave retardation of the incident light. In most instances, it is easier and more convenient to modulate the retardation element between (+) and (-) 90° phase changes than it is to rotate the linear analyzer by ±90°.

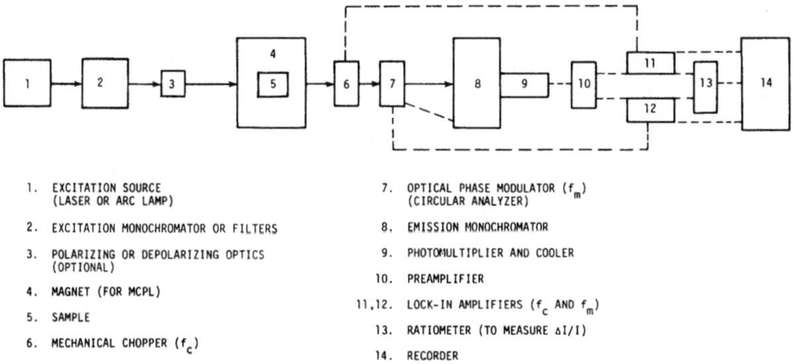

Figure 4: Block diagram depicting the basic components of a CPL spectrophotometer.

4.2. Instrumentation

A block diagram depicting the basic components of a general-purpose, steady-state CPL instrument is shown in Figure 4. These components include: (a) excitation source (continuous output); (b) excitation monochromator; (c) excitation beam polarizer (or depolarizer); (d) sample cell; (e) mechanical beam chopper; (f) modulated circular analyzer; (g) emission monochromator; (h) emission detector; (i) phase-sensitive signal amplification electronics; and, (j) recorder. In experiments where unpolarized excitation light is desired, the excitation beam polarizer may either be removed or be replaced by a depolarizing element (as may be required, for example, when a laser excitation source is used).

The mechanical beam chopper (operated at a chopping frequency f_c) modulates the total emission output of the sample and provides the reference signal for one of the two phase-sensitive lock-in amplifiers in the system. The circular analyzer is comprised of an optical phase modulator (operated at a frequency f_m) and a stationary linear analyzer. The optical phase modulator alternately advances and retards the electric vector of the emitted light by 90° which has the effect of converting the circularly polarized components of the light into mutually orthogonal linearly polarized components. The modulated (at f_m) optical signal emerging from the linear analyzer is, then, alternately proportional to the number of left and right circularly polarized photons emitted by the sample. The second phase-sensitive, lock-in amplifier in the system is referenced to the f_m frequency of the optical phase modulator.

The electronics of the CPL instrumentation system are designed to amplify the electrical signals supplied by the photomultiplier detector, to discriminate between the signal variables which carry differential circular polarization information and those which are proportional to total emission intensity, and to record any two of the following: (1) CPL intensity (ΔI) in arbitrary units; (2) total luminescence intensity (I) in arbitrary units; and, (3) $\Delta I/I$ in absolute units (following instrumentation calibration). The magnitude and phase of the f_m signal carry the CPL intensity and sign information, and the f_c signal carries the total luminescence intensity information.

As noted previously, in all of the CPL experiments reported to date ΔI and I have been measured in arbitrary intensity units. However, in most cases absolute values have been reported for $\Delta I/I$ (the degree of circular polarization). Absolute values of $\Delta I/I$ may be obtained by simply introducing light into the circular analyzer/detector train of the instrument which has a known degree of circular polarization (or ellipticity), and then determining the proportionality constant in the relationship,

$$(\Delta I/I)_{\text{measured}} = C\,(\Delta I/I)_{\text{known}} \quad .$$

C then becomes the instrument calibration factor and may be used to convert measured values of ($\Delta I/I$) into actual (absolute) values of the degree of circular polarization. This calibration factor will, in general, be wavelength-dependent and it is necesary to know C as a function of (emission) wavelength for doing highly accurate measurements. Perhaps the most accurate and ingenius (but simple) calibration procedure has been developed and discussed by Steinberg and Gafni.[31]

Detailed accounts of CPL instrumentation design and construction can be found in references 10, 31, and 32.

REFERENCES

1. L.D. Barron and A.D. Buckingham, Annu. Rev. Phys. Chem., 26, 381 (1975).

2. (a) D.H. Turner, I. Tinoco, and M. Maestre, J. Am. Chem. Soc., 96, 4340 (1974);

(b) D.H. Turner, I. Tinoco, and M. Maestre, Biochemistry, 14, 3794 (1975);

(c) I. Tinoco and D.H. Turner, J. Am. Chem. Soc., 98, 6453 (1976);

(d) I. Tinoco, B. Ehrenberg, and I.Z. Steinberg, J. Chem. Phys., 66, 916 (1977);

(e) D.H. Tuner, in "Methods in Enzymology", Vol. XLIX, Part G, edited by C.H.W. Hirs and S.N. Timasheff, Academic Press, New York, N.Y., 1978, Section II.8., pp. 199-214.

3. C.H. Emeis and L.J. Oosterhoff, Chem. Phys. Lett., 1, 129 (1967).

4. C.A. Emeis, Ph.D. Thesis, The University of Leiden, The Netherlands, 1968.

5. H.P.J.M. Dekkers, C.A. Emeis, and L.J. Oosterhoff, J. Am. Chem. Soc., 91, 4589 (1969).

6. C.A. Emeis and L.J. Oosterhoff, J. Chem. Phys., 54, 4809 (1971).

7. H.P.J.M. Dekkers and L.E. Closs, J. Am. Chem. Soc., 98, 2210 (1976).

8. H. Wynberg, H. Numan, and H.P.J.M. Dekkers, J. Am. Chem. Soc., 99, 3870 (1977).

9. I.Z. Steinberg, in "Biochemical Fluorescence: Concepts", Vol. I, edited by R.F. Chen and H. Edelhoch, Marcel-Dekker, New York, N.Y., 1975, Chapter 3.

10. I.Z. Steinberg, in "Methods in Enzymology", Vol. XLIX, Part G, edited by C.H.W. Hirs and S.N. Timasheff, Academic Press, New York, N.Y., 1975, Section II.7., pp. 179-199.

11. I.Z. Steinberg, J. Schlessinger, and A. Gafni, in "Peptides, Polypeptides, and Proteins", edited by E.R. Blout, F.A. Bovey, M. Goodman, and N. Lotan, Wiley-Interscience, New York, N.Y. 1974, pp. 351-369.

12. J. Snir and J.A. Schellman, J. Phys. Chem., 78, 387 (1974).

13. I.Z. Steinberg and B. Ehrenberg, J. Chem. Phys., 61, 3382 (1974).

14. J.P. Riehl and F.S. Richardson, J. Chem. Phys., 65, 1011 (1976).

15. F.S. Richardson and J.P. Riehl, Chem. Rev., 77, 773 (1977).

16. J.P. Riehl and F.S. Richardson, J. Chem. Phys., 66, 1988 (1977).

17. P. Wahl, in "Biochemical Fluorescence: Concepts", Vol. I, edited by R.F. Chen and H. Edelhoch, Marcel-Dekker, New York, N.Y., 1975, Chapter 1.

18. A. Gafni and I.Z. Steinberg, Photochem. Photobiol., 15, 93 (1972).

19. J. Schlessinger and A. Warshel, Chem. Phys. Lett., 28, 380 (1974).

20. A Gafni, H. Hardt, J. Schlessinger, and I.Z. Steinberg, Biochim. Biophys. Acta, 387, 256 (1975).

21. C.K. Luk and F.S. Richardson, J. Am. Chem. Soc., 96, 2006 (1974).

22. J. Schlessinger, A. Gafni, and I.Z. Steinberg, J. Am. Chem. Soc., 96, 7396 (1974).

23. J. Schlessinger and I.Z. Steinberg, Proc. Natl. Acad. Sci. U.S.A., 69, 769 (1972).

24. J. Schlessinger, I.Z. Steinberg, and I. Pecht, J. Mol. Biol., 87, 725 (1974).

25. C.K. Luk and F.S. Richardson, J. Am. Chem. Soc., 97, 6666 (1975).

26. (a) H.G. Brittain and F.S. Richardson, Inorg. Chem., 15, 1507 (1976);

(b) H.G. Brittain and F.S. Richardson, J. Am. Chem. Soc., 98, 5858 (1976);

(c) H.G. Brittain and F.S. Richardson, J. Am. Chem. Soc., 99, 65 (1977);

26. (d) H.G. Brittain and F.S. Richardson, Bioinorg. Chem., 7, 233 (1977).

27. A. Gafni and I.Z. Steinberg, Biochemistry, 13, 800 (1974).

28. H.G. Brittain, F.S. Richardson, and R.B. Martin, J. Am. Chem. Soc., 98, 8255 (1976).

29. J. Schlessinger, R. Roche, and I.Z. Steinberg, Biochemistry, 14, 255 (1975).

30. A. Gafni, J. Schlessinger, and I.Z. Steinberg, Isr. J. Chem., 11, 423 (1973).

31. I.Z. Steinberg and A. Gafni, Rev. Sci. Instrum., 43, 409 (1976).

32. R.A. Shatwell and A.J. McCaffery, J. Phys. E, 7, 297 (1974).

RAMAN OPTICAL ACTIVITY

L. D. Barron

Chemistry Department, The University,
Glasgow G12 8QQ, U. K.

ABSTRACT. Optical activity associated with the vibrations of chiral molecules can be measured by means of a small difference in the intensity of Raman scattering in right and left circularly polarized incident light. This technique enables complete vibrational optical activity spectra from about 50 to 4000 cm^{-1} to be measured routinely, although the largest effects occur at low frequency. The basic theory of Raman optical activity is developed, together with a 'bond-polarizability' theory which gives the sign and magnitude of the effects generated by idealized normal modes of model chiral structures. The experimental method is described briefly, and some typical Raman optical activity spectra presented. Correlations pointed out include features in methyl and trifluoromethyl asymmetric deformations, carbon-methyl deformations, methyl torsions, carbonyl deformations, methyl rocking in single methyl groups and in isopropyl group, skeletal vibrations and carbon-hydrogen and carbon-deuterium deformations.

1. INTRODUCTION

1.1 Vibrational Optical Activity

It has been appreciated for some time that the measurement of optical activity associated with molecular vibrations could provide a wealth of delicate stereochemical information. But only in the last few years, thanks to developments in infrared, optical and electronic technology, have the formidable technical difficulties been overcome and vibrational optical activity spectra been obtained using both infrared and Raman techniques.

The significance of vibrational optical activity becomes

Stephen F. Mason (ed.), Optical Activity and Chiral Discrimination. 219–262.

apparant when it is compared with conventional electronic optical activity in the form of optical rotatory dispersion and circular dichroism of visible and near-ultraviolet radiation. These conventional techniques have proved most valuable in stereochemical studies, but since the electronic transition frequencies of most structural units in a molecule occur in inaccessible regions of the far ultraviolet, they are restricted to probing limited regions of molecules, in particular chromophores and their immediate intramolecular environments, and cannot be used at all when a molecule lacks a chromophore (although optical rotation measurements at transparent frequencies can still be of value). But since a vibrational spectrum contains bands from most parts of a molecule, measurement of some form of vibrational optical activity could provide much more information.

The obvious method of measuring vibrational optical activity is by extending optical rotatory dispersion and circular dichroism into the infrared. But in addition to the technical difficulties in manipulating polarized infrared radiaiton, there is a fundamental physical difficulty inherent in this approach: optical activity is proportional to the frequency of the exciting light, and infrared frequencies are several orders of magnitude smaller than visible and near-ultraviolet frequencies. This snag could be side-stepped if a manifestation of optical activity existed in Raman scattering because the Raman effect provides complete vibrational spectra using visible exciting light. And indeed Raman optical activity has been discovered recently in the form of a small difference in the intensity of Raman scattered light in right and left circularly polarized incident light. At the time of writing the two techniques appear to be complementary: infrared circular dichroism gives better spectra in the carbon-hydrogen stretching frequency region, but cannot penetrate below about 2000 cm^{-1}; whereas Raman optical activity provides complete spectra down to about 50 cm^{-1}. In fact it is the region below 2000 cm^{-1} that carries most stereochemical information since skeletal deformation and torsion modes of vibration contribute here. On the other hand, infrared circular dichroism is more favourable for studying dilute non-aqueous solutions since transparent Raman spectroscopy reqires concentrated solutions or neat liquids.

In addition to the study of low-frequency vibrations, another application of Raman optical activity not accessible to the infrared approach at present is the study of biological molecules in aqueous media. Optical activity measurements are important in biochemistry and biophysics since they are sensitive to the delicate stereochemical features that determine biological function. In this respect vibrational optical activity should be particularly useful. However, the infrared approach is hampered by the difficulties that prevent widespread application of conventional infrared spectroscopy to biological molecules, namely, that infrared radiation is absorbed strongly by water, the natural biological medium, and that the complexity of biological molecules can lead to thousands of modes of vibration. Raman spectroscopy does not suffer from the first limitation since water does not absorb visible wavelengths and has no Raman bands. The second limitation can be overcome by using the resonance Raman technique in which the laser frequency falls within an electronic absorption band, usually localized at a site of biological function, resulting in a tremendous enhancement of the Raman intensities of those few vibrations coupled to the electronic transition {1}. Thus resonance Raman optical activity could provide unique stereochemical information at sites of biological function, even in molecules on surfaces in particulate samples such as membranes in highly scattering suspensions. However, apart from the magnetic experiment mentioned below, this type of biological observation has not yet been realized. But with anticipated developments in optical and electronic technology, particularly with regard to tunable ultraviolet lasers, this application could become important.

1.2 Rayleigh and Raman optical activity

When a light wave encounters an obstacle, bound charges are set into oscillation and secondary waves are scattered in all directions. Within a medium the obstacles responsible for light scattering can be gross inclusions of foreign matter such as impurities in crystals, dust particles in the atmosphere or colloidal matter suspended in fluids. But light scattering also occurs in transparent materials completely free of contaminants, and this was shown by Lord Rayleigh to originate in the individual molecules {2}.

Polarization studies have always been an important feature in molecular light scattering. Lord Rayleigh had shown that the light scattered at right angles from a spherical molecule should be completely linearly polarized perpendicular to the scattering plane; the observed imperfection in the polarization of light scattered from dust-free gases was subsequently ascribed to a lack of spherical symmetry in the optical properties of the molecules. But it was not until 1923 that the first article appeared discussing possible new effects in light scattering from optically active molecules {3}. In it, Gans considered the additional contribution to Rayleigh scattering from the optical activity tensor alone and claimed to have observed its effect on the depolarization ratio; but de Mallemann {4} pointed out that the anomalies in the depolarization ratio originated in optical rotation of the incident and scattered beams. Gans also omitted the scattering contributions arising from interference between the polarizability and the optical activity tensors: this interference is central to the discussion of Rayleigh and Raman optical activity since, as shown below, it leads to a circular polarization dependence of the scattered intensity and to a circularly polarized component in the scattered light. Shortly after the discovery of the Raman effect, Bhagavantam and Venkataswaran {5} found differences in the relative intensities of some of the vibrational Raman lines of two enantiomers in unpolarized incident light, but these were attributed subsequently to impurities. Although he had no theory, Kastler {6} thought that, since optically active molecules respond differently to right and left circularly polarized light, a difference might exist between the Raman spectra of optically active molecules in right and left circularly polarized incident light, and he attempted unsuccessfully to observe it. Perrin {7} alluded to the existence of additional polarization effects in light scattered from optically active molecules, and Atkins and Barron {8} showed explicitly that interference between the molecular polarizability and optical activity tensors leads to a dependence of the scattered intensity on the degree of circular polarization of the incident light and to a circularly polarized component in the scattered light. Barron and Buckingham {9} developed subsequently the theory of the Rayleigh and Raman circular intensity difference (c.i.d.) defined by

$$\Delta_\alpha = (I_\alpha^R - I_\alpha^L)/(I_\alpha^R + I_\alpha^L), \tag{1.1}$$

where I_α^R and I_α^L are the scattered intensities with α-polarization in right and left circularly polarized incident light, this being the appropriate experimental quantity. The first reported Raman c.i.d. spectra by Bosnich et al. {10} and Diem et al. {11} originated in experimental artifacts, but the spectra reported subsequently by Barron et al. {12-14} in simple molecules such as α-phenylethylamine and α-pinene have now been confirmed as genuine by Hug et al. {15}, Diem et al. {16} and Boucher et al. {17}.

Since all molecules can show optical rotation and circular dichroism in a magnetic field, it is not surprising that all molecules in a strong magnetic field parallel to the incident light beam should show Rayleigh and Raman optical activity {18}. Magnetic c.i.d.s have not been observed so far at transparent wavelengths, but have been observed in the resonance Raman spectrum of ferrocytochrome c, a haem protein {19}. More surprisingly, although there is no simple electrical analogue of magnetic optical rotation and circular dichroism, Rayleigh and Raman optical activity should be shown by any fluid in a static electric field perpendicular to both the incident and scattered directions {20}, but this has not yet been observed.

This article concentrates on Raman optical activity at transparent wavelengths in chiral molecules, since this is the most important aspect for stereochemical studies. More details of other aspects such as magnetic and resonance Raman optical activity can be found in earlier reviews {21-23}.

2. THE MOLECULAR POLARIZABILITY AND OPTICAL ACTIVITY TENSORS

2.1 Multipole moments induced by a light wave

The response of a molecule to applied electric and magnetic fields can be characterized by property tensors that relate components of the applied fields to components of the electric and magnetic multipole moments induced in the molecule. In addition to the familiar polarizability tensor that gives the electric dipole moment induced by an electric field that can be either static or oscillating, we require optical activity tensors that are defined only for oscillating electric and magnetic fields and give rise to only oscillating electric and magnetic multipole moments.

In cartesian tensor notation, the electric dipole, magnetic dipole and traceless electric quadrupole moments are defined, in SI, by

$$\mu_\alpha = \sum_i e_i r_{i\alpha}, \tag{2.1a}$$

$$m_\alpha = \sum_i (e_i/2m_i)\,\varepsilon_{\alpha\beta\gamma} r_{i\beta} p_{i\gamma}, \tag{2.1b}$$

$$\Theta_{\alpha\beta} = \tfrac{1}{2}\sum_i e_i (3r_{i\alpha}r_{i\beta} - r_i^2\delta_{\alpha\beta}), \tag{2.1c}$$

where particle i at $\vec{r}_i$ has charge e_i, mass m_i and momentum $\vec{p}_i$. The Greek subscripts denote vector or tensor components and can be equal to x, y or z. A repeated Greek subscript denotes a summation over all three cartesian components: this is the tensor equivalent of a scalar product so that, for example,

$$a_\alpha b_\alpha = a_x b_x + a_y b_y + a_z b_z \equiv \vec{a}\cdot\vec{b}.$$

The unit alternating tensor $\varepsilon_{\alpha\beta\gamma}$ is defined such that $\varepsilon_{\alpha\beta\gamma} r_\beta p_\gamma$ is the α-component of the vector product $\vec{r}\times\vec{p}$; that is, $\varepsilon_{\alpha\beta\gamma}$ is equal to +1 or -1 if α,β,γ is an even or odd permutation of x,y,z and is zero if any two subscripts are the same. The scalar product of a vector with a vector product would thus be written

$$a_\alpha \varepsilon_{\alpha\beta\gamma} b_\beta c_\gamma = a_x(b_y c_z - b_z c_y) + a_y(b_z c_x - b_x c_z) + a_z(b_x c_y - b_y c_x) \equiv \vec{a}\cdot(\vec{b}\times\vec{c}).$$

The electric and magnetic field vectors of a plane-wave light beam of angular frequency ω travelling in the direction of the unit vector $\vec{n}$ with angular velocity c are, in complex notation,

$$\tilde{E}_\alpha = E_\alpha^{(0)} \exp[-i\omega(t - n_\beta r_\beta/c)], \tag{2.2a}$$

$$\tilde{B}_\alpha = \frac{1}{c}\varepsilon_{\alpha\beta\gamma} n_\beta \tilde{E}_\gamma. \tag{2.2b}$$

A tilde denotes a complex quantity. The choice of sign in the exponent of (2.2a) is arbitrary since only the real part has physical significance. We also require the electric field gradient tensor

$$\tilde{E}_{\alpha\beta} \equiv \nabla_\alpha \tilde{E}_\beta = \frac{i\omega}{c} n_\alpha \tilde{E}_\beta. \tag{2.2c}$$

Quantum-mechanical expressions for the molecular polarizability and optical activity tensors are obtained by taking expectation values of the operator equivalents of the multipole moments (2.1) using molecular wavefunctions perturbed by the light wave, and identifying the tensors in the resulting series. The periodically-perturbed molecular eigenfunctions $\psi_{n'}$, are obtained by solving the time-dependent Schrödinger equation

$$\left(i\hbar\frac{\partial}{\partial t} - H\right)\Psi_{n'} = V\Psi_{n'}, \tag{2.3}$$

where H is the unperturbed molecular Hamiltonian and V is the Hamiltonian describing the interaction of the molecule with the light wave. It is convenient to write V as a series displaying the coupling of the molecular multipole moment operators with appropriate real fields of the light wave {24}:

$$V = -\mu_\alpha(E_\alpha)_0 - m_\alpha(B_\alpha)_0 - \tfrac{1}{3}\Theta_{\alpha\beta}(E_{\alpha\beta})_0 + \cdots, \tag{2.4}$$

where a subscript zero means that the corresponding field is to be taken at the molecular origin; for example

$$(E_\alpha)_0 = \tfrac{1}{2}[(\tilde{E}_\alpha)_0 + (\tilde{E}_\alpha^*)_0] = \tfrac{1}{2}[\tilde{E}_\alpha^{(0)}\exp(-i\omega t) + \tilde{E}_\alpha^{(0)*}\exp(i\omega t)].$$

We find the following real oscillating induced electric dipole, magnetic dipole and electric quadrupole moments of the molecule in the nth quantum state {25,23}:

$$\mu_\alpha = \langle n'|\mu_\alpha|n'\rangle = \alpha_{\alpha\beta}(E_\beta)_0 + \frac{1}{\omega}\alpha'_{\alpha\beta}(\dot{E}_\beta)_0 + G_{\alpha\beta}(B_\beta)_0$$
$$+ \frac{1}{\omega}G'_{\alpha\beta}(\dot{B}_\beta)_0 + \frac{1}{3}A_{\alpha\beta\gamma}(E_{\beta\gamma})_0 + \frac{1}{3\omega}A'_{\alpha\beta\gamma}(\dot{E}_{\beta\gamma})_0 + \cdots, \tag{2.5a}$$

$$m_\alpha = \langle n'|m_\alpha|n'\rangle = G_{\beta\alpha}(E_\beta)_0 - \frac{1}{\omega}G'_{\beta\alpha}(\dot{E}_\beta)_0 + \cdots, \tag{2.5b}$$

$$\Theta_{\alpha\beta} = \langle n'|\Theta_{\alpha\beta}|n'\rangle = A_{\gamma\alpha\beta}(E_\gamma)_0 - \frac{1}{\omega}A'_{\gamma\alpha\beta}(\dot{E}_\gamma)_0 + \cdots, \tag{2.5c}$$

where

$$\alpha_{\alpha\beta} = \frac{2}{\hbar}\sum_{j\neq n}\frac{\omega_{jn}}{\omega_{jn}^2-\omega^2}\mathrm{Re}(\langle n|\mu_\alpha|j\rangle\langle j|\mu_\beta|n\rangle) = \alpha_{\beta\alpha}, \tag{2.6a}$$

$$\alpha'_{\alpha\beta} = -\frac{2}{\hbar}\sum_{j\neq n}\frac{\omega}{\omega_{jn}^2-\omega^2}\mathrm{Im}(\langle n|\mu_\alpha|j\rangle\langle j|\mu_\beta|n\rangle) = -\alpha'_{\beta\alpha}, \tag{2.6b}$$

$$G_{\alpha\beta} = \frac{2}{\hbar}\sum_{j\neq n}\frac{\omega_{jn}}{\omega_{jn}^2-\omega^2}\mathrm{Re}(\langle n|\mu_\alpha|j\rangle\langle j|m_\beta|n\rangle), \tag{2.6c}$$

$$G'_{\alpha\beta} = -\frac{2}{\hbar}\sum_{j\neq n}\frac{\omega}{\omega_{jn}^2-\omega^2}\mathrm{Im}(\langle n|\mu_\alpha|j\rangle\langle j|m_\beta|n\rangle), \tag{2.6d}$$

$$A_{\alpha\beta\gamma} = \frac{2}{\hbar}\sum_{j\neq n} \frac{\omega_{jn}}{\omega_{jn}^2 - \omega^2} \mathrm{Re}(\langle n|\mu_\alpha|j\rangle\langle j|\Theta_{\beta\gamma}|n\rangle), \tag{2.6e}$$

$$A'_{\alpha\beta\gamma} = -\frac{2}{\hbar}\sum_{j\neq n} \frac{\omega}{\omega_{jn}^2 - \omega^2} \mathrm{Im}(\langle n|\mu_\alpha|j\rangle\langle j|\Theta_{\beta\gamma}|n\rangle), \tag{2.6f}$$

where $\omega_{jn} = \omega_j - \omega_n$. Notice that $\alpha_{\alpha\beta}$ is symmetric with respect to interchange of the tensor subscripts, whereas $\alpha'_{\alpha\beta}$ is antisymmetric. This follows from the fact that μ_α is a Hermitian operator so that

$$\langle n|\mu_\alpha|j\rangle\langle j|\mu_\beta|n\rangle = \langle n|\mu_\beta|j\rangle^*\langle j|\mu_\alpha|n\rangle^*,$$

$$\mathrm{Re}(\langle n|\mu_\alpha|j\rangle\langle j|\mu_\beta|n\rangle) = \mathrm{Re}(\langle n|\mu_\beta|j\rangle\langle j|\mu_\alpha|n\rangle),$$

$$\mathrm{Im}(\langle n|\mu_\alpha|j\rangle\langle j|\mu_\beta|n\rangle) = -\mathrm{Im}(\langle n|\mu_\beta|j\rangle\langle j|\mu_\alpha|n\rangle).$$

No analogous separation into symmetric and antisymmetric parts exists for the remaining tensors since each involves products of two different multipole transition moments.

It is convenient to present these results in a complex representation. Introducing the complex tensors

$$\tilde{\alpha}_{\alpha\beta} = \alpha_{\alpha\beta} - i\alpha'_{\alpha\beta} = \tilde{\alpha}^*_{\beta\alpha}, \tag{2.7a}$$

$$\tilde{G}_{\alpha\beta} = G_{\alpha\beta} - iG'_{\alpha\beta}, \tag{2.7b}$$

$$\tilde{A}_{\alpha\beta\gamma} = A_{\alpha\beta\gamma} - iA'_{\alpha\beta\gamma} = \tilde{A}_{\alpha\gamma\beta}, \tag{2.7c}$$

where the minus sign arises from our choice of a negative exponent in the complex light wave (2.2), the amplitudes of the corresponding complex moments are

$$\tilde{\mu}^{(0)}_\alpha = \tilde{\alpha}_{\alpha\beta}\tilde{E}^{(0)}_\beta + \tilde{G}_{\alpha\beta}\tilde{B}^{(0)}_\beta + \tfrac{1}{3}\tilde{A}_{\alpha\beta\gamma}\tilde{E}^{(0)}_{\beta\gamma} + \cdots$$

$$= \left(\tilde{\alpha}_{\alpha\beta} + \tfrac{1}{c}\varepsilon_{\gamma\delta\beta}n_\delta\tilde{G}_{\alpha\gamma} + \tfrac{i\omega}{3c}n_\delta\tilde{A}_{\alpha\delta\beta} + \cdots\right)\tilde{E}^{(0)}_\beta, \tag{2.8a}$$

$$\tilde{m}^{(0)}_\alpha = \tilde{G}^*_{\beta\alpha}\tilde{E}^{(0)}_\beta + \cdots, \tag{2.8b}$$

$$\Theta^{(0)}_{\alpha\beta} = \tilde{A}^*_{\gamma\alpha\beta}\tilde{E}^{(0)}_\gamma + \cdots. \tag{2.8c}$$

The symmetric polarizability $\alpha_{\alpha\beta}$ provides the major contribution to light scattering and refraction and, since the real part of a product of electric dipole transition moments is specified, it is supported by all molecules. In the antisymmetric polarizability $\alpha'_{\alpha\beta}$, the imaginary part of a product of electric dipole transition moments is specified: this means that, since μ_α is real, $\alpha'_{\alpha\beta}$ is supported only by systems for which the only good wavefunctions are complex, implying time-antisymmetry, as in a magnetic field {26}. The Faraday effect, which is linear in $\alpha'_{\alpha\beta}$, requires a magnetic field fixed in space {27}, whereas antisymmetric

Rayleigh scattering, which depends on α'^2, can arise from internal magnetic fields {28,29}.

The scalar part of the tensor $G'_{\alpha\beta}$ is familiar from the Rosenfeld equation for optical rotation in an isotropic sample {30},

$$\phi = -\tfrac{1}{3}\omega\mu_o LNG'_{\alpha\alpha} = -\tfrac{1}{3}\omega\mu_o LN(G'_{xx}+G'_{yy}+G'_{zz}), \tag{2.9}$$

where μ_o is the permeability of free space, L is the path length and N is the number density of molecules. Optical rotation in isotropic samples is supported only by molecules belonging to the proper rotation groups C_n, D_n, O, T and I (the chiral point groups). This is because $G'_{\alpha\alpha}$ contains the scalar product of an electric dipole and a magnetic dipole transition moment between the same initial and intermediate states. The polar vector $\vec{\mu}$ (which changes sign under inversion) and the axial vector $\vec{m}$ (which does not change sign under inversion) only transform the same under proper rotations and are distinguished by improper rotations. The same components of $\vec{\mu}$ and $\vec{m}$ therefore only span the same irreducible representations, and so can connect the same initial and intermediate states, in chiral molecules. Although $A_{\alpha\beta\gamma}$ does not contribute to optical rotation in isotropic samples, it provides a contribution of the same order as $G'_{\alpha\beta}$ in oriented samples. Thus the optical rotation of a light beam propagating along z in a completely oriented sample is given by the Buckingham-Dunn equation {31}:

$$\phi = -\tfrac{1}{2}\omega\mu_o LN[G'_{xx}+G'_{yy}+\tfrac{1}{3}\omega(A_{xyz}-A_{yxz})]. \tag{2.10}$$

This additional electric dipole-electric quadrupole optical activity generated by $A_{\alpha\beta\gamma}$ guarantees the origin-invariance of (2.10) and has been identified in the circular dichroism spectra of uniaxial crystals of Co^{3+} and Rh^{3+} complexes {32}. The symmetry requirements for optical rotation in oriented samples are not quite as severe as in isotropic samples. Although a centre of inversion must not be present, optical rotation is generated for certain directions of propagation in collections of molecules with reflection planes or rotation-reflection axes {33,34}. The remaining tensors $G_{\alpha\beta}$ and $A'_{\alpha\beta\gamma}$ are only non-zero in systems that are time-antisymmetric, and contribute to exotic phenomena such as gyrotropic birefringence {35}.

The polarizability and optical activity tensors (2.6) are strictly valid only when the frequency of the incident light is far from one of the natural transition frequencies ω_{jn} of the molecule. The generalization to absorbing frequencies is not required here and is given elsewhere {25,23}.

2.2 Vibrational Raman transition tensors

The exposition so far has derived the multipole moments that are oscillating with the same frequency as, and have a definite phase relation to, the inducing light wave. Radiation from such moments is responsible for Rayleigh scattering. But the Raman components of the scattered waves have frequencies different from, and are unrelated in phase to, the incident wave, so the polarizability and optical activity tensors must be replaced by corresponding transition tensors which take account of the different initial and final molecular states. In place of the expectation values of the multipole moment operators, real transition moments between initial and final molecular states n' and m', perturbed by the light wave, are introduced {28,36}. For example, the transition electric dipole moment is

$$(\mu_\alpha)_{mn} = \langle m'|\mu_\alpha|n'\rangle + \langle m'|\mu_\alpha|n'\rangle^* . \tag{2.11}$$

It is found that {37,23}

$$(\mu_\alpha)_{mn} = (\alpha_{\alpha\beta})_{mn} E_\beta(\omega-\omega_{mn}) + \frac{1}{\omega-\omega_{mn}}(\alpha'_{\alpha\beta})_{mn}\dot{E}_\beta(\omega-\omega_{mn}), \tag{2.12}$$

where, for simplicity, we have retained only pure electric dipole terms. When $\omega_m > \omega_n$, $(\omega-\omega_{mn}) < \omega$ and (2.12) is responsible for Stokes Raman scattering, whereas when $\omega_m < \omega_n$, $(\omega-\omega_{mn}) > \omega$ and (2.12) is responsible for anti-Stokes Raman scattering. The transition polarizabilities are

$$(\alpha_{\alpha\beta})_{mn} = \frac{1}{2\hbar}\sum_{j\neq n,m}\frac{1}{(\omega_{jn}-\omega)(\omega_{jm}+\omega)}$$

$$\times[(\omega_{jn}+\omega_{jm})\,\mathrm{Re}(\langle m|\mu_\alpha|j\rangle\langle j|\mu_\beta|n\rangle+\langle m|\mu_\beta|j\rangle\langle j|\mu_\alpha|n\rangle)$$

$$+(2\omega+\omega_{nm})\,\mathrm{Re}(\langle m|\mu_\alpha|j\rangle\langle j|\mu_\beta|n\rangle-\langle m|\mu_\beta|j\rangle\langle j|\mu_\alpha|n\rangle)], \tag{2.13a}$$

$$(\alpha'_{\alpha\beta})_{mn} = -\frac{1}{2\hbar}\sum_{j\neq n,m}\frac{1}{(\omega_{jn}-\omega)(\omega_{jm}+\omega)}$$

$$\times[(\omega_{jn}+\omega_{jm})\,\mathrm{Im}(\langle m|\mu_\alpha|j\rangle\langle j|\mu_\beta|n\rangle+\langle m|\mu_\beta|j\rangle\langle j|\mu_\alpha|n\rangle)$$

$$+(2\omega+\omega_{nm})\,\mathrm{Im}(\langle m|\mu_\alpha|j\rangle\langle j|\mu_\beta|n\rangle-\langle m|\mu_\beta|j\rangle\langle j|\mu_\alpha|n\rangle)], \tag{2.13b}$$

and may be regarded as the real and imaginary parts of the complex transition polarizability

$$(\tilde{\alpha}_{\alpha\beta})_{mn} = (\alpha_{\alpha\beta})_{mn} - i(\alpha'_{\alpha\beta})_{mn} . \tag{2.13c}$$

When n = m (Rayleigh scattering), $(\tilde{\alpha}_{\alpha\beta})_{mn}$ reduces to $\tilde{\alpha}_{\alpha\beta}$, as required; the real part of the complex transition polarizability is then purely symmetric and the imaginary part purely antisymmetric. But when n ≠ m (Raman scattering), both real and imaginary parts of the complex transition polarizability can contain symmetric and antisymmetric parts, although the additional parts only contribute to resonance Raman scattering and disappear at transparent frequencies {28,37}. Similar expressions are obtained for the transition optical activity tensors $(\tilde{G}_{\alpha\beta})_{mn}$ and $(\tilde{A}_{\alpha\beta\gamma})_{mn}$, with μ_β replaced by m_β and $\Theta_{\beta\gamma}$, respectively, but now there is no meaningful separation into symmetric and antisymmetric parts when n = m.

The vibrational Raman transition tensors can be developed in two distinct ways: Placzek's polarizability theory {28,36}, which considers the dependence of the ground state electronic polarizability on the normal coordinates of vibration; and the vibronic expansion theory {38,37}, which considers the vibrational perturbation of the ground and excited electronic states. The polarizability theory is best for transparent frequencies, but the vibronic expansion theory is best for the resonance situation since ground state theories depend on a formal sum over all excited states. We use here only the polarizability theory since this article is restricted to Raman optical activity at transparent frequencies. The adiabatic approximation is first invoked to write a molecular eigenstate as a product of electronic, vibrational and rotational parts:

$$|j_{evr}\rangle = |j_e j_v j_r\rangle. \tag{2.14}$$

At transparent frequencies the Raman transition tensors can be written to a good approximation as, for example,

$$(\alpha_{\alpha\beta})_{m_{evr}n_{evr}} = (\alpha_{\alpha'\beta'})_{m_{ev}n_{ev}}\langle m_r|l_{\alpha\alpha'}l_{\beta\beta'}|n_r\rangle, \tag{2.15}$$

where $(\alpha_{\alpha\beta})_{mevnev}$ is simply (2.13a) with all rotational states and energies removed, and $l_{\alpha\alpha'}$ is the direction cosine between the space-fixed axis α and the molecule-fixed axis α'. Since we are concerned here only with vibrational Raman scattering from fluids, the rotational states are henceforth dropped: the Raman transition tensors are written in space-fixed coordinates and isotropic averages taken of intensity expressions. For vibrational Raman scattering we need only consider the vibrational transition polarizability $(\alpha_{\alpha\beta})_{m_v n_v}$ in which both m_v and n_v belong to the lowest electronic level $m_e = n_e = 0$. By neglecting the 'ionic' part of the vibrational transition polarizability describing scattering through virtual excited vibrational states alone, the molecule remaining in the ground electronic state, and invoking the closure theorem with respect to the complete set of vibrational states in the remaining 'electronic' part, the real vibrational transition polarizability can be written {36}

$$(\alpha_{\alpha\beta})_{m_v n_v} = \langle m_v|\alpha_{\alpha\beta}(Q)|n_v\rangle = (\alpha_{\beta\alpha})_{m_v n_v}, \tag{2.16a}$$

where $\alpha_{\alpha\beta}(Q)$ is (2.13a) with only electronic states and energies retained, the states and operators depending parametrically on the normal vibrational coordinate Q. Similarly for the imaginary vibrational transition polarizability:

$$(\alpha'_{\alpha\beta})_{m_v n_v} = \langle m_v | \alpha'_{\alpha\beta}(Q) | n_v \rangle = -(\alpha'_{\beta\alpha})_{m_v n_v}. \tag{2.16b}$$

In this approximation, the antisymmetric part of the real transition polarizability, and the symmetric part of the imaginary transition polarizability, vanish. Similar developments are possible for the vibrational transition optical activity tensors, leading to

$$(G'_{\alpha\beta})_{m_v n_v} = \langle m_v | G'_{\alpha\beta}(Q) | n_v \rangle, \tag{2.16c}$$

$$(A_{\alpha\beta\gamma})_{m_v n_v} = \langle m_v | A_{\alpha\beta\gamma}(Q) | n_v \rangle = (A_{\alpha\gamma\beta})_{m_v n_v}. \tag{2.16d}$$

Finally, the electronic polarizability and optical activity tensors are expanded as Taylor series in the normal vibrational coordinates; for example

$$(\alpha_{\alpha\beta})_{m_v n_v} = (\alpha_{\alpha\beta})_0 \delta_{m_v n_v} + \sum_p \left(\frac{\partial \alpha_{\alpha\beta}}{\partial Q_p}\right)_0 \langle m_v | Q_p | n_v \rangle + \cdots, \tag{2.17}$$

where a subscript zero indicates that the function is taken at the equilibrium nuclear configuration. The first term describes Rayleigh scattering and the second vibrational Raman scattering with the selection rule $\Delta v_p = \pm 1$ for a simple harmonic oscillator.

3. THE RAYLEIGH AND RAMAN CIRCULAR INTENSITY DIFFERENCES

3.1 Molecular scattering of circularly polarized light

We consider the origin of scattered light to be the radiation of electromagnetic waves by the oscillating electric and magnetic multipole moments (2.8) induced in the molecule by the incident light wave. In fact Rayleigh scattering, because it is coherent, cannot occur in a perfectly homogeneous medium on account of destructive interference: only forward-scattering survives and gives rise to refraction through interference with the transmitted component of the incident wave. Rayleigh scattering is actually observed only because local density fluctuations prevent the destructive interference from being complete at any instant. The total Rayleigh scattering power per molecule decreases with increasing density, the isotropic contribution being suppressed more than the anisotropic contribution. Vibrational Raman scattering, on the other hand, is completely incoherent, the total Raman intensity from N molecules being approximately N times that from a single molecule at all sample densities. Consequently, the polarization properties of scattered light derived below apply to Rayleigh scattering only from an ideal gas, but apply to vibrational Raman scattering from any sample.

We shall only derive explicitly the c.i.d.s for light scattered at 90^0 at transparent frequencies, since at present this is the most important experimental situation. More general calculations are presented elsewhere {22}.

A plane wave travelling along z can be written as a sum of two coherent waves linearly polarized along x and y:

$$\vec{E} = E_x\vec{i} + E_y\vec{j}, \tag{3.1}$$

where $\vec{i}$ and $\vec{j}$ are unit vectors along x and y ($\vec{k}$ is the unit vector along z and $\vec{i}\times\vec{j} = \vec{k}$ since right-hand axes are used). The polarization of $\vec{E}$ is determined by the relative phases and magnitudes of E_x and E_y: for circular polarization, E_x and E_y are equal in magnitude and $\pm\frac{1}{2}\pi$ out of phase. The traditional convention for right (R) or left (L) circular polarization is a clockwise or anticlockwise rotation of E when viewed by an observer looking towards the source of the wave. Hence from (2.2a) and (3.1),

$$\begin{aligned}\tilde{\vec{E}}^{R}_{L} &= 2^{-\frac{1}{2}}E^{(0)}[\vec{i} + \exp(\mp i\pi/2)\vec{j}]\exp[-i\omega(t - \vec{k}.\vec{r}/c)] \\ &= 2^{-\frac{1}{2}}E^{(0)}(\vec{i} \mp i\vec{j})\exp[-i\omega(t - \vec{k}.\vec{r}/c)].\end{aligned} \tag{3.2}$$

Notice that the sign of $i\vec{j}$ in (3.2) is determined by the choice of sign in the exponent of (2.2a)

The molecule is taken at the origin of the space-fixed coordinate system x,y,z in an incident plane wave light beam travelling along z. For scattering at 90^0 we require the following expression for the complex electric vector radiated into the y direction at a distance from the origin much greater than the wavelength {39}:

$$\begin{aligned}\tilde{E}_\alpha = \frac{\omega^2\mu_0}{4\pi y}\Big[&(\tilde{\mu}^{(0)}_\alpha - j_\alpha\tilde{\mu}^{(0)}_y) - \frac{1}{c}\varepsilon_{\alpha y\beta}\tilde{m}^{(0)}_\beta \\ &- \frac{i\omega}{3c}(\tilde{\Theta}^{(0)}_{\alpha y} - j_\alpha\tilde{\Theta}^{(0)}_{yy}) + \cdots\Big]\exp[-i\omega(t - y/c)].\end{aligned} \tag{3.3}$$

The terms in $j_\alpha\tilde{\mu}^{(0)}_y$ and $j_\alpha\tilde{\Theta}^{(0)}_{yy}$ guarantee that the radiated wave is transverse. The intensities of the components of this scattered wave perpendicular (I_x) and parallel (I_z) to the scattering plane yz (Figure 1) are

$$I_x = \frac{1}{2\mu_0 c}\tilde{E}_x\tilde{E}^*_x, \qquad I_z = \frac{1}{2\mu_0 c}\tilde{E}_z\tilde{E}_z. \tag{3.4}$$

Using (3.2), (3.3), (3.4) and (2.8), the differences in the scattered intensity components in right and left circularly polarized incident light are

$$\begin{aligned}I^R_x - I^L_x = \frac{\omega^4\mu_0 E^{(0)2}}{16\pi^2c^2y^2}\Big[&\mathrm{Im}\,(c\tilde{\alpha}_{xy}\tilde{\alpha}^*_{xx} + \tilde{\alpha}_{xy}\tilde{G}^*_{xy} \\ &+ \tilde{\alpha}_{xx}\tilde{G}^*_{xx} + \tilde{\alpha}_{xy}\tilde{G}^*_{xz} - \tilde{\alpha}_{xx}\tilde{G}^*_{yz}) \\ &+ \tfrac{1}{3}\omega\,\mathrm{Re}(\tilde{\alpha}_{xx}\tilde{A}^*_{xzy} - \tilde{\alpha}_{xy}\tilde{A}^*_{xzx} + \tilde{\alpha}_{xy}\tilde{A}^*_{xxy} - \tilde{\alpha}_{xx}\tilde{A}^*_{yxy}) + \cdots\Big],\end{aligned} \tag{3.5a}$$

$$I_z^R - I_z^L = \frac{\omega^4 \mu_0 E^{(0)2}}{16\pi^2 c^2 y^2}\Big[\mathrm{Im}\big(c\tilde{\alpha}_{zy}\tilde{\alpha}_{zx}^* + \tilde{\alpha}_{zy}\tilde{G}_{zy}^* + \tilde{\alpha}_{zx}\tilde{G}_{zx}^* - \tilde{\alpha}_{zy}\tilde{G}_{xx}^* + \tilde{\alpha}_{zx}\tilde{G}_{yx}^*\big) + \tfrac{1}{3}\omega\,\mathrm{Re}\big(\tilde{\alpha}_{zx}\tilde{A}_{zzy}^* - \tilde{\alpha}_{zy}\tilde{A}_{zzx}^* + \tilde{\alpha}_{zy}\tilde{A}_{xzy}^* - \tilde{\alpha}_{zx}\tilde{A}_{yzy}^*\big) + \cdots\Big]. \tag{3.5b}$$

The corresponding sums of the scattered intensity components in right and left circularly polarized incident light are

$$I_x^R + I_x^L = \frac{\omega^4 \mu_0 E^{(0)2}}{32\pi^2 c y^2}\left(\tilde{\alpha}_{xx}\tilde{\alpha}_{xx}^* + \tilde{\alpha}_{xy}\tilde{\alpha}_{xy}^* + \cdots\right), \tag{3.6a}$$

$$I_z^R + I_z^L = \frac{\omega^4 \mu_0 E^{(0)2}}{32\pi^2 c y^2}\left(\tilde{\alpha}_{zx}\tilde{\alpha}_{zx}^* + \tilde{\alpha}_{zy}\tilde{\alpha}_{zy}^* + \cdots\right). \tag{3.6b}$$

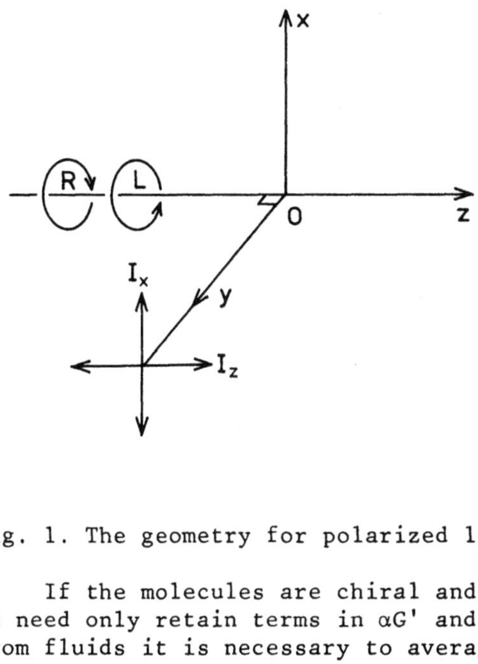

Fig. 1. The geometry for polarized light scattering at 90^0.

If the molecules are chiral and no magnetic fields are present, we need only retain terms in $\alpha G'$ and αA in (3.5). For scattering from fluids it is necessary to average (3.5) and (3.6) over all orientations of the molecule making use of the equations {40}

$$\langle i_\alpha i_\beta i_\gamma i_\delta \rangle = \tfrac{1}{15}\left(\delta_{\alpha\beta}\delta_{\gamma\delta} + \delta_{\alpha\gamma}\delta_{\beta\delta} + \delta_{\alpha\delta}\delta_{\beta\gamma}\right), \tag{3.7a}$$

$$\langle i_\alpha i_\beta j_\gamma j_\delta \rangle = \tfrac{1}{30}\left(4\delta_{\alpha\beta}\delta_{\gamma\delta} - \delta_{\alpha\gamma}\delta_{\beta\delta} - \delta_{\alpha\delta}\delta_{\beta\gamma}\right), \tag{3.7b}$$

$$\langle i_\alpha j_\beta k_\gamma k_\delta k_\epsilon \rangle = \tfrac{1}{30}\left(\varepsilon_{\alpha\beta\gamma}\delta_{\delta\epsilon} + \varepsilon_{\alpha\beta\delta}\delta_{\gamma\epsilon} + \varepsilon_{\alpha\beta\epsilon}\delta_{\gamma\delta}\right), \tag{3.7c}$$

where $i_\alpha, j_\alpha, k_\alpha$ are direction cosines between the space-fixed x,y,z axes and a molecule-fixed α axis, and the angular brackets denote an unweighted average. From (1.1), (3.5),(3.6) and (3.7), we find the following c.i.d. components for scattering at 90^0 at transparent frequencies from an isotropic sample {9}:

$$\Delta_x = \frac{2(7\alpha_{\alpha\beta}G'_{\alpha\beta} + \alpha_{\alpha\alpha}G'_{\beta\beta} + \frac{1}{3}\omega\alpha_{\alpha\beta}\varepsilon_{\alpha\gamma\delta}A_{\gamma\delta\beta})}{c(7\alpha_{\lambda\mu}\alpha_{\lambda\mu} + \alpha_{\lambda\lambda}\alpha_{\mu\mu})}, \tag{3.8a}$$

$$\Delta_z = \frac{4(3\alpha_{\alpha\beta}G'_{\alpha\beta} - \alpha_{\alpha\alpha}G'_{\beta\beta} - \frac{1}{3}\omega\alpha_{\alpha\beta}\varepsilon_{\alpha\gamma\delta}A_{\gamma\delta\beta})}{2c(3\alpha_{\lambda\mu}\alpha_{\lambda\mu} - \alpha_{\lambda\lambda}\alpha_{\mu\mu})}. \tag{3.8b}$$

We refer to Δ_x and Δ_z as the polarized and depolarized c.i.d.s; the c.i.d. with no analyzer in the scattered beam is obtained by adding the numerators and denominators. The same expressions apply to Raman scattering if the polarizability and optical activity tensors are replaced by corresponding transition tensors. Notice that although $A_{\alpha\beta\gamma}$ does not contribute to optical rotation in isotropic samples, it still contributes to Rayleigh and Raman c.i.d.

So Rayleigh (or Raman) optical activity in isotropic samples originates in interference between the scattered waves generated through the same components of the polarizability and optical activity tensors (or the transition polarizability and transition optical activity tensors). Using expansions of the form (2.17) in accordance with the polarizability theory, the Raman intensity and optical activity in the fundamental transition associated with the normal coordinate Q_p are therefore determined by

$$\langle p_0|\alpha_{\alpha\beta}|p_1\rangle\langle p_1|\alpha_{\alpha\beta}|p_0\rangle = \frac{\hbar}{2\omega_p}\left(\frac{\partial\alpha_{\alpha\beta}}{\partial Q_p}\right)_0^*\left(\frac{\partial\alpha_{\alpha\beta}}{\partial Q_p}\right)_0, \tag{3.9a}$$

$$\langle p_0|\alpha_{\alpha\beta}|p_1\rangle\langle p_1|G'_{\alpha\beta}|p_0\rangle = \frac{\hbar}{2\omega_p}\left(\frac{\partial\alpha_{\alpha\beta}}{\partial Q_p}\right)_0^*\left(\frac{\partial G'_{\alpha\beta}}{\partial Q_p}\right)_0, \tag{3.9b}$$

$$\langle p_0|\alpha_{\alpha\beta}|p_1\rangle\langle p_1|\varepsilon_{\alpha\gamma\delta}A_{\gamma\delta\beta}|p_0\rangle = \frac{\hbar}{2\omega_p}\left(\frac{\partial\alpha_{\alpha\beta}}{\partial Q_p}\right)_0^*\varepsilon_{\alpha\gamma\delta}\left(\frac{\partial A_{\gamma\delta\beta}}{\partial Q_p}\right)_0, \tag{3.9c}$$

the factor $\hbar/2\omega_p$ being the value of $|\langle p_1|Q_p|p_0\rangle|^2$.

Consequently, the symmetry requirements for natural Rayleigh optical activity are that the same components of $\alpha_{\alpha\beta}$ and $G'_{\alpha\beta}$ span the totally symmetric representation; and for natural vibrational Raman optical activity the same components of $\alpha_{\alpha\beta}$ and $G'_{\alpha\beta}$ must span the irreducible representation of the particular normal coordinate of vibration. This can only happen in the chiral point groups in which $\alpha_{\alpha\beta}$, which transforms like $\mu_\alpha\mu_\beta$ (a second rank polar tensor), and $G'_{\alpha\beta}$, which transforms like $\mu_\alpha m_\beta$ (a second rank axial tensor), have identical transformation properties. Furthermore, although $A_{\alpha\beta\gamma}$ does not transform like $G'_{\alpha\beta}$, the second rank axial tensor $\varepsilon_{\alpha\gamma\delta}A_{\gamma\delta\beta}$ that combines with $\alpha_{\alpha\beta}$ in the expressions for optically active scattering has transformation properties identical with $G'_{\alpha\beta}$. Consequently, all the Raman-active vibrations in a chiral molecule should show Raman optical activity. In the chiral point groups the assignment of components of $G'_{\alpha\beta}$ among the irreducible representations is trivial since it transforms the same as $\alpha_{\alpha\beta}$. But it is useful to know the distribution of components of $G'_{\alpha\beta}$ in the achiral point groups (for example, to determine symmetry rules for Raman optical activity induced in local achiral group vibrations by chiral intramolecular perturbations). One simple method for non-degenerate irreducible representations is to inspect the characters of the irreducible representations of the corresponding components of $\alpha_{\alpha\beta}$ and find the irreducible representations whose characters are the same except for a sign change under improper rotations. For example, in C_{2v}, A_1 is spanned by $\alpha_{XX}, \alpha_{YY}, \alpha_{ZZ}$, so A_2 is spanned by $G'_{XX}, G'_{YY}, G'_{ZZ}$; A_2 is spanned by α_{XY}, so A_1 is spanned by $G'_{XY}+G'_{YX}$, etc.

4. THE GENERATION OF RAYLEIGH AND RAMAN OPTICAL ACTIVITY WITHIN CHIRAL MOLECULES

4.1 Optical activity in transmitted and scattered light

A complete theory of the generation of Rayleigh and Raman optical activity within chiral molecules is difficult and will take some time to evolve. The electronic optical activity tensors $G'_{\alpha\beta}$ and $A_{\alpha\beta\gamma}$ (and also the polarizability $\alpha_{\alpha\beta}$) must be calculated; and for Raman optical activity at transparent frequencies we require the variation of these tensors with the normal vibrational coordinates, in accordance with (3.9).

Considerable physical insight into conventional electronic optical rotation and circular dichroism has been obtained from two limiting models {41}. The 'inherently chiral chromophore' model applies when electronic states are delocalized significantly over a chiral nuclear framework so that no selection rules restrict the corresponding electronic transitions induced by the light wave: electric dipole, magnetic dipole and electric quadrupole transitions are fully allowed between all states and the electronic optical activity tensors are maximized. At the opposite extreme, 'coupling' models apply when all groups within a molecule are inherently achiral and no electron exchange exists between them: any optical activity is assumed to arise from perturbations of local group electronic states by the chiral intramolecular environment. Two types of coupling model can be distinguished. The 'static coupling' or 'one electron' theory of Condon et al. {42} emphasises the perturbations due to the electrostatic fields of other groups. The 'dynamic coupling' or 'coupled oscillator' theory, which was brought to fruition by Kirkwood {43}, emphasises the perturbations due to the electromagnetic fields radiated by other groups under the influence of the light wave. The static and dynamic coupling models can make comparable contributions in the same molecule {44}.

These models are usually applied to optical rotation and circular dichroism in isotropic collections of molecules and so only generate the mean optical activity $G'_{\alpha\alpha}$. To deal with Rayleigh and Raman optical activity, they must be extended to generate general components of $G'_{\alpha\beta}$ and $A_{\alpha\beta\gamma}$, taking care to include the origin-dependent parts. As Barron and Buckingham have pointed out {45}, these origin-dependent parts give rise to a mechanism for Rayleigh and Raman optical activity that has no counterpart in optical rotation and circular dichroism. The latter are birefringence phenomena, and therefore originate in interference between transmitted and forward-scattered waves. Thus in the Kirkwood model, the optical rotation generated by a chiral structure comprising two achiral groups involves dynamic coupling: only forward-scattered waves that have been deflected from one group to the other have sampled the chirality and can generate optical rotation on combining with the transmitted wave at the detector (Figure 2a). But the transmitted wave is not important in Rayleigh and Raman scattering, so interference between two waves independently scattered from the two groups provides chiral information (Figure 2b). Dynamic coupling is not required, although it can make higher-order contributions. Both groups must be anisotropically polarizable for

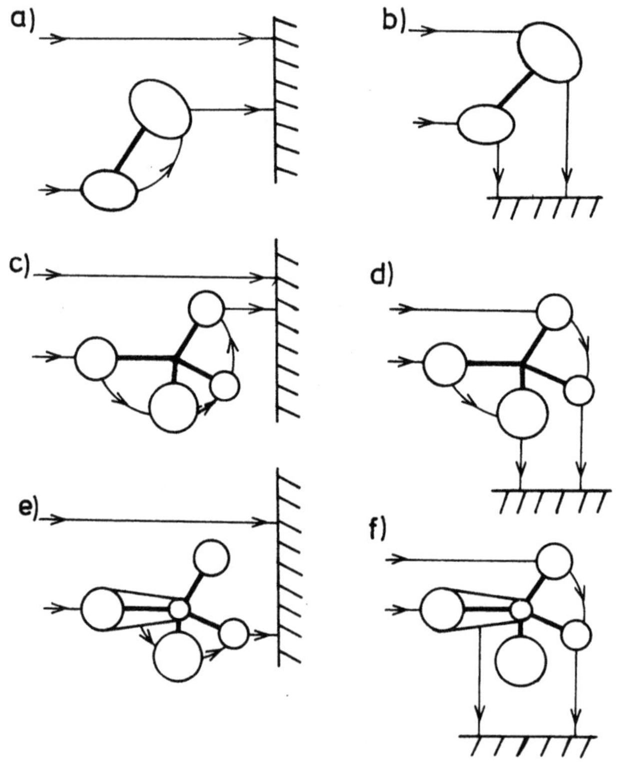

Fig. 2. The generation of optical rotation (first column) and Rayleigh and Raman optical activity (second column).

the complete structure to be chiral.

This picture can be extended to a chiral tetrahedral structure such as CHFClBr. Since a pair of dynamically-coupled spherical atoms constitutes a single anisotropically polarizable group, the Born-Boys model of optical rotation, which considers just the four ligand atoms, requires dynamic coupling extending over all four atoms (Figure 2c) {44}; whereas Rayleigh and Raman optical activity require interference between waves scattered independently from two pairs of dynamically coupled atoms (Figure 2d) {21,23}, or from one atom and three other dynamically coupled atoms. If the central carbon atom is included, the carbon-ligand bonds constitute anisotropic groups so less dynamic coupling is required (Figures 2e and 2f). Each diagram represents just one possible scattering sequence: any explicit calculation would sum over all permutations.

The development of (3.9b) and (3.9c), on which Raman optical activity depends, can proceed in two distinct ways.

I. We can consider a chiral electronic distribution, generated through electron delocalization over a chral nuclear framework or through static and dynamic coupling of two or more localized achiral electronic distributions that together constitute a chiral structure, modulated by achiral modes of vibration. This would apply to the situation where a normal coordinate is dominated by an internal coordinate localized on an achiral group (so that a group frequency approximation obtains) in which electronic chirality is induced. An early theory developed symmetry rules for Raman optical activity within this model {46}.

II. Alternatively, we can consider an achiral electronic distribution modulated by chiral modes of vibration. A normal mode of vibration can be considered chiral in zeroth order if it contains significant contributions from internal coordinates distributed over a chiral nuclear framework. Here we require interference of waves scattered independently from two or more groups that together constitute a chiral structure: a chiral electronic distribution in the usual sense is not generated since static or dynamic coupling is not required, but the normal mode of vibration must embrace the complete structure. In addition, we shall see that there is the possibility of 'vibrational chirality' contributions originating in the mixing of two or more internal coordinates, localized on the same group, that are orthogonal in the parent achiral local group symmetry.

In practice it might be difficult to disentangle contributions I and II since they are intimately connected: static and dynamic coupling will exist in all chiral molecules, and every normal mode in a completely asymmetric (C_1) molecule will contain contributions from every internal coordinate. Contribution II is emphasised in the discussion below since the appearance of typical Raman c.i.d. spectra indicates that vibrational chirality effects dominate. Notice that the processes described by Figures 2d and 2f constitute 'mixed' type I and type II mechanisms.

4.2 The two-group model of Rayleigh optical activity

The Rayleigh optical activity generated by a chiral molecule consisting of two neutral achiral groups 1 and 2 is now developed in detail since this provides much of the background required for a consideration of the Raman optical activity. It is in fact possible to derive the two-group Rayleigh c.i.d.s by developing explicitly the physical picture outlined in the previous section and illustrated in Figure 2b: we simply compute the intensity components at the detector arising from waves radiated independently by the oscillating multipole moments induced in the two groups by the incident light wave. This approach, which is given elsewhere {23}, has the advantage that it is not necessary to invoke the optical activity tensors $G'_{\alpha\beta}$ and $A_{\alpha\beta\gamma}$ and gives results that are valid for all separations of the two groups. However, the approach that is most easily incorporated into general theories, and which is used here, invokes the origin-dependence of $G'_{\alpha\beta}$ and $A_{\alpha\beta\gamma}$ {45}. Although the results are valid only for groups with separations much less than the wavelength of the incident light, this is not a significant restriction in practice (except for long-chain polymers).

Under an origin shift from $\vec{0}$ to a point $\vec{0}+\vec{a}$, where $\vec{a}$ is some constant vector, the polarizability and optical activity tensors change to {47}

$$\alpha_{\alpha\beta} \rightarrow \alpha_{\alpha\beta}, \tag{4.1a}$$

$$G'_{\alpha\beta} \rightarrow G'_{\alpha\beta} + \tfrac{1}{2}\omega\varepsilon_{\beta\gamma\delta}a_\gamma\alpha_{\alpha\delta}, \tag{4.1b}$$

$$A_{\alpha\beta\gamma} \rightarrow A_{\alpha\beta\gamma} - \tfrac{3}{2}a_\beta\alpha_{\alpha\gamma} - \tfrac{3}{2}a_\gamma\alpha_{\alpha\beta} + a_\delta\alpha_{\alpha\delta}\delta_{\beta\gamma}. \tag{4.1c}$$

We assume no electron exchange between the groups and write the polarizability and optical activity tensors of the molecule as the sum of the corresponding group tensors. The group tensors must be referred to a fixed origin within the molecule, which we choose to be the local origin in group 1:

$$\alpha_{\alpha\beta} = \alpha_{1\alpha\beta} + \alpha_{2\alpha\beta} + \text{coupling terms}, \tag{4.2a}$$

$$G'_{\alpha\beta} = G'_{1\alpha\beta} + G'_{2\alpha\beta} - \tfrac{1}{2}\omega\varepsilon_{\beta\gamma\delta}R_{21\gamma}\alpha_{2\alpha\delta} + \text{coupling terms}, \tag{4.2b}$$

$$\begin{aligned} A_{\alpha\beta\gamma} = {} & A_{1\alpha\beta\gamma} + A_{2\alpha\beta\gamma} + \tfrac{3}{2}R_{21\beta}\alpha_{2\alpha\gamma} + \tfrac{3}{2}R_{21\gamma}\alpha_{2\alpha\beta} \\ & - R_{21\delta}\alpha_{2\alpha\delta}\delta_{\beta\gamma} + \text{coupling terms}, \end{aligned} \tag{4.2c}$$

where $\alpha_{i\alpha\beta}$, $G'_{i\alpha\beta}$ and $A_{i\alpha\beta\gamma}$ are tensors referred to a local origin in the ith group, and $\vec{R}_{21} = \vec{R}_2 - \vec{R}_1$ is the vector from the origin in 1 to that in 2. Even though all components of $G'_{i\alpha\beta}$ and $A_{i\alpha\beta\gamma}$ may be zero, the origin-dependent parts may not be. Also, although the groups are assumed to be achiral in the usual sense, for certain symmetries (such as C_{2v}) there are non-zero components of the optical activity tensors that can contribute to Rayleigh optical activity. Using (4.2), the relevant polarizability-optical activity products in the c.i.d. components (3.8) can be approximated by

$$\alpha_{\alpha\beta}G'_{\alpha\beta} = -\tfrac{1}{2}\omega\varepsilon_{\beta\gamma\delta}R_{21\gamma}\alpha_{1\alpha\beta}\alpha_{2\delta\alpha} + \alpha_{1\alpha\beta}G'_{2\alpha\beta} + \alpha_{2\alpha\beta}G'_{1\alpha\beta}, \tag{4.3a}$$

$$\tfrac{1}{3}\omega\alpha_{\alpha\beta}\varepsilon_{\alpha\gamma\delta}A_{\gamma\delta\beta} = -\tfrac{1}{2}\omega\varepsilon_{\beta\gamma\delta}R_{21\gamma}\alpha_{1\alpha\beta}\alpha_{2\delta\alpha} + \tfrac{1}{3}\omega\alpha_{1\alpha\beta}\varepsilon_{\alpha\gamma\delta}A_{2\gamma\delta\beta} + \tfrac{1}{3}\omega\alpha_{2\alpha\beta}\varepsilon_{\alpha\gamma\delta}A_{1\gamma\delta\beta}, \tag{4.3b}$$

$$\alpha_{\alpha\alpha}G'_{\beta\beta} = 0, \tag{4.3c}$$

where the coupling terms of higher order have been negelected.

If both groups have threefold or higher proper rotation axes (which we take to be the 3 axis), equations (4.3) can be given a tractable form. If the groups are achiral, they cannot belong to the proper rotation point groups, and for the remaining point groups the components of the second rank axial tensors $G'_{\alpha\beta}$ and $\varepsilon_{\alpha\gamma\delta}A_{\gamma\delta\beta}$ are either zero or have $G'_{12} = -G'_{21}$ and $\varepsilon_{1\gamma\delta}A_{\gamma\delta 2} = -\varepsilon_{2\gamma\delta}A_{\gamma\delta 1}$ {48}. The terms $\alpha_{i_{\alpha\beta}}G'_{j\alpha\beta}$ and $\alpha_{i_{\alpha\beta}}\varepsilon_{\alpha\gamma\delta}A_{j\gamma\delta\beta}$ in (4.3) are then zero because $\alpha_{\alpha\beta} = \alpha_{\beta\alpha}$. If the principal axes 1,2,3 of the ith group are associated with unit vectors $\vec{s}_i, \vec{t}_i, \vec{u}_i$, its polarizability tensor may be written

$$\alpha_{i_{\alpha\beta}} = \alpha_i(1-\kappa_i)\delta_{\alpha\beta} + 3\alpha_i\kappa_i u_{i_\alpha}u_{i_\beta}, \tag{4.4}$$

where

$$\alpha = \tfrac{1}{3}(\alpha_{11}+\alpha_{22}+\alpha_{33}) = \tfrac{1}{3}(2\alpha_{11}+\alpha_{33}), \tag{4.5a}$$

$$\kappa = (\alpha_{33}-\alpha_{11})/3\alpha \tag{4.5b}$$

are the mean polarizability and the dimensionless polarizability anisotropy. Then

$$\varepsilon_{\beta\gamma\delta}R_{21\gamma}\alpha_{1\alpha\beta}\alpha_{2\delta\alpha} = 9\varepsilon_{\beta\gamma\delta}R_{21\gamma}\alpha_1\alpha_2\kappa_1\kappa_2 u_{1\alpha}u_{2\alpha}u_{1\beta}u_{2\delta}. \tag{4.6}$$

For the simplest chiral pair where the principal axes of the two groups are in parallel planes (Figure 3), this becomes

$$\varepsilon_{\beta\gamma\delta}R_{21\gamma}\alpha_{1\alpha\beta}\alpha_{2\delta\alpha} = -\tfrac{9}{2}R_{21}\alpha_1\alpha_2\kappa_1\kappa_2\sin 2\theta. \tag{4.7}$$

Using (4.7) in (4.3), the combinations of polarizability-optical activity products required in the c.i.d. components (3.8) are

$$3\alpha_{\alpha\beta}G'_{\alpha\beta} - \alpha_{\alpha\alpha}G'_{\beta\beta} - \tfrac{1}{3}\omega\alpha_{\alpha\beta}\varepsilon_{\alpha\gamma\delta}A_{\gamma\delta\beta} = \tfrac{9}{2}\omega R_{21}\alpha_1\alpha_2\kappa_1\kappa_2\sin 2\theta, \tag{4.8a}$$

$$7\alpha_{\alpha\beta}G'_{\alpha\beta} + \alpha_{\alpha\alpha}G'_{\beta\beta} + \tfrac{1}{3}\omega\alpha_{\alpha\beta}\varepsilon_{\alpha\gamma\delta}A_{\gamma\delta\beta} = 18\omega R_{21}\alpha_1\alpha_2\kappa_1\kappa_2\sin 2\theta. \tag{4.8b}$$

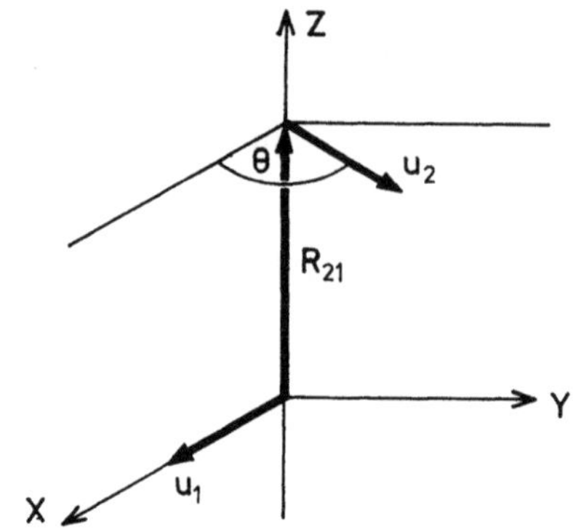

Fig. 3. A simple chiral two-group structure.

We also require combinations of polarizability-polarizability products in the c.i.d. components (3.8). Using (4.2a) to write

$$\alpha_{\alpha\beta}\alpha_{\alpha\beta} = \alpha_{1\alpha\beta}\alpha_{1\alpha\beta} + \alpha_{2\alpha\beta}\alpha_{2\alpha\beta} + 2\alpha_{1\alpha\beta}\alpha_{2\alpha\beta},$$

and using (4.4) for the polarizability tensors of the axially-symmetric groups, we find

$$3\alpha_{\alpha\beta}\alpha_{\alpha\beta} - \alpha_{\alpha\alpha}\alpha_{\beta\beta} = 18(\alpha_1^2\kappa_1^2 + \alpha_2^2\kappa_2^2) + 9\alpha_1\alpha_2\kappa_1\kappa_2(1+3\cos 2\theta), \quad (4.9a)$$

$$7\alpha_{\alpha\beta}\alpha_{\alpha\beta} + \alpha_{\alpha\alpha}\alpha_{\beta\beta} = 30(\alpha_1^2 + \alpha_2^2 + 2\alpha_1\alpha_2) + 42(\alpha_1^2\kappa_1^2 + \alpha_2^2\kappa_2^2) + 21\alpha_1\alpha_2\kappa_1\kappa_2(1+3\cos 2\theta). \quad (4.9b)$$

If groups 1 and 2 are identical, these results give the following Rayleigh c.i.d. components {45}:

$$\Delta_x = \frac{24\pi R_{21}\kappa^2 \sin 2\theta}{\lambda[40+7\kappa^2(5+3\cos 2\theta)]}, \quad (4.10a)$$

$$\Delta_z = \frac{2\pi R_{21} \sin 2\theta}{\lambda(5+3\cos 2\theta)}, \quad (4.10b)$$

where κ is a group anisotropy. Positive Δ values correspond to the absolute configuration of Figure 3. Notice that, provided $\lambda >> R_{21}$, Rayleigh c.i.d. increases with increasing separation of the two groups, whereas optical rotation decreases since it depends on dynamic coupling. Stone has extended the calculation to a two-group structure of more general geometry {49} (see also {23}).

In the case of a twisted biphenyl with $\theta = \frac{1}{4}\pi$ and $\lambda = 500$ nm, (4.10) gives $\Delta_x \simeq 0.6\times10^{-4}$ and $\Delta_z \simeq 1.3\times10^{-3}$, so the polarized Rayleigh c.i.d. is expected to be an order of magnitude smaller than the depolarized Rayleigh c.i.d. The two-group model can be extended to more complicated structures by summing over all chiral pairs, and a calculation on hexahelicene {50} gives $\Delta_x \simeq 0.4\times10^{-4}$ and $\Delta_z \simeq 0.6\times10^{-3}$. Thus the Rayleigh c.i.d. of hexahelicene should be rather smaller than that of a twisted biphenyl, in contrast with the specific rotation, which is considerably larger. This is because the sum over all the pairwise I^R-I^L contributions in the c.i.d. numerator is weighted by a sum over all the corresponding pairwise I^R+I^L contributions in the denominator, but no such weighting occurs in optical rotation. So a molecule with a large specific rotation does not necessarily show a large Rayleigh c.i.d.

4.3 A bond-polarizability theory of Raman optical activity

In an earlier review {23}, expressions for the Raman optical activity generated by idealized modes of a two-group structure were developed, together with a naive perturbation approach which indicated how conservative Raman c.i.d. couplets could be generated from the coupling of orthogonal internal coordinates localized on the same group. Here we develop instead a general theory, based on the 'bond-polarizability' theory of Raman intensities {51}, which embraces these earlier results and enables the Raman optical activity associated with any normal mode to be calculated analytically provided that a normal coordinate analysis is available.

Internal vibrational coordinates S_q, such as local bond stretches and angle bendings, can be written as a sum over the set of normal coordinates {52}:

$$S_q = \sum_{p=1}^{3N-6} L_{qp} Q_p , \tag{4.11}$$

the $\vec{L}$-matrix elements being determined from a normal coordinate analysis. In the bond-polarizability theory, the variation of the molecular polarizability tensor with a normal coordinate of vibration, which determines the Raman intensity, is calculated by way of the variation of the molecular polarizability tensor with local internal coordinates:

$$\left(\frac{\partial\alpha_{\alpha\beta}}{\partial Q_p}\right)_0 = \sum_q \left(\frac{\partial\alpha_{\alpha\beta}}{\partial S_q}\frac{\partial S_q}{\partial Q_p}\right)_0 = \sum_q \left(\frac{\partial\alpha_{\alpha\beta}}{\partial S_q}\right)_0 L_{qp} . \tag{4.12}$$

The molecular polarizability tensor is written as a sum of local bond polarizabilities, and the local bond polarizability derivatives with respect to internal coordinates evaluated.

The extension to Raman optical activity involves writing the polarizability and optical activity tensors as sums of corresponding bond tensors, taking account of the origin-dependence of the optical activity tensors:

$$\alpha_{\alpha\beta} = \sum_i \alpha_{i\,\alpha\beta}\,, \tag{4.13a}$$

$$G'_{\alpha\beta} = \sum_i G'_{i\,\alpha\beta} - \tfrac{1}{2}\omega\varepsilon_{\beta\gamma\delta}\sum_i R_{i\gamma}\alpha_{i\,\alpha\delta}\,, \tag{4.13b}$$

$$A_{\alpha\beta\gamma} = \sum_i A_{i\,\alpha\beta\gamma} + \tfrac{3}{2}\sum_i R_{i\beta}\alpha_{i\,\alpha\gamma} + \tfrac{3}{2}\sum_i R_{i\gamma}\alpha_{i\,\alpha\beta} - \sum_i R_{i\delta}\alpha_{i\,\alpha\delta}\delta_{\beta\gamma}\,, \tag{4.13c}$$

where $\alpha_{i_{\alpha\beta}}$ etc. are the intrinsic tensors of bond (or group) i referred to a local origin within i, and $\vec{R}_i$ is the vector from the molecular origin to the local bond origin. In neglecting static and dynamic coupling contributions, we are effectively limiting the applicability to the type II situation described above. Using (4.12) and (4.13), the polarizability theory products (3.9) that determine the Raman intensity and optical activity in fundamental transitions of a particular normal coordinate Q_p can be written

$$\langle p_0|\alpha_{\alpha\beta}|p_1\rangle\langle p_1|\alpha_{\alpha\beta}|p_0\rangle = \frac{\hbar}{2\omega_p}\Big[\sum_i\sum_q\Big(\frac{\partial\alpha_{i\,\alpha\beta}}{\partial S_q}\Big)_0 L_{qp}\Big]\Big[\sum_j\sum_r\Big(\frac{\partial\alpha_{j\,\alpha\beta}}{\partial S_r}\Big)_0 L_{rp}\Big]\,, \tag{4.14a}$$

$$\begin{aligned}\langle p_0|\alpha_{\alpha\beta}|p_1\rangle\langle p_1|G'_{\alpha\beta}|p_0\rangle = \\ -\frac{\hbar}{4\omega_p}\varepsilon_{\beta\gamma\delta}\Big\{\sum_{i<j}(R_{ji\gamma})_0\Big[\sum_q\Big(\frac{\partial\alpha_{i\,\alpha\beta}}{\partial S_q}\Big)_0 L_{qp}\Big]\Big[\sum_r\Big(\frac{\partial\alpha_{j\,\delta\alpha}}{\partial S_r}\Big)_0 L_{rp}\Big] \\ +\Big[\sum_i\sum_q\Big(\frac{\partial\alpha_{i\,\alpha\beta}}{\partial S_q}\Big)_0 L_{qp}\Big]\Big[\sum_j(\alpha_{j\,\delta\alpha})_0\sum_r\Big(\frac{\partial R_{j\gamma}}{\partial S_r}\Big)_0 L_{rp}\Big]\Big\} \\ +\frac{\hbar}{2\omega_p}\Big[\sum_i\sum_q\Big(\frac{\partial\alpha_{i\,\alpha\beta}}{\partial S_q}\Big)_0 L_{qp}\Big]\Big[\sum_j\sum_r\Big(\frac{\partial G'_{j\,\alpha\beta}}{\partial S_r}\Big)_0 L_{rp}\Big]\,,\end{aligned} \tag{4.14b}$$

$$\begin{aligned}\tfrac{1}{3}\omega\langle p_0|\alpha_{\alpha\beta}|p_1\rangle\langle p_1|\varepsilon_{\alpha\gamma\delta}A_{\gamma\delta\beta}|p_0\rangle = \\ -\frac{\hbar}{4\omega_p}\varepsilon_{\beta\gamma\delta}\Big\{\sum_{i<j}(R_{ji\gamma})_0\Big[\sum_q\Big(\frac{\partial\alpha_{i\,\alpha\beta}}{\partial S_q}\Big)_0 L_{qp}\Big]\Big[\sum_r\Big(\frac{\partial\alpha_{j\,\delta\alpha}}{\partial S_r}\Big)_0 L_{rp}\Big] \\ +\Big[\sum_i\sum_q\Big(\frac{\partial\alpha_{i\,\alpha\beta}}{\partial S_q}\Big)_0 L_{qp}\Big]\Big[\sum_j(\alpha_{j\,\delta\alpha})_0\sum_r\Big(\frac{\partial R_{j\gamma}}{\partial S_r}\Big)_0 L_{rp}\Big]\Big\} \\ +\frac{\hbar}{6\omega_p}\Big[\sum_i\sum_q\Big(\frac{\partial\alpha_{i\,\alpha\beta}}{\partial S_q}\Big)_0 L_{qp}\Big]\Big[\varepsilon_{\alpha\gamma\delta}\sum_j\sum_r\Big(\frac{\partial A_{j\,\gamma\delta\beta}}{\partial S_r}\Big)_0 L_{rp}\Big]\,.\end{aligned} \tag{4.14c}$$

We also require the products

$$\langle p_0|\alpha_{\alpha\alpha}|p_1\rangle\langle p_1|\alpha_{\beta\beta}|p_0\rangle = \frac{\hbar}{2\omega_p}\Big[\sum_i\sum_q\Big(\frac{\partial\alpha_{i\,\alpha\alpha}}{\partial S_q}\Big)_0 L_{qp}\Big]\Big[\sum_j\sum_r\Big(\frac{\partial\alpha_{j\,\beta\beta}}{\partial S_r}\Big)_0 L_{rp}\Big]\,, \tag{4.14d}$$

$$\langle p_0|\alpha_{\alpha\alpha}|p_1\rangle\langle p_1|G'_{\beta\beta}|p_0\rangle = \frac{\hbar}{2\omega_p}\Big[\sum_i\sum_q\Big(\frac{\partial\alpha_{i\,\alpha\alpha}}{\partial S_q}\Big)_0 L_{qp}\Big]\Big[\sum_j\sum_r\Big(\frac{\partial G'_{j\,\beta\beta}}{\partial S_r}\Big)_0 L_{rp}\Big]\,. \tag{4.14e}$$

We shall illustrate the application of these results by considering idealized normal modes, containing just one or two internal coordinates, of some model chiral structures.

4.4 The two-group model of Raman optical activity

Consider first the simple two-group structure of Figure 3. Since this has a twofold proper rotation axis, pairs of equivalent internal coordinates associated with the two groups, such as local bond stretchings or angle deformations, will always contribute to normal modes in symmetric and antisymmetric combinations. Consider two idealized normal coordinates containing just symmetric and antisymmetric combinations, respectively, of two equivalent internal coordinates localized on groups 1 and 2:

$$Q_+ = N_+(S_1+S_2)\,, \qquad Q_- = N_-(S_1-S_2)\,, \tag{4.15a}$$

where $N_+ = N_- = N$ is a constant. Consequently,

$$S_1 = \frac{1}{2N}(Q_+ + Q_-) \qquad S_2 = \frac{1}{2N}(Q_+ - Q_-)\,, \tag{4.15b}$$

so that $L_{1+} = L_{1-} = L_{2+} = 1/2N$, $L_{2-} = -1/2N$. Equations (4.14) become

$$\langle \pm_0|\alpha_{\alpha\beta}|\pm_1\rangle\langle \pm_1|\alpha_{\alpha\beta}|\pm_0\rangle = \left(\frac{\hbar}{2\omega_\pm}\right)\left(\frac{1}{4N^2}\right)\Bigg\{\left(\frac{\partial\alpha_{1\alpha\beta}}{\partial S_1}\right)^*_0\left(\frac{\partial\alpha_{1\alpha\beta}}{\partial S_1}\right)_0$$

$$+\left(\frac{\partial\alpha_{2\alpha\beta}}{\partial S_2}\right)^*_0\left(\frac{\partial\alpha_{2\alpha\beta}}{\partial S_2}\right)_0 \pm\left[\left(\frac{\partial\alpha_{1\alpha\beta}}{\partial S_1}\right)^*_0\left(\frac{\partial\alpha_{2\alpha\beta}}{\partial S_2}\right)_0+\left(\frac{\partial\alpha_{2\alpha\beta}}{\partial S_2}\right)^*_0\left(\frac{\partial\alpha_{1\alpha\beta}}{\partial S_1}\right)_0\right]\Bigg\}\,, \tag{4.16a}$$

$$\langle \pm_0|\alpha_{\alpha\beta}|\pm_1\rangle\langle \pm_1|G'_{\alpha\beta}|\pm_0\rangle = \tfrac{1}{3}\omega\langle \pm_0|\alpha_{\alpha\beta}|\pm_1\rangle\langle \pm_1|\varepsilon_{\alpha\gamma\delta}A_{\gamma\delta\beta}|\pm_0\rangle =$$

$$\mp\left(\frac{\hbar\omega}{4\omega_\pm}\right)\left(\frac{1}{4N^2}\right)\varepsilon_{\beta\gamma\delta}(R_{21\gamma})_0\left(\frac{\partial\alpha_{1\alpha\beta}}{\partial S_1}\right)^*_0\left(\frac{\partial\alpha_{2\alpha\beta}}{\partial S_2}\right)_0\,, \tag{4.16b}$$

$$\langle \pm_0|\alpha_{\alpha\alpha}|\pm_1\rangle\langle \pm_1|G'_{\beta\beta}|\pm_0\rangle = 0. \tag{4.16c}$$

In writing (4.16b) and (4.16c) we have neglected terms in $\alpha_i G'_i$, $\alpha_i A_i$, $\alpha_i G'_j$ and $\alpha_i A_j$. These are discussed later. If the two groups have threefold or higher proper rotation axes, we can use (4.4) for the group polarizability tensors. For totally symmetric local group internal coordinates (so that the relative orientation of the two groups does not change) we find after a little trigonometry the following Raman c.i.d. components:

$$\Delta_x^+ = \frac{24\pi R_{21}[\partial(\alpha_i\kappa_i)/\partial S_i]_0^2 \sin 2\theta}{\lambda\{40[\partial\alpha_i/\partial S_i]_0^2 + 7[\partial(\alpha_i\kappa_i)/\partial S_i]_0^2(5+3\cos 2\theta)\}}\,, \tag{4.17a}$$

$$\Delta_x^- = -\frac{8\pi R_{21}\sin 2\theta}{7\lambda(1-\cos 2\theta)}\,, \tag{4.17b}$$

$$\Delta_z^+ = \frac{2\pi R_{21}\sin 2\theta}{\lambda(5+3\cos 2\theta)}, \quad (4.17c)$$

$$\Delta_z^- = -\frac{2\pi R_{21}\sin 2\theta}{3\lambda(1-\cos 2\theta)}. \quad (4.17d)$$

Notice that only Δ_x^+ depends on the derivatives of the group polarizability tensor components with respect to the group internal coordinate. These derivatives are usually difficult to evaluate, so empirical values, transferable from one molecule to another, are often used {51}. Although the dimensionless c.i.d.s have different magnitudes with opposite signs for the symmetric and antisymmetric bands, $I_z^R - I_z^L$ for the two bands have equal magnitudes and opposite signs. The calculation of the c.i.d.s for the idealized normal modes corresponding to the symmetric and antisymmetric combinations of the angle bendings between the axes of the groups and the connecting bond is more complicated and is not given here.

Next consider an idealized normal coordinate containing just the internal coordinate of torsion S_t between groups 1 and 2 in the two-group structure:

$$Q_t = N_t S_t. \quad (4.18)$$

Taking θ to be the equilibrium value of the torsion angle, as shown in Figure 3, we can write the general torsion angle at some instant during the torsion vibration as $\theta+\Delta\theta$, and so identify S_t with $\Delta\theta$. If $\vec{I},\vec{J},\vec{K}$ are unit vectors along the internal molecular X,Y,Z axes, the principal axes of the two groups for some general torsion angle can be written

$$u_{1\alpha} = I_\alpha \cos(\tfrac{1}{2}\Delta\theta) - J_\alpha \sin(\tfrac{1}{2}\Delta\theta), \quad (4.19a)$$

$$u_{2\alpha} = I_\alpha \cos(\theta+\tfrac{1}{2}\Delta\theta) + J_\alpha \sin(\theta+\tfrac{1}{2}\Delta\theta). \quad (4.19b)$$

Using these expressions in group polarizability tensors of the form (4.4), we obtain the derivatives

$$\left(\frac{\partial\alpha_{1\alpha\beta}}{\partial\Delta\theta}\right)_0 = -\tfrac{3}{2}\alpha_1\kappa_1(I_\alpha J_\beta + J_\alpha I_\beta), \quad (4.20a)$$

$$\left(\frac{\partial\alpha_{2\alpha\beta}}{\partial\Delta\theta}\right)_0 = -\tfrac{3}{2}\alpha_2\kappa_2[(I_\alpha I_\beta - J_\alpha J_\beta)\sin 2\theta - (I_\alpha J_\beta + J_\alpha I_\beta)\cos 2\theta]. \quad (4.20b)$$

These provide the following products:

$$\left(\frac{\partial\alpha_{1\alpha\beta}}{\partial\Delta\theta}\right)_0\left(\frac{\partial\alpha_{1\alpha\beta}}{\partial\Delta\theta}\right)_0 = \tfrac{9}{2}\alpha_1^2\kappa_1^2, \quad (4.21a)$$

$$\left(\frac{\partial\alpha_{1\alpha\beta}}{\partial\Delta\theta}\right)_0\left(\frac{\partial\alpha_{2\alpha\beta}}{\partial\Delta\theta}\right)_0 = -\tfrac{9}{2}\alpha_1\alpha_2\kappa_1\kappa_2\cos 2\theta, \quad (4.21b)$$

$$\left(\frac{\partial \alpha_{1\alpha\beta}}{\partial \Delta\theta}\right)_0 \varepsilon_{\beta\gamma\delta}(R_{21\gamma})_0 \left(\frac{\partial \alpha_{2\delta\alpha}}{\partial \Delta\theta}\right)_0 = \frac{9}{2}\alpha_1\alpha_2 K_1 K_2 R_{21} \sin 2\theta, \tag{4.21c}$$

$$\left(\frac{\partial \alpha_{1\alpha\alpha}}{\partial \Delta\theta}\right)_0 \left(\frac{\partial \alpha_{2\beta\beta}}{\partial \Delta\theta}\right)_0 = 0 . \tag{4.21d}$$

When used in (4.14), these lead to the following Raman c.i.d. components:

$$\Delta_x = -\frac{8\pi R_{21} \sin 2\theta}{7\lambda(1-\cos 2\theta)}, \tag{4.22a}$$

$$\Delta_z = -\frac{2\pi R_{21} \sin 2\theta}{3\lambda(1-\cos 2\theta)} . \tag{4.22b}$$

4.5 Methyl torsion Raman optical activity

In the examples considered so far, the second terms of (4.14b) and (4.14c), involving the change in the position vector $\vec{R}_j$ of a group relative to the molecule-fixed origin, have made no contribution. Normal coordinates containing contributions from changes in either the length of $\vec{R}_j$, or its orientation, or both, will activate this term. The evaluation of this term is simplified if the molecule-fixed axes are chosen with origin at the centre of mass and with zero net vibrational angular momentum at any instant about each axis. These axes, which are translating and rotating with the molecule as a whole, coincide with the principal inertial axes. It might be thought that this term would generate Raman optical activity associated with the stretching of the connecting bond in the two-group model, but the appropriate terms reduce to zero, and in fact static or dynamic coupling between the two groups is required.

To illustrate the application of this term, we now turn to another type of model structure supporting methyl torsion vibrations. Organic molecules containing methyl groups often show large Raman optical activity at low frequency (between about 100 and 300 cm^{-1}) some of which probably originates in methyl torsion vibrations {53-55}. If the hindered internal rotation of the methyl group relative to the rest of the molecule is to be treated as a normal mode of vibration, both the top and the framework must move relative to the 'principal internal' axes. If I_i and I_{Me} are the moments of inertia of the frame and the methyl group about the torsion axis, the requirement that the torsion vibration generates zero net angular momentum can be expressed as

$$I_i \dot{\tau}_i = -I_{Me} \dot{\tau}_{Me} , \tag{4.23}$$

where $\tau_i = \tau_i^{(0)} + \Delta\tau_i$ and $\tau_{Me} = \tau_{Me}^{(0)} + \Delta\tau_{Me}$ specify the instantaneous orientations of the frame and the methyl group relative to a principal internal axis, perpendicular to the torsion axis, that remains stationary during the torsion vibration. $\tau_i^{(0)}$ and $\tau_{Me}^{(0)}$ are the corresponding equilibrium orientations, and

$$\Delta\tau = \Delta\tau_i - \Delta\tau_{Me} \tag{4.24}$$

is the internal coordinate of torsion, the two 'intrinsic' displacements satisfying

$$I_i \Delta\tau_i = -I_{Me}\Delta\tau_{Me} . \tag{4.25}$$

Thus it is the change in orientation of $\vec{R}_i$ relative to a principal internal axis that activates the second term of (4.14b) here, and if the threefold axis of the methyl group coincides with a principal inertial axis of the molecule, the methyl group itself makes no contribution: it is the oscillations in space of the chiral framework, compensating the oscillations of the methyl group, that generate the Raman intensity and optical activity (and also the infrared intensity and optical activity). So the tail wags the dog!

Our basic model consists of an anisotropic, intrinsically achiral, group i with a principal axis of polarizability along the unit vector $\vec{u}_i$ which is oriented relative to the threefold axis of the methyl group such that the anisotropic group and the three-

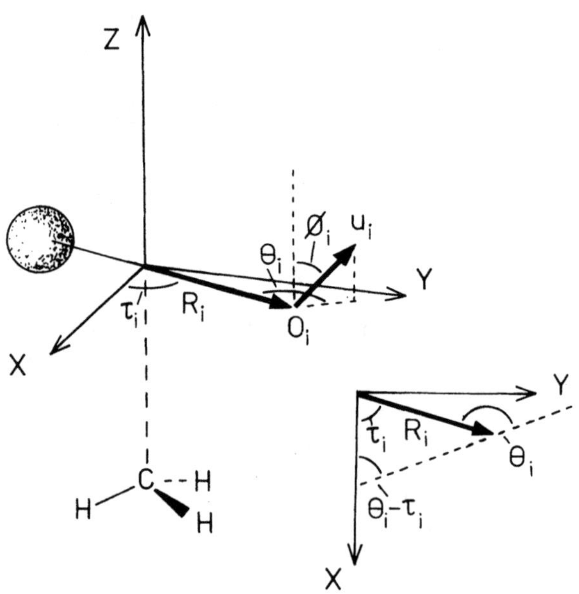

Fig. 4. The methyl torsion model based on a hindered single-bladed propellor.

fold axis constitute a chiral structure. Group i is balanced dynamically by a spherical group so that, assuming the existence of a hindering potential, torsional oscillations are executed about the threefold axis of the methyl group (Figure 4). If group i were a benzene ring with $\vec{u}_i$ along the sixfold axis, the structure would have the appearance of a single-bladed propellor.

For an idealized normal coordinate containing just the internal coordinate of torsion $\Delta\tau$,

$$Q_t = N_t \Delta\tau , \tag{4.26}$$

and we can use (4.24) and (4.25) to write

$$Q_t = N_t\left(\frac{I}{I_{Me}}\right)\Delta\tau_i = -N_t\left(\frac{I}{I_i}\right)\Delta\tau_{Me} , \tag{4.27}$$

where $I = I_i + I_{Me}$, from which $L_{it} = (I_{Me}/IN_t)$ and $L_{Met} = -(I_i/IN_t)$. Writing the polarizability tensor of the molecule as the sum of tensors intrinsic to group i and the methyl group, these $\vec{L}$-matrix elements multiply $(\partial\alpha_{i\alpha\beta}/\partial\Delta\tau_i)_o$ and $(\partial\alpha_{Me\alpha\beta}/\partial\Delta\tau_{Me})_o$ in the second terms of (4.14b) and (4.14c). In fact

$$\left(\frac{\partial\alpha_{Me\alpha\beta}}{\partial\Delta\tau_{Me}}\right)_o = 0$$

because the physical properties of an object with a threefold or higher proper rotation axis are isotropic with respect to rotations about that axis, so the methyl group makes no intrinsic contribution.

Referring to Figure 5, the unit vector along the principal axis of group i at some instant during the torsion vibration can be written in terms of the unit vectors $\vec{I},\vec{J},\vec{K}$ along the principal internal axes X,Y,Z:

$$\begin{aligned} u_{i\alpha} = &-I_\alpha \sin\phi_i \cos[\theta_i - (\tau_i^{(o)} + \Delta\tau_i)] \\ &+ J_\alpha \sin\phi_i \sin[\theta_i - (\tau_i^{(o)} + \Delta\tau_i)] + K_\alpha \cos\phi_i . \end{aligned} \tag{4.28}$$

If the principal axis of group i is a threefold or higher proper rotation axis, we can again write its polarizability tensor in the form (4.4), and using (4.28) we obtain

$$\begin{aligned} \alpha_{i\alpha\beta} = {}& \alpha_i(1-\kappa_i)\delta_{\alpha\beta} + 3\alpha_i\kappa_i\{I_\alpha I_\beta \sin^2\phi_i \cos^2[\theta_i - (\tau_i^{(o)} + \Delta\tau_i)] \\ &+ J_\alpha J_\beta \sin^2\phi_i \sin^2[\theta_i - (\tau_i^{(o)} + \Delta\tau_i)] + K_\alpha K_\beta \cos^2\phi_i \\ &- \tfrac{1}{2}(I_\alpha J_\beta + J_\alpha I_\beta)\sin^2\phi_i \sin 2[\theta_i - (\tau_i^{(o)} + \Delta\tau_i)] \\ &- \tfrac{1}{2}(I_\alpha K_\beta + K_\alpha I_\beta)\sin 2\phi_i \cos[\theta_i - (\tau_i^{(o)} + \Delta\tau_i)] \\ &+ \tfrac{1}{2}(J_\alpha K_\beta + K_\alpha J_\beta)\sin 2\phi_i \sin[\theta_i - (\tau_i^{(o)} + \Delta\tau_i)]\}. \end{aligned} \tag{4.29}$$

We also require

$$R_{i\gamma} = R_i[I_\gamma \cos(\tau_i^{(0)} + \Delta\tau_i) + J_\gamma \sin(\tau_i^{(0)} + \Delta\tau_i)]. \qquad (4.30)$$

The partial derivatives of (4.29) and (4.30) with respect to the internal coordinate of torsion are

$$\left(\frac{\partial \alpha_{i\alpha\beta}}{\partial \Delta\tau_i}\right)_0 = 3\alpha_i\kappa_i[(I_\alpha I_\beta - J_\alpha J_\beta)\sin^2\phi_i \sin 2(\theta_i - \tau_i^{(0)})$$

$$+ (I_\alpha J_\beta + J_\alpha I_\beta)\sin^2\phi_i \cos 2(\theta_i - \tau_i^{(0)}) - \tfrac{1}{2}(I_\alpha K_\beta + K_\alpha I_\beta)\sin 2\phi_i \sin(\theta_i - \tau_i^{(0)})$$

$$- \tfrac{1}{2}(J_\alpha K_\beta + K_\alpha J_\beta)\sin 2\phi_i \cos(\theta_i - \tau_i^{(0)})], \qquad (4.31)$$

$$\left(\frac{\partial R_{i\gamma}}{\partial \Delta\tau_i}\right)_0 = R_i(-I_\gamma \sin\tau_i^{(0)} + J_\gamma \cos\tau_i^{(0)}). \qquad (4.32)$$

These provide the following products:

$$\left(\frac{\partial \alpha_{i\alpha\beta}}{\partial \Delta\tau_i}\right)_0 \left(\frac{\partial \alpha_{i\alpha\beta}}{\partial \Delta\tau_i}\right)_0 = 9\alpha_i^2\kappa_i^2(1 - \cos 2\phi_i), \qquad (4.33a)$$

$$\left(\frac{\partial \alpha_{i\alpha\alpha}}{\partial \Delta\tau_i}\right)_0 \left(\frac{\partial \alpha_{i\beta\beta}}{\partial \Delta\tau_i}\right)_0 = 0, \qquad (4.33b)$$

$$\left(\frac{\partial \alpha_{i\alpha\beta}}{\partial \Delta\tau_i}\right)_0 \varepsilon_{\beta\gamma\delta}(\alpha_{i\delta\alpha})_0 \left(\frac{\partial R_{i\gamma}}{\partial \Delta\tau_i}\right)_0 = -\frac{9}{2}R_i\alpha_i^2\kappa_i^2 \sin 2\phi_i \sin\theta_i. \qquad (4.33c)$$

When used in (4.14a) and the second terms of (4.14b) and (4.14c), these generate the following Raman c.i.d. components:

$$\Delta_x = \frac{8\pi R_i \sin 2\phi_i \sin\theta_i}{7\lambda(1 - \cos 2\phi_i)}, \qquad (4.34a)$$

$$\Delta_z = \frac{2\pi R_i \sin 2\phi_i \sin\theta_i}{3\lambda(1 - \cos 2\phi_i)}. \qquad (4.34b)$$

These expressions were first derived by Barron and Buckingham {54}. The numerators reduce to zero if $\theta_i = 0^0$ or 180^0 or if $\phi_i = 0^0$ or 90^0.

These c.i.d.s are not exclusive to the methyl group; the same results would be obtained for the oscillation of a single-bladed propellor about any well-defined torsion axis. An intriguing extension is to molecules containing two adjacent methyl groups since the in-phase and out-of-phase combinations of the two methyl torsions can generate compensating torsion vibrations of the rest of the molecule about a well-defined symmetry axis. For example, in the bridged biphenyl below, the in-phase combination generates a torsion of the rest of the molecule (which has the appearance of a double-bladed propellor) about its C_2 axis, and the associated

Raman c.i.d. is easily calculated {54}.

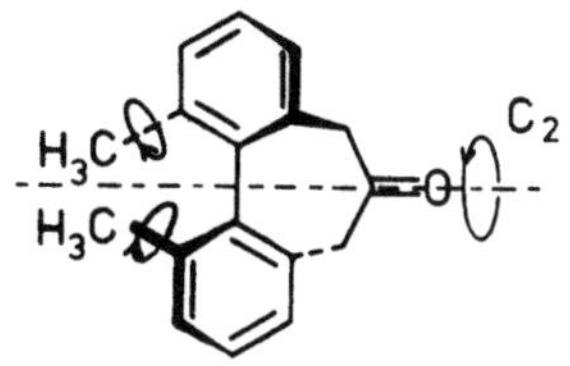

Notice that the second terms of (4.14b) and (4.14c) can also deal with normal modes of vibration in which the methyl torsion is mixed with other internal coordinates.

4.6 Intrinsic group optical activity tensors

In the models developed so far we have neglected the terms in $\alpha_i G'_j$ and $\alpha_i A_j$ in the general polarizability theory expressions (4.14). These can make significant contributions and, although they cannot be calculated explicitly at present, we can make some general statements about them. Consider two idealized normal modes containing symmetric and antisymmetric combinations of two internal coordinates which are, in general, non-equivalent:

$$Q_+ = N_1 S_1 + N_2 S_2 \, , \quad Q_- = N_2 S_1 - N_1 S_2 \, . \tag{4.35a}$$

The inverse expressions are

$$S_1 = \frac{1}{N_1^2 + N_2^2}(N_1 Q_+ + N_2 Q_-), \quad S_2 = \frac{1}{N_1^2 + N_2^2}(N_2 Q_+ - N_1 Q_-), \tag{4.35b}$$

so that the $\vec{L}$-matrix elements are $L_{1+} = N_1/(N_1^2+N_2^2)$, $L_{1-} = N_2/(N_1^2+N_2^2)$, $L_{2+} = N_2/(N_1^2+N_2^2)$ and $L_{2-} = -N_1/(N_1^2+N_2^2)$.

If S_1 and S_2 are localized on the same group i, the required contribution to (4.14b) is

$$\langle \pm_0 | \alpha_{\alpha\beta} | \pm_1 \rangle \langle \pm_1 | G'_{\alpha\beta} | \pm_0 \rangle =$$
$$\pm\left(\frac{\hbar}{2\omega_\pm}\right)\frac{N_1 N_2}{(N_1^2+N_2^2)^2}\left[\left(\frac{\partial \alpha_{i\alpha\beta}}{\partial S_1}\right)_0\left(\frac{\partial G'_{i\alpha\beta}}{\partial S_2}\right)_0 + \left(\frac{\partial \alpha_{i\alpha\beta}}{\partial S_2}\right)_0\left(\frac{\partial G'_{i\alpha\beta}}{\partial S_1}\right)_0\right] \tag{4.36}$$

with an analogous $\alpha_i A_j$ contribution to (4.14c). Terms in $(\partial\alpha_{i\alpha\beta}/\partial S_1)_0(\partial G'_{i\alpha\beta}/\partial S_1)_0$ and $(\partial\alpha_{i\alpha\beta}/\partial S_2)_0(\partial G'_{i\alpha\beta}/\partial S_2)_0$ are zero because the group i is assumed to be intrinsically achiral. A possible example is the carbonyl group in molecules such as 3-methylcyclohexanone. The in-plane and out-of-plane bending modes belong to symmetry species B_1 and B_2 in the local C_{2v} symmetry: B_1 is spanned by α_{XZ} and G'_{YZ}, and B_2 is spanned by α_{YZ} and G'_{XZ}. The skeletal chirality will lead to normal modes of vibration containing symmetric and antisymmetric combinations of the two locally orthogonal deformations, generating equal and opposite Raman optical activities. The absolute signs depend on N_1 and N_2, given by the normal coordinate analysis, and on

$$\left(\frac{\partial \alpha_{xz}}{\partial S_{B_1}}\right)_0 \left(\frac{\partial G'_{xz}}{\partial S_{B_2}}\right)_0 + \left(\frac{\partial \alpha_{yz}}{\partial S_{B_2}}\right)_0 \left(\frac{\partial G'_{yz}}{\partial S_{B_1}}\right)_0 ,$$

which is an intrinsic property of the carbonyl group. Another possible example is the mixing of the two degenerate asymmetric deformations in a group such as methyl or trifluoromethyl attached to a chiral carbon atom.

If S_1 and S_2 are localized on two different achiral groups 1 and 2 that together constitute a chiral structure, the required contribution to (4.14b) is

$$\langle \pm_0|\alpha_{\alpha\beta}|\pm_1\rangle\langle\pm_1|G'_{\alpha\beta}|\pm_0\rangle =$$
$$\pm\left(\frac{\hbar}{2\omega_\pm}\right)\frac{N_1 N_2}{(N_1^2+N_2^2)^2}\left[\left(\frac{\partial \alpha_{1\alpha\beta}}{\partial S_1}\right)_0\left(\frac{\partial G'_{2\alpha\beta}}{\partial S_2}\right)_0 + \left(\frac{\partial \alpha_{2\alpha\beta}}{\partial S_2}\right)_0\left(\frac{\partial G'_{1\alpha\beta}}{\partial S_1}\right)_0\right] \quad (4.37)$$

with an analogous $\alpha_i A_j$ contribution to (4.14c). The two groups, and also S_1 and S_2, may be equivalent, in which case the normal modes (4.35) reduce to (4.15). If the two groups have threefold or higher proper rotation axes, the terms in $\alpha_i G'_j$ are zero if S_1 and S_2 are totally symmetric (because no components of $G'_{\alpha\beta}$ span the totally symmetric irreducible representation), but are not necessarily zero if S_1 and S_2 are non-totally symmetric. If the two groups have less than a threefold proper rotation axis, terms in $\alpha_i G'_j$ can be non-zero even if S_1 and S_2 are totally symmetric. The discussion can be extended readily to non-equivalent modes on equivalent or non-equivalent groups.

4.7 Electronic coupling models of Raman optical activity

The discussion so far has concentrated on 'vibrational chirality' contributions to Raman optical activity; the type II mechanisms discussed in section 4.1. The type I mechanisms, involving 'electronic chirality' can be developed by adding to the molecular polarizability and optical activity tensors (4.13) appropriate electronic coupling terms, evaluating their variations with internal vibrational coordinates, and calculating products of the form (4.14). For example, for a two-group structure, dynamic coupling contributions would be calculated by adding the terms {44,56}

$$\frac{1}{4\pi\epsilon_0}\left(-\tfrac{1}{2}\omega\varepsilon_{\beta\gamma\delta}R_{21\gamma}\alpha_{2\delta\epsilon}T_{21\epsilon\lambda}\alpha_{1\lambda\alpha} + G'_{1\gamma\beta}T_{21\gamma\delta}\alpha_{2\delta\alpha} + G'_{2\gamma\beta}T_{21\gamma\delta}\alpha_{1\delta\alpha}\right) \quad (4.38a)$$

to (4.13b), with analogous terms added to (4.13a) and (4.13c), where

$$T_{ij\alpha\beta} = (3R_{ij\alpha} R_{ij\beta} - R_{ij}^{2}\delta_{\alpha\beta}) R_{ij}^{-5} \tag{4.38b}$$

is the dipole-dipole coupling tensor. However, we shall not develop the electronic coupling contributions explicitly here.

4.8 Comparison with infrared circular dichroism

A theory of infrared circular dichroism has been developed in which, given a normal coordinate analysis in terms of cartesian atomic displacement coordinates, the sign and magnitude of the infrared rotatory strength associated with fundamental transitions of all the normal modes can be calculated analytically if fixed partial charges are assigned to all the atoms {57}. There is no simple Raman analogue of this model because the molecular polarizability, expressed as a sum of 'fixed atom polarizabilities', has no dependence on the molecular geometry. Such a dependence can be introduced by allowing for dynamic and static coupling between the atoms, but at present the bond-polarizability theory outlined above appears to be preferable. There is, however, an infrared analogue of the bond-polarizability theory, namely, the 'bond-dipole' theory {51}, in which the variations of bond dipoles with internal vibrational coordinates are calculated. This leads to the following infrared dipole and rotational strengths:

$$\mathrm{Re}(\langle p_0|\mu_\alpha|p_1\rangle\langle p_1|\mu_\alpha|p_0\rangle) = \frac{\hbar}{2\omega_p}\left[\sum_i\sum_q\left(\frac{\partial\mu_{i\alpha}}{\partial S_q}\right)_0 L_{qp}\right]\left[\sum_j\sum_t\left(\frac{\partial\mu_{j\alpha}}{\partial S_t}\right)_0 L_{tp}\right], \tag{4.39a}$$

$$\begin{aligned}
\mathrm{Im}(\langle p_0|\mu_\alpha|p_1\rangle\langle p_1|m_\alpha|p_0\rangle) = {} \\
&\frac{\hbar}{4}\varepsilon_{\alpha\beta\gamma}\Big\{\sum_{i<j}(R_{ji\beta})_0\left[\sum_q\left(\frac{\partial\mu_{i\alpha}}{\partial S_q}\right)_0 L_{qp}\right]\left[\sum_t\left(\frac{\partial\mu_{j\gamma}}{\partial S_t}\right)_0 L_{tp}\right] \\
&+\left[\sum_i\sum_q\left(\frac{\partial\mu_{i\alpha}}{\partial S_q}\right)_0 L_{qp}\right]\left[\sum_j(\mu_{j\gamma})_0\sum_t\left(\frac{\partial R_{j\beta}}{\partial S_t}\right)_0 L_{tp}\right]\Big\} \\
&+\frac{\hbar}{2\omega_p}\left[\sum_i\sum_q\left(\frac{\partial\mu_{i\alpha}}{\partial S_q}\right)_0 L_{qp}\right]\left[\sum_j\sum_t\left(\frac{\partial m_{j\alpha}}{\partial S_t}\right)_0 L_{tp}\right].
\end{aligned} \tag{4.39b}$$

In comparing these infrared expressions with their Raman counterparts (4.14), the important point to notice is that the 'two-group' term in (4.14b) is a function of ω/ω_p, the ratio of the exciting visible frequency to the vibrational frequency, whereas the corresponding term in (4.39b) has no such factor (because now the exciting infrared frequency equals the vibrational frequency). Consequently the Raman circular intensity difference, I^R-I^L, tends to larger values with decreasing vibrational frequency, whereas the infrared circular absorption difference, $\varepsilon^L-\varepsilon^R$, is comparable at high and low vibrational frequencies. The experimentally significant quantities are the dimensionless c.i.d. Δ (1.1) in the Raman case, and the dimensionless dissymmetry factor {58}

$$g = \frac{\epsilon^L - \epsilon^R}{\frac{1}{2}(\epsilon^L + \epsilon^R)} = \frac{4\,\mathrm{Im}(\langle p_0|\mu_\alpha|p_1\rangle\langle p_1|m_\alpha|p_0\rangle)}{c\,\mathrm{Re}(\langle p_0|\mu_\alpha|p_1\rangle\langle p_1|\mu_\alpha|p_0\rangle)} \qquad (4.40)$$

in the infrared case (a c appears in the denominator in SI). Consequently, the Raman Δ values are larger than the infrared g values by λ_p/λ, the ratio of the wavelength of the vibrational transition to the exciting Raman laser wavelength. So the Raman approach to vibrational optical activity, because it uses visible exciting light, has a natural advantage over the infrared approach, particularly at low frequency. For example, taking λ_p = 500,000 Å (corresponding to ω_p = 200 cm^{-1}), and λ = 5,000 Å, the Raman experiment is 10^2 more favourable.

Applying the bond-dipole expressions (4.39) to the model situations considered previously for the Raman case gives the following infrared dissymmetry factors:
For coupled equivalent totally symmetric internal group coordinates in a chiral two-group structure, c.f. (4.17),

$$g^+ = -\frac{2\pi R_{21}\sin\theta}{\lambda_+(1+\cos\theta)}, \qquad g^- = \frac{2\pi R_{21}\sin\theta}{\lambda_-(1-\cos\theta)}. \qquad (4.41)$$

For the torsion between the two groups, c.f. (4.22),

$$g = \frac{2\pi R_{21}\sin\theta}{\lambda_p(1-\cos\theta)}. \qquad (4.42)$$

For the methyl torsion in the single-bladed propellor, c.f. (4.34),

$$g = -\frac{4\pi R_i \sin 2\phi_i \sin\theta_i}{\lambda_p(1-\cos 2\phi_i)}. \qquad (4.43)$$

The opposite signs arise from the definitions of Δ and g (so perhaps the Raman circular intensity difference should be redefined as I^L-I^R). A further point, which is not apparent from these expressions because the relevant factors have cancelled, is that infrared optical activity generated by deformation coordinates requires the groups to have permanent electric dipole moments. This is more restrictive than the corresponding requirement for Raman optical activity since all non-spherical groups have a polarizability anisotropy.

5. EXPERIMENTAL

5.1 Measurement of Raman c.i.d.

The arrangement currently used in Glasgow for measuring Raman c.i.d. is shown in Figure 5. A beam from an argon-ion laser is focused into the sample. The light scattered at 90^0 is processed by a triple monochromator and the Raman components detected by a photomultiplier, which produces discrete voltage pulses from individual photon strikes. These pass through a discriminator which gives equal weight to all pulses that exceed a certain voltage threshold; this enables much of the noise from the photomultiplier to be eliminated and gives the scattered intensity as a 'photon' count rate. The polarization of the incident beam is modulated directly between right and left circular at 90 Hz using a KDP crystal driven by a square-wave alternating voltage synchronized with two matched pulse counters. While the modulator performs the first half of its cycle and produces right circular polarization, pulses are channeled into counter A; while the modulator performs the second half of its cycle and produces left circular polarization, pulses are channeled into counter B. After accumulating a statistically significant number of counts, the sum of the two counters $A+B = I^R+I^L$ and the difference $A-B = I^R-I^L$ are read. The statistical significance of N counts is determined by the standard deviation $N^{\frac{1}{2}}$. Thus on 10^8 counts, the 'statistical noise' is $N^{\frac{1}{2}}/N = 10^{-4}$, an order of magnitude smaller than the larger c.i.d.s expected. On typical Raman bands at low resolution, 10^8 counts can be collected in less than a minute using 4 W laser power at 488 nm, and well-defined I^R-I^L spectra can be recorded automatically at a scan rate of 1 cm^{-1} min^{-1}.

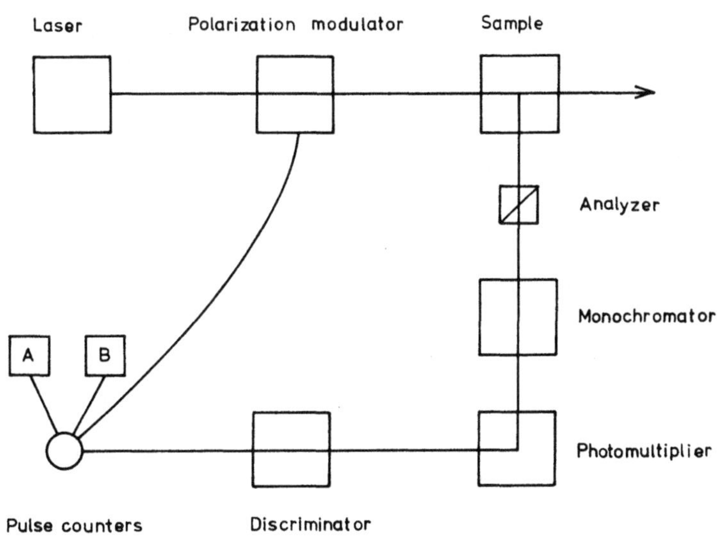

Fig. 5. Dual synchronous photon counting system.

Imperfections in the modulation system and the optical train can generate large artifacts in the I^R-I^L spectrum. The origin of these artifacts is discussed in detail in {21}. The artifacts are minimized by using square-wave modulation to switch the polarization of the laser beam between pure right and pure left circular. The extent of the problem depends on the polarization characteristics of the particular Raman band, strongly polarized Raman bands being affected most. Also Δ_x, the polarized c.i.d., is much more susceptible than Δ_z, the depolarized c.i.d. On account of these problems, only depolarized c.i.d.s are measured routinely at present, and then only on depolarized and weakly polarized Raman bands. This is not restrictive in practice, however, since only one or two Raman bands in a chiral molecule are strongly polarized, and the model calculations above indicate that depolarized c.i.d.s should be easier to interpret than polarized c.i.d.s.

Only neat liquids and strong solutions have been studied so far. Crystals and powders present greater difficulties: crystals must be oriented carefully to eliminate artifacts, and multiple scattering in powders depolarizes the incident light.

It should be mentioned that measurement of the degree of circular polarization of the scattered light using incident light of fixed polarization could give information identical with that from c.i.d. measurements {8,22}. However, this approach has many problems and is not discussed here.

5.2 Typical Raman c.i.d. spectra

The Raman c.i.d. spectra of a large number of chiral molecules have been obtained, and a few typical examples are shown here. It is usually worthwhile to run only depolarized c.i.d. spectra, and then only from 0 to 2,000 cm^{-1}. Polarized c.i.d.s are more susceptible to artifacts and harder to interpret, and the region above 2,000 cm^{-1} contains mainly C-H and O-H stretching modes which give marginal effects. It is best to present I^R-I^L and I^R+I^L separately because the background must be subtracted from I^R+I^L before calculating Δ. Another reason, which emerged from the theory above, is that equal and opposite I^R-I^L values in adjacent bands indicate that the corresponding modes contain symmetric and antisymmetric combinations of the same pair of internal coordinates: this diagnostic feature is lost if Δ values are presented directly because the Δ values in two such bands are usually very different on account of different I^R+I^L values. The I^R+I^L spectra are presented on a linear scale, whereas the I^R-I^L spectra are presented on a scale that is linear within each decade range but logarithmic between decade ranges. This enables the exponent in the I^R-I^L photon count to be recorded, although the spectrum takes on a 'stretched out' appearance in which the statistical noise level looks much greater than it really is since, on all but the weakest bands, only values within the $\pm 10^4$ range are significant. S and W denote strongly and weakly polarized bands; all other bands are effectively depolarized. All the spectra were recorded with a slit width of 10 cm^{-1} at a scan speed of 1 cm^{-1} min^{-1} using a 4 W argon-ion laser beam of wavelength 488 nm focused into samples that were either neat liquids or near-saturated solutions.

None of the theories of the generation of Raman optical activity discussed above has yet been applied in detail to give a definitive interpretation of any observed feature, due mainly to the difficulty in obtaining sufficiently simple model compounds, so discussion of the spectra is confined to general remarks.

Figure 6 shows the depolarized Raman circular intensity sum and difference spectra of the two neat optical isomers of α-pinene.

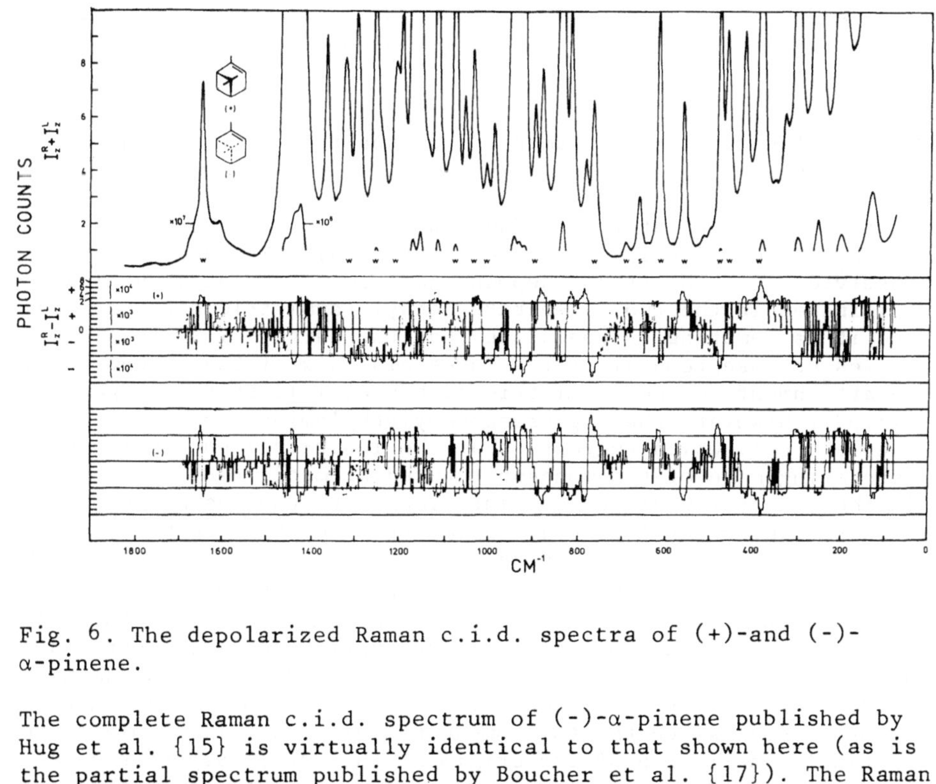

Fig. 6. The depolarized Raman c.i.d. spectra of (+)-and (-)-α-pinene.

The complete Raman c.i.d. spectrum of (-)-α-pinene published by Hug et al. {15} is virtually identical to that shown here (as is the partial spectrum published by Boucher et al. {17}). The Raman c.i.d. features between about 380 and 900 cm^{-1} are assigned to skeletal vibrations of pinane-type molecules and to out-of-plane olefinic hydrogen deformations {59}.

Figure 7 shows the spectra of neat (-)-α-phenylethylamine. The dominant features are the c.i.d. couplets centred at about 1450 and 340 cm^{-1}. It is gratifying that the same features, but with opposite signs, appear in the spectrum of (+)-α-phenylethanol {60} since the two molecules have opposite absolute configurations

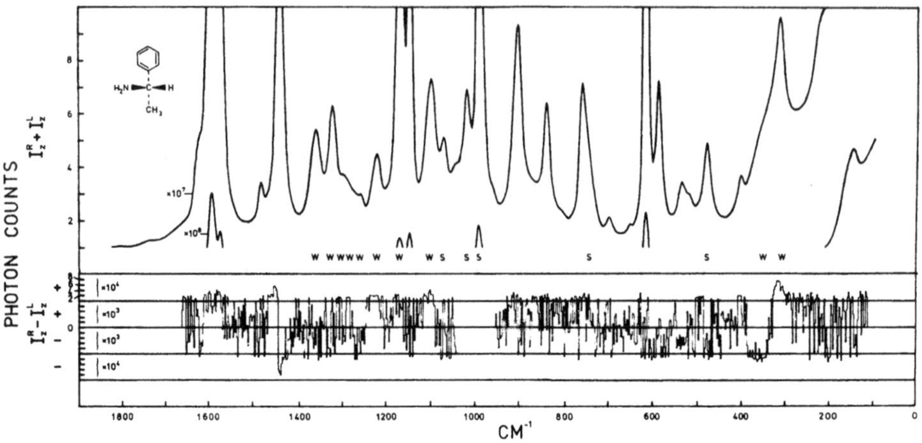

Fig. 7. The depolarized Raman c.i.d. spectrum of (-)-α-phenylethylamine.

in the R,S nomenclature. In fact the couplets at about 340 cm^{-1} in these two molecules were the first genuine Raman c.i.d.s to be observed {12}. The strong depolarized Raman band at about 1450 cm^{-1} originates in the two degenerate methyl asymmetric deformations, and it is tempting to speculate that the large c.i.d. couplet arises because removal of the threefold axis of the methyl group on attachment to the asymmetric carbon atom lifts this degeneracy {15,53}. However, the α-p-bromophenylethylamine shown no such couplet {60}, which suggests that the lower-frequency semicircle stretching mode of the monosubstituted aromatic ring is involved in the generation of the couplet: the corresponding Raman band occurs right under the methyl asymmetric deformation band, but p-substitution shifts it by about 40 cm^{-1} to lower frequency.

Figure 8 shows the spectra of neat (+)-α-phenyltrifluoromethylethanol. There are a number of large effects, but the only feature we shall discuss is the conservative couplet centred at about 510 cm^{-1} associated with two adjacent depolarized Raman bands. This region is appropriate for trifluoromethyl asymmetric deformations: since the two components are now well-resolved, the original description of the origin of the methyl asymmetric deformation c.i.d. couplet, which proved to be over-simplified, may well be valid for the trifluoromethyl asymmetric deformations {60}.

Figure 9 shows the spectra of neat (+)-3-methylcyclohexanone. A large negative c.i.d. appears in the broad, weak Raman band at about 250 cm^{-1} which almost certainly originates in the methyl torsion {53}. However, the fact that there is an equal and opposite c.i.d. at lower frequency indicates that the effect is generated through the coupling of the methyl torsion with a skeletal def-

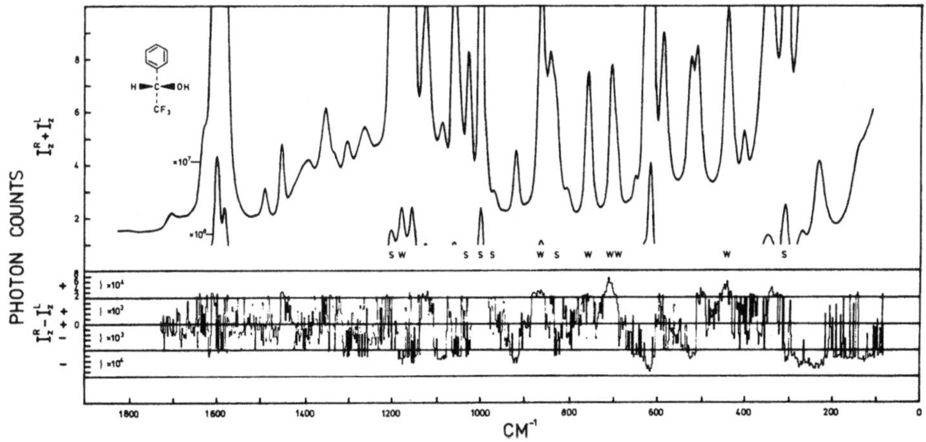

Fig. 8. The depolarized Raman c.i.d. spectrum of (+)-α-trifluoromethylethanol.

ormation rather than through the inertial mechanism described in section 4.5. A striking feature is the large conservative c.i.d. couplets which could originate in the mechanism outlined in section 4.6 in which orthogonal internal coordinates of the same group contribute to symmetric and antisymmetric normal coordinates on account of the low symmetry. Thus the couplet at about 400 cm^{-1} might be generated by in-plane and out-of-plane deformations of

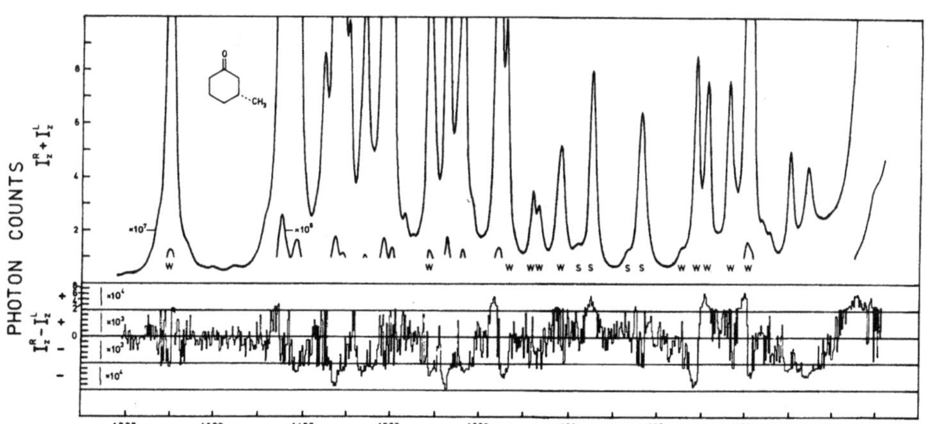

Fig. 9. The depolarized Raman c.i.d. spectrum of (+)-3-methylcyclohexanone.

the C-CH_3 group, and that at about 500 cm^{-1} by corresponding de formations of the carbonyl group. There is also the possibility of a two-group contribution involving a carbonyl deformation and a C-CH_3 deformation. The conservative couplet at about 960 cm^{-1} might involve the two orthogonal methyl rock modes. The corresponding 3-methylcyclopentanone shows much smaller effects {23}, probably on account of the near-planarity of the five-membered ring (the assignment of the band at 280 cm^{-1} to the methyl torsion is probably wrong; it is more likely to be a C-CH_3 deformation).

Camphor shows a large c.i.d. couplet centred at about 500 cm^{-1} that could again originate in the carbonyl deformations {61}. Bromocamphor shows several new features, including a couplet at about 1,300 cm^{-1}, that could originate in C-H deformations {61}.

Figure 10 shows the spectra of (-)-menthol in methanol solution. The complicated c.i.d. structure between about 1,100 and 1,400 cm^{-1} is repeated almost exactly (in sign and magnitude) in (-)-menthylamine and (-)-menthyl chloride {55}. Definite band assignments in this region are difficult due to the large number of CH and CH_2 deformations and C-C stretches which can contribute. However, the rocking of the two isopropyl groups gives rise to good group frequencies in this region, being responsible for the two Raman bands at about 1,170 and 1,140 cm^{-1} which are associated with a c.i.d. couplet {55}. The weak depolarized Raman band at about 260 cm^{-1} showing a large positive c.i.d. might originate in the torsion of the ring methyl group.

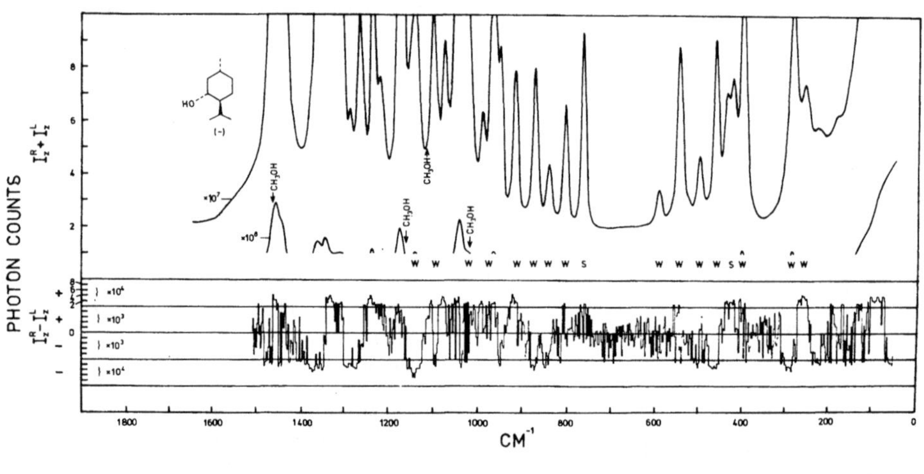

Fig. 10. The depolarized Raman c.i.d. spectrum of (-)-menthol.

Figure 11 shows the spectra of (+)-carvone. The large positive c.i.d. at about 250 cm^{-1}, associated with a broad weak depolarized Raman band, might originate in the torsion of the isoprop-

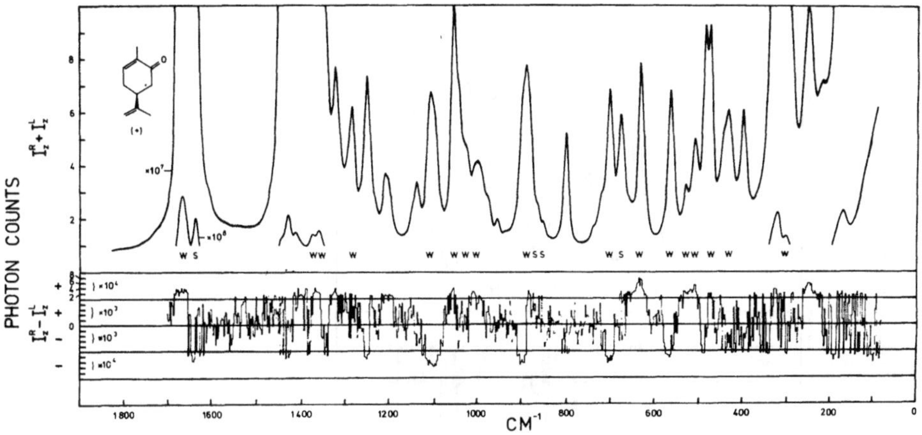

Fig. 11. The depolarized Raman c.i.d. spectrum of (+)-carvone.

enyl methyl group since (-)-limonene shows a large negative effect at the same frequency; this correlates with the opposite absolute configurations of the structural features common to both molecules {53,55}. But it is also possible that C-CH_3 deformations of the ring olefinic methyl group are responsible: without deuteration, methyl trosion assignments are not certain. However, a large 'inertial' contribution, as discussed in section 4.5, is expected in the isopropenyl methyl torsion, and the absence of any compensating c.i.d. in the vicinity is evidence that the effect is not generated through coupling with some other internal coordinate. The c.i.d. features between about 550 and 720 cm^{-1} might originate in carbonyl deformations since they are not shown by limonene. The small positive c.i.d. in the strong, weakly polarized, Raman band at 885 cm^{-1} has been assigned to the vinylidene CH_2 out-of-plane in-phase vibration {55}.

Figure 12 shows the low-frequency spectra of (S)-(-)-6,6'-dinitro-2,2'-dimethyl biphenyl in benzene solution. The large positive c.i.d. at about 270 cm^{-1} has the correct sign and magnitude to originate in the oscillations of the double-bladed propellor about the C_2 axis that compensate the in-phase combination of the two methyl torsions {54}. But this is probably fortuitous: the negative-positive-negative c.i.d. pattern between about 200 and 300 cm^{-1} most likely originates in complicated normal coordinates containing contributions from methyl torsions together with C-CH_3 and C-NO_2 deformations.

Figure 13 shows the spectra of the two optical isomers of tartaric acid in aqueous solution. The simplicity of the spectra indicates that one of the three possible staggered conformations

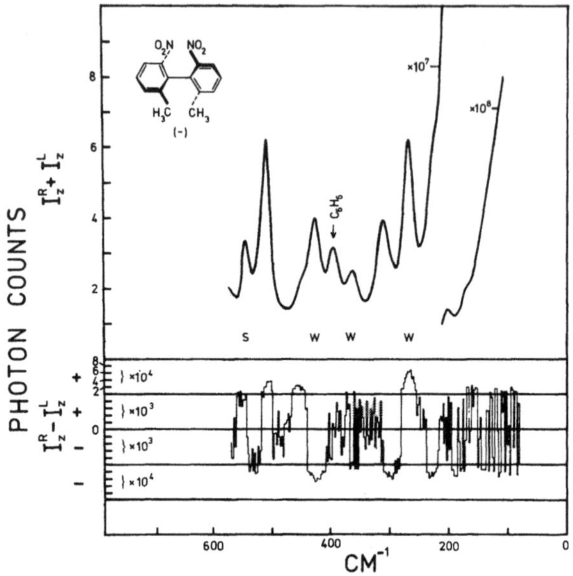

Fig. 12 . Part of the depolarized Raman c.i.d. spectrum of (S)-(-)-6,6'-dinitro-2,2'-dimethyl biphenyl.

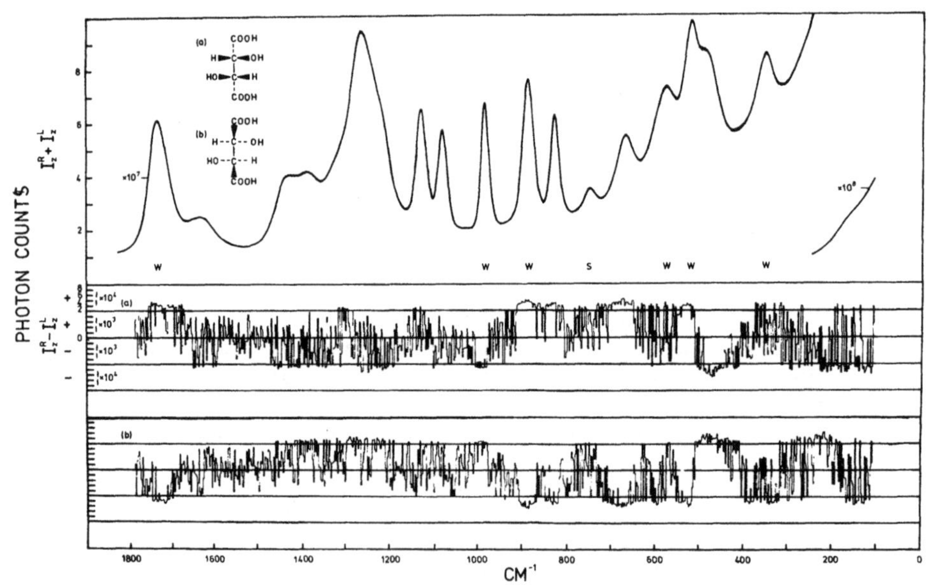

Fig. 13. The depolarized Raman c.i.d. spectra of (+)- and (-)-tartaric acid.

dominates at room temperature. A striking feature is the large c.i.d. couplet centred at about 500 cm^{-1}. Similar couplets appear in this region in dimethyl tartrate and 2,3-butanediol {62}. The couplets might originate in the symmetric and antisymmetric combinations of the two in-plane C-C-O deformations in the twisted O-C-C-O unit. If so, knowing the absolute configuration, it would be possible to distinguish between the three possible staggered conformers (and vice-versa). However, the interpretation might be more complicated because C-C-C-C deformations and out-of-plane hydrogen bonded O-H deformations also occur in this region.

The only significant effect in the c.i.d. spectrum of (S)-(+)-α-D-benzyl alcohol (Figure14) is a couplet associated with two weakly polarized Raman bands at about 860 and 950 cm^{-1}. These

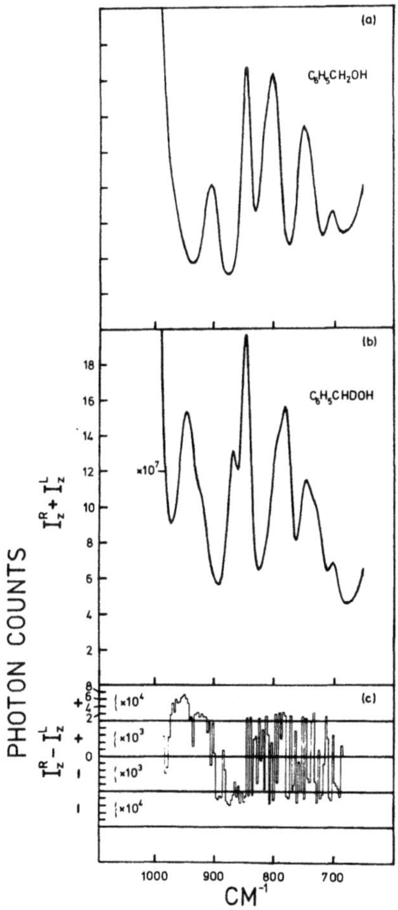

Fig. 14. Part of the depolarized Raman spectrum of benzyl alcohol, and the c.i.d. spectrum of (S)-(+)-α-D-benzyl alcohol.

bands are not present in benzyl alcohol itself, and are assigned to C-D deformations {63}. As indicated in section 4.6, this type of couplet is expected to arise from the coupling of two orthogonal deformation modes on account of the low symmetry. In another example of Raman optical activity due to isotopic substitution, rich c.i.d. structure was observed between about 800 and 1,300 cm^{-1} in (1S)-4,4-dideuteroadamantan-2-one {64}, this being the region for CH_2 and CD_2 deformations. Molecules that owe their chirality to isotopic substitution are important in stereochemical studies, but the determination of their absolute configurations can be difficult on account of very small (if any) electronic optical rotations and Cotton effects. On the other hand, in the case of deuterium substitution, the vibrational chirality of the skeleton in the region of the deuterium atom is expected to be large, and this is born out by the above Raman c.i.d. results. So vibrational optical activity measurements could provide reliable correlations of absolute configurations in these molecules.

REFERENCES

1. T.G. Spiro, Acc. Chem. Res., **7**, 339 (1974).
2. Lord Rayleigh, Phil. Mag., **41**, 107 (1871).
3. R. Gans, Z. Phys., **17**, 353 (1923).
4. R. de Mallemenn, C. R. Acad. Sci. Paris, **181**, 371 (1925).
5. S. Bhagavantam and S. Venkateswaran, Nature, **125**, 237 (1930).
6. A. Kastler, C. R. Acad. Sci. Paris, **191**, 565 (1930).
7. F. Perrin, J. Chem. Phys., **10**, 415 (1942).
8. P.W. Atkins and L.D. Barron, Mol. Phys., **16**, 453 (1969).
9. L.D. Barron and A.D. Buckingham, Mol. Phys., **20**, 1111 (1971).
10. B. Bosnich, M. Moskovits and G.A. Ozin, J. Amer. Chem. Soc., **94**, 4750 (1972).
11. M. Diem, J.L. Fry and D.F. Burow, J. Amer. Chem. Soc., **95**, 253 (1973).
12. L.D. Barron, M.P. Bogaard and A.D. Buckingham, J. Amer. Chem. Soc., **95**, 603 (1973).
13. L.D. Barron, M.P. Bogaard and A.D. Buckingham, Nature, **241**, 113 (1973).
14. L.D. Barron and A.D. Buckingham, J. C. S. Chem. Comm., 152 (1974).
15. W. Hug, S. Kint, G.F. Bailey and J.R. Scherer, J. Amer. Chem. Soc., **97**, 5589 (1975).
16. M. Diem, M.J. Diem, B.A. Hudgens, J.L. Fry and D.F. Burow, J. C. S. Chem. Comm., 1028 (1976).
17. H. Boucher, T.R. Brocki, M. Moskovits and B. Bosnich, J. Amer. Chem. Soc., **99**, 6870 (1977).
18. L.D. Barron and A.D. Buckingham, Mol. Phys., **23**, 145 (1972).
19. L.D. Barron, Nature, **257**, 372 (1975).
20. A.D. Buckingham and R.E. Raab, Proc. Roy. Soc. A, **345**, 365 (1975).

21. L.D. Barron and A.D. Buckingham, Ann. Rev. Phys. Chem., 26, 381 (1975).
22. L.D. Barron, in: Molecular Spectroscopy, Vol. 4, ed. by R.F. Barrow, D.A. Long and J. Sheriden, The Chemical Society, London, 1976.
23. L.D. Barron, in: Advances in Infrared and Raman Spectroscopy, Vol. 4, ed. by R.J.H. Clark and R.E. Hester, Heyden, London, 1978.
24. L.D. Barron and C.G. Gray, J. Phys. A, 6, 59 (1973).
25. A.D. Buckingham, Adv. Chem. Phys., 12, 107 (1967).
26. V.B. Berestetskii, E.M. Lifshitz and L.P. Pitaevskii, Relativistic Quantum Theory, Part 1, Pergamon Press, Oxford, 1971.
27. A.D. Buckingham and P.J. Stephens, Ann. Rev. Phys. Chem., 17, 399 (1966).
28. G. Placzek, in: Handbuch der Radiologie, Vol. 6, Part 2, ed. by E. Marx, Akademische Verlagsgesellschaft, Berlin, 1934.
29. C.M. Penney, J. Opt. Soc. Amer., 59, 34 (1969).
30. L. Rosenfeld, Z. Phys., 52, 161 (1928).
31. A.D. Buckingham and M.B. Dunn, J. Chem. Soc. A, 1988 (1971).
32. L.D. Barron, Mol. Phys., 21, 241 (1971).
33. L.D. Landau and E.M. Lifshitz, Electrodynamics of Continuous Media, Pergamon Press, Oxford, 1960.
34. L.D. Barron, Nature, 238, 17 (1972).
35. R. Hornreich and S. Shtrikman, Phys. Rev., 171, 1065 (1968).
36. M. Born and K. Huang, Dynamical Theory of Crystal Lattices, Oxford University Press, Oxford, 1954.
37. L.D. Barron, Mol. Phys., 31, 129 (1976).
38. A.C. Albrecht, J. Chem. Phys., 34, 1476 (1961).
39. L.D. Landau and E.M. Lifshitz, The Classical Theory of Fields, Pergamon Press, Oxford, 1971.
40. A.D. Buckingham and J.A. Pople, Proc. Phys. Soc. A, 68, 905 (1955).
41. J.A. Schellman, Acc. Chem. Res., 1, 144 (1968).
42. E.U. Condon, W. Alter and H. Eyring, J. Chem. Phys., 5, 753 (1937).
43. J.G. Kirkwood, J. Chem. Phys., 5, 479 (1937).
44. A.D. Buckingham and P.J. Stiles, Acc. Chem. Res., 7, 258 (1974).
45. L.D. Barron and A.D. Buckingham, J. Amer. Chem. Soc., 96, 4769 (1974).
46. L.D. Barron, J. Chem. Soc. A, 2899 (1971).
47. A.D. Buckingham and H.C. Longuet-Higgins, Mol. Phys., 14, 63 (1968).
48. R.R. Birss, Symmetry and Magnetism, North Holland, Amsterdam, 1966.
49. A.J. Stone, Mol. Phys. 29, 1461 (1975); 33, 293 (1977).
50. L.D. Barron, J. Amer. Chem. Soc., 96, 6761 (1974).
51. L.M. Sverdlov, M.A. Kovner and E.P. Krainov, Vibrational Spectra of Polyatomic Molecules, Israel program for Scientific Translations, Jerusalem, 1974.
52. E.B. Wilson, J.C. Decius and P.C. Cross, Molecular Vibrations,

McGraw-Hill, New York, 1955.
53. L.D. Barron, Nature, 255, 458 (1975).
54. L.D. Barron and A.D. Buckingham, to be published.
55. L.D. Barron and B.P. Clark, to be published.
56. L.D. Barron, J. C. S. Faraday II, 71, 293 (1975).
57. J.A. Schellman, J. Chem. Phys., 58, 2882 (1973).
58. S.F. Mason, Quart. Rev. Chem. Soc., 17, 20 (1963).
59. L.D. Barron and B.P. Clark, to be published.
60. L.D. Barron, J. C. S. Perkin II, 1790 (1977).
61. L.D. Barron, J. C. S. Perkin II, 1074 (1977).
62. L.D. Barron, Tetrahedron, 34, 607 (1978).
63. L.D. Barron, J. C. S. Chem. Comm., 305 (1977).
64. L.D. Barron, H. Numan and H. Wynberg, J. C. S. Chem. Comm., 259 (1978).

VIBRATIONAL CIRCULAR DICHROISM: THE EXPERIMENTAL VIEWPOINT

P. J. STEPHENS AND R. CLARK

Department of Chemistry, University of Southern California, Los Angeles, California

INTRODUCTION

Chirality in molecular structure manifests itself in the optical phenomena collectively referred to as optical activity, the common feature of which is the differentiation in behavior of left and right circularly polarized light. In turn, optical activity can be used to probe molecular chirality. The optical rotation and circular dichroism associated with the electronic transitions of chiral molecules and commonly observed in the visible and ultraviolet regions of the spectrum have been studied for many years to this end. Recently, as a result of developments in instrumentation for the measurement of circular dichroism in the infrared spectral domain, the observability of circular dichroism in vibrational transitions has been demonstrated, thereby making available a new technique for the study of chiral molecules.

In these lectures we shall review expermental studies of vibrational circular dichroism (VCD) carried out so far. Considerable attention will be devoted to the measurement of very small circular dichroism in the medium infrared - a sine qua non for the study of VCD.

HISTORY

Prior to the 1970's, attempts to observe anomalous dispersion of optical rotation through vibrational transitions were plagued by instrumental artefacts. Positive reports by Gutowsky (2) and by Gunthard et al.(5) were not confirmed (4,7,8).

In the early 1970's, interest in vibrational optical activity re-

Stephen F. Mason (ed.), Optical Activity and Chiral Discrimination. 263–287.

vived, with the focus now being on the circular dichroism (CD) phenomenon. In George Holzwarth's laboratory at the University of Chicago a medium infrared (IR) CD instrument was constructed which led to the first definitive, published (VCD) spectra. In 1973, Holzwarth's group reported CD in vibrational transitions associated with the H_2O and D_2O molecules in the chiral crystalline lattices of α-$ZnSO_4.6H_2O$, α-$ZnSO_4.6D_2O$ and α-$ZnSeO_4.6H_2O$ (15). Following this, in 1974, the CD of the C-H stretch fundamental in $C_6H_5.CH(OH).CF_3$ was reported, as was the existence of CD in the C-D stretch of $C(CH_3)_3.CHDCl$ (17). These latter measurements were the first to be successfully carried out on chiral molecules in the isotropic liquid phase.

In our laboratory at USC a near IR CD instrument had been constructed in the late 1960's by George Osborne and Jack Cheng for the study of magnetic circular dichroism due to metal ions in ionic crystals (34). In 1973, extension of this instrument to the medium IR spectral region was initiated by Larry Nafie, Tim Keiderling and Jack Cheng. By late 1974, Holzwarth's data on $C_6H_5CH(OH).CF_3$ had been confirmed and extended (19). During 1975, a large number of molecules were examined and a survey of the VCD data obtained was published in 1976 (22). This work established that VCD could be observed with good signal-to-noise ratio in the hydrogenic stretching region (2500-4000 cm^{-1}) in a wide variety of chiral molecules.

Exploitation and development of these two instruments has continued. Holzwarth's instrument is now at the University of Minnesota in the laboratories of Al Moscowitz and John Overend. The VCD in the C-H stretch of tartaric acid obtained with this instrument was published in 1976 (24). At USC, Tim Keiderling showed in 1976 that the VCD of overtone and combination bands could be observed as easily as that of fundamentals (23). He also carried out an extensive VCD study of the C-H and O-H stretch transitions of dialkyl tartrate esters and their deuterated analogues, a preliminary report of which was published in 1977 (26) Since 1977, extensive modifications have been undertaken by Roy Clark, with the aim of increasing instrumental sensitivity, particularly at frequencies below 2000 cm^{-1}, in which range VCD has not previously been published. Inter alia, this work has led to the first VCD spectra of fundamental carbonyl and ethylenic stretch transitions, as reported below.

Since their departure from USC, Larry Nafie and Tim Keiderling have built new VCD instruments at Syracuse University and the Chicago Circle campus of the University of Illinois, respectively. To date, VCD in the C-H stretch modes of alanine and alanyl peptides has been published by Nafie and his group (30, 33)

INSTRUMENTATION

The basic technique used by all modern CD instruments was developed by Legrand and Grosjean (37) around 1960 and is illustrated schematically in Figure 1.

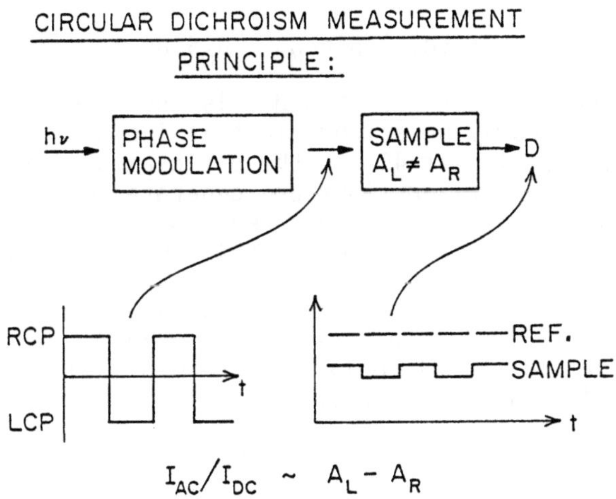

FIGURE 1

Monochromatic light is phase modulated, the polarization of the light oscillating between left and right circular polarizations at a modulation frequency ω_M. The light intensity is identical on each half of the modulation cycle and is, therefore, time-independent. Passage of the light beam through an absorbing sample then creates an oscillating, time-dependent intensity if, and only if, the sample absorbs left and right circular polarizations differently, i.e. if there is circular dichroism. Detection of this AC signal using phase sensitive electronics then completes the measurement of the CD.

The practical arrangement of a CD instrument is illustrated in Figure 2.

$$L \rightarrow MC \rightarrow P \rightarrow \overset{h\nu}{M} \rightarrow S \rightarrow D \rightarrow \text{Electronics}$$

FIGURE 2. The components of a CD instrument.

Light from a source L is monochromated by monochromator MC and subsequently linearly polarized by polarizer P. Phase modulation is then carried out by a modulator M consisting of a material possessing a time-dependent linear birefringence. For the sake of definiteness, we suppose here that this birefringence varies sinusoidally with time:

$$\Delta n = (\Delta n)_0 \mathrm{Sin}\,\omega_M t \qquad -(1)$$

Then the phase lag (retardation) between linearly polarized waves propagating along the optic axes of M is

$$\delta = 2\pi(\Delta n)_0 \frac{\ell}{\lambda} \mathrm{Sin}\,\omega_M t \equiv \delta_0 \mathrm{Sin}\,\omega_M t \qquad -(2)$$

where ℓ is the pathlength of M and λ is the wavelength of the light. When linearly polarized light is incident on M, with the plane of polarization at 45° to the optic axes of M, right and left circularly polarized light are generated when $\delta = \pm \frac{\pi}{2}$; if δ_0 is arranged to be $\frac{\pi}{2}$, the light beam polarization passes sinusoidally between left and right circular polarizations.

Subsequent to the modulator the light passes through the sample S to a detector D. If S possesses different absorbances for left and right circularly polarized light (A_L and A_R, respectively) so that the circular dichroism

$$\Delta A = A_L - A_R \qquad -(3)$$

is non-zero, then the intensity at D contains an AC component at the frequency ω_M. More precisely (34), the ratio of the intensity oscillating at ω_M to the time-independent, DC intensity is

$$\frac{I_{AC}^{\omega_M}}{I_{DC}} = 2J_1(\delta_0)\tanh\left[\frac{\ell n 10}{2}(\Delta A)\right]\mathrm{Sin}\,\omega_M t \qquad -(4)$$

where J_1 is the first Bessel function. When ΔA is $\lesssim 0.1$ (which, in practice, is very infrequently not the case), equation 4 simplifies to

$$\frac{I_{AC}^{\omega_M}}{I_{DC}} = [2.30\ J_1(\delta_0)\mathrm{Sin}\,\omega_M t]\,\Delta A \qquad -(5)$$

The CD, ΔA, is therefore obtained by separating the detector signal into its DC component and the AC component having the same frequency and phase as the modulator, followed by the division of the DC signal into the AC signal. The resultant is proportional to ΔA. If δ_0 is maintained constant as the wavelength of the light varies the proportionality constant is independent of wavelength. The absolute magnitude of ΔA is then obtained either by an optical or electronic calibration procedure or, more commonly, by the use of a secondary standard. Note that it is not essential for δ_0 to be $\pi/2$. The optimum value, at which $J_1(\delta_0)$ is a maximum, in fact corresponds to $\delta_0 = 0.58\pi$- slightly greater than quarter-wave ($\pi/2$) retardation.

The heart of the CD instrument is the modulator. The first generation of visible-ultraviolet (UV) CD instruments were based on electro-optic modulators constructed from potassium dihydrogen phosphate (KDP) and related materials. Here, an alternating electric field applied to an appropriately cut crystal creates the alternating linear birefringence. While having made possible a large volume of CD experimentation, these modulators have three serious drawbacks. First, due to the specific material used for their construction their wavelength range is limited and excludes the majority of the IR and vacuum UV regions. Second, the need for very large voltages in order to generate the necessary birefringences leads to electrical breakdown problems which are difficult to control. Third, the use of an intrinsically anisotropic (uniaxial) crystal leads to a limited angular aperture.

These disadvantages do not occur with photoelastic modulators (PEM). The PEM is based on the stress-optic effect. An isotropic transparent optical element is alternately stressed and expanded by coupling to a piezoelectric device, which is driven electrically. Various PEM designs have been developed, the earliest being those of Badoz and Billardon (38), Kemp (39), Schnatterly (40), and Mollenauer (41).

As a result of the wider variety of materials that can be used, PEM s can be constructed with much wider transmission range than KDP-type EOM s. Because of the intrinsic isotropy of the element, the angular aperture is much greater. The main limitation of the PEM is that, although electrical breakdown cannot occur, mechanical breakdown (i.e. fracture) can. The magnitude of the retardation, δ_0, achieved by a PEM is determined by the birefringence, $(\Delta n)_0$, and by λ (according to equation 2). In turn, $(\Delta n)_0$ is proportional to the stress applied and to the stress-optic coefficient of the material. At a given wavelength, the retardation that can be achieved is limited by the fracture point of the material; only if fracture occurs at stresses greater than required to make $\delta_0 \sim \pi/2$ is the PEM optimal for CD measurement. Since for many transparent materials $(\Delta n)_0$ is nearly independent of wavelength, the stresses required to yield a given δ_0 increase approximately linearly with wavelength. As wavelength increases, therefore, a PEM is either limited by a transmission cut-off, or by its stress-optic and fracture properties. Unlike an EOM, a PEM is also restricted to a fixed frequency and to sinusoidal phase modulation.

As a consequence of their generally superior characteristics, PEM s have increasingly replaced EOM s in visible-near UV CD instrumentation. In addition, they have made possible the development of IR CD instrumentation capable of the measurement of VCD - to which subject we now turn.

The optical lay-out of the instrument in our laboratory, as it was in 1973 (34), is shown in Figure 3. This instrument was built to operate in the 1-3μ range. Light from a tungsten-halogen lamp passed through a Spex 1500 monochromator. Linear polarization was achieved, using either a Glan-Taylor calcite polarizer (at $\lambda \gtrsim 2\mu$) or a gold-grid polarizer (at $\lesssim 2\mu$). An infrasil fused quartz Morvue PEM provided phase modulation at $\sim$50kHz. The detector was a 77°K InAs intrinsic semiconductor.

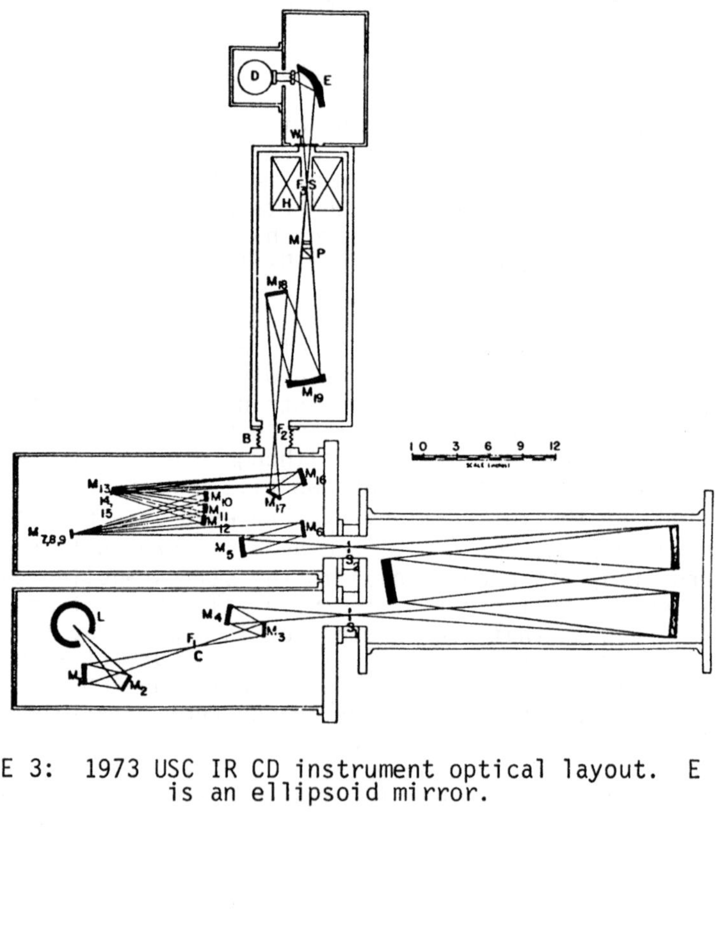

FIGURE 3: 1973 USC IR CD instrument optical layout. E is an ellipsoid mirror.

The first modifications of this instrument carried out in order to look for VCD consisted of replacing the InAs detector by a 77°K InSb device whose band-gap (and hence detection cut-off) lies at $\sim 5.5\,\mu$, permitting operation through the hydrogenic stretching region (~ 2.5-$4\,\mu$; 2500-4000 cm^{-1}). In addition, a Morvue CaF_2 PEM was incorporated to allow transmission throughout the InSb detection range (which fused quartz does not). These steps proved inadequate, however, and preliminary VCD experiments failed.

The Morvue CaF_2 modulator does not allow quarter-wave peak retardation at wavelengths greater than $\sim 2\,\mu$ and the CD sensitivity in experiments between 2.5 and $4\,\mu$ is consequently less than optimal. We therefore sought alternative materials whose stress-optic coefficient, fracture point and transmission range allowed $\lambda/4$ retardation to much longer wavelengths. After several materials were examined and a variety of modulators constructed by Jack Cheng, ZnSe was the material of choice. Tests with several units constructed showed (36) that $\lambda/4$ peak retardation could be obtained up to wavelengths greater than $10\,\mu$.

A second major difficulty with the early instrumentation was the existence of substantial "artefact" signals. It is easy to show that if the CD sample S is replaced by an element introducing a phase change (for example, a birefringent plate) followed by a linear polarizer, an AC signal is again received by the detector with identical phase and frequency to a true CD signal. Artefact signals of this type occur in our instrument principally due to the ellipsoid mirror used to focus the beam onto the detector combined with the polarization sensitivity of the detector.

Such artefact signals would not exist if the optics subsequent to the sample were completely insensitive to the polarization state of the light beam. In principle, therefore, a device which scrambles polarization, creating a completely unpolarized beam, placed immediately after the sample, would eliminate artefacts. It occurred to Jack Cheng to use a second PEM for this purpose and a study of this idea proved it to be feasible if the "scrambler" PEM is incoherent with the modulating PEM and adjusted to provide a retardation level such that $J_0(\delta_0) = 0$, where J_0 is the zeroth Bessel function (35). This condition occurs when $\delta_0 = 0.76\pi$, a somewhat greater level of retardation than required for the optimum CD measurement.

With the incorporation of two ZnSe PEM s, one to phase modulate the light beam (tuned to make $J_1(\delta_0)$ maximal) and one to eliminate artefacts due to the ellipsoid and detector (tuned to make $J_0(\delta_0) = 0$), it proved possible, finally, to observe VCD.

Since Holzwarth's spectrum of $C_6H_5.CH(OH).CF_3$ (Figure 4) had just appeared (17), the first studies focussed on this molecule. In the fall of 1974, Holzwarth's experiment was reproduced. Indeed, the signal-to-noise ratio turned out to be sufficiently greater that a more extended spectrum was obtained (Figure 5), clearly exhibiting the complex bandshape of the C-H VCD stretch and including the O-H stretch VCD (19).

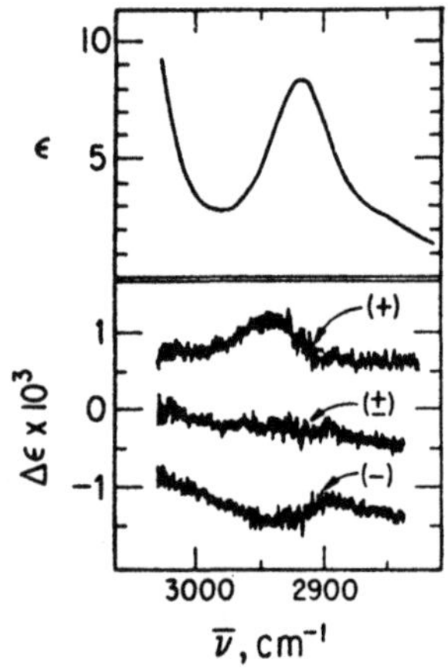

Infrared absorption (upper frame) and CD spectra (lower frame) of 2,2,2-trifluoro-1-phenylethanol, neat, with monochromator spectral bandpass 20 cm^{-1} (full width at half-height). The CD spectra show the raw data exactly as recorded. A fivefold reduction in bandpass causes only 8% change in the absorption band half-width. The CD zero has been displaced between (+), (±), and (−) measurements; the actual curves are indistinguishable at 3030 cm^{-1}.

FIGURE 4: The first VCD spectrum of $C_6H_5.CH(OH).CF_3$.

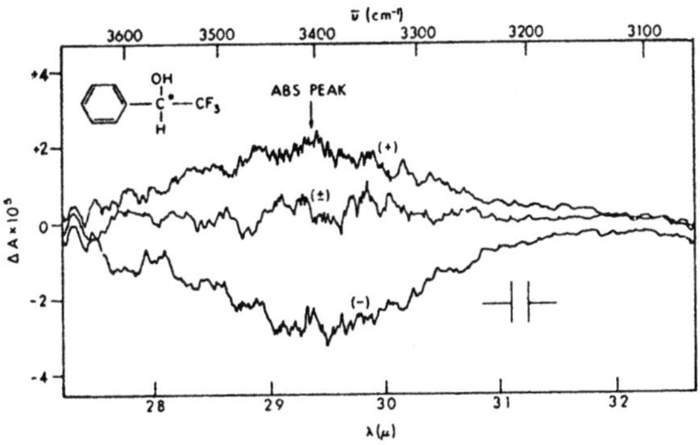

The CD spectrum of the O–H stretching band of liquid 2,2,2-trifluoro-1-phenylethanol. $\Delta A = A_L - A_R$. The spectral band pass, shown in the figure, is 15 cm^{-1} and the time constant is 10 sec. The pathlength is 20 μ giving an absorbance (A) of 1.0 at the absorption maximum, indicated by an arrow.

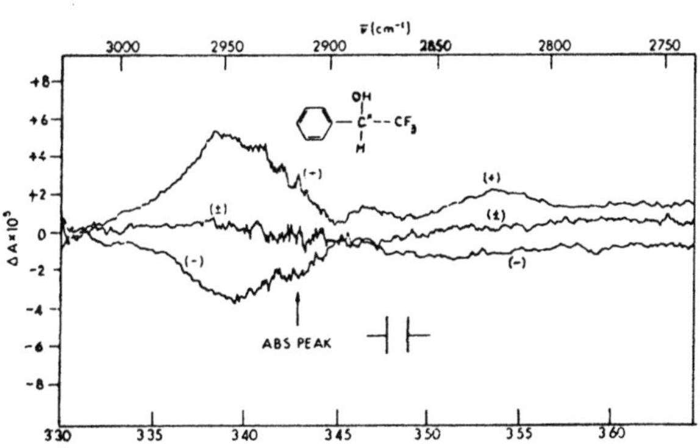

The CD spectrum of the C*–H stretching band of liquid 2,2,2-trifluoro-1-phenylethanol. $\Delta A = A_L - A_R$. The spectral band pass, shown in the figure, is 10 cm^{-1} and the time constant is 10 sec. The pathlength is 100 μ giving an absorbance A of 1.0 at the absorption peak, indicated by an arrow.

FIGURE 5: The first USC VCD spectra of $C_6H_5.CH(OH).CF_3$.

In the ideal CD experiment a racemic (dl) sample produces a flat baseline while opposite (d and l) optical isomers produce exactly mirror image spectra. Despite the vast improvement produced by the scrambler, subsequent experiments showed that artefacts could still be present, leading to non-ideal CD experiments. Two additional modifications further improved instrument performance. The grid polarizer was replaced by a Glan-Foucault $LiIO_3$ polarizer (whose long wavelength limit is ∿5μ (with superior polarizing properties. Secondly, improved electronics were devised to eliminate artefacts arising from unwanted pick-up of various types. With these developments it proved to be possible to obtain VCD spectra for a wide variety of molecules, with good signal-to-noise ratio. The first results of a general survey study were published in 1976 (22) and demonstrate the quality of spectra obtainable at that time.

The most important limitation of the instrumentation described above was the deterioration of performance with increasing wavelength above ∿4μ. This could be attributed to a variety of factors. Particularly important were the decreasing photon flux from the black-body source, made worse by the increasing absorption of the quartz envelope, and the loss of detector sensitivity at ∿5.5μ. To improve the photon flux we have switched to a 300W Varian Xe arc lamp, equipped with a sapphire window. The increase in color temperature (3000° to 6000° K) and improved window transmission lead to substantially enhanced output, particularly at wavelengths greater than 4μ. At the InSb cut-off we have added a 77°K PbSnTe detector, whose detectivity peaks at ∿9μ.

In addition, we have taken advantage of new PEMs developed by Professor J. C. Kemp and now being marketed by Hinds International Inc.. These use an octagonal optical element, either crystalline CaF_2 or amorphous ZnSe, driven by two transducers. In contrast to Kemp's earlier PEM design, this CaF_2 modulator provides greater than λ/4 retardation to its transmission limit (∿8μ). We are now using a Hinds CaF_2 PEM as principal modulator and a Hinds ZnSe PEM as a scrambler. Their advantages over the earlier ZnSe PEMs include larger aperture, lower reflection losses, higher optical quality and greater efficiency (lower power dissipation).

These latest modifications have led to improved sensitivity, improved baselines, and an extension of the upper wavelength limit to ∿6μ. As a result, VCD measurements in the 2000-2500 cm^{-1} region are much easier; this is important for the study of C-D stretch modes, for example. And below 2000 cm^{-1} VCD signals are now observable for the first time (as discussed below). Of especial importance is the detection of VCD in carbonyl and ethylenic stretching modes - non-hydrogenic fundamentals for which VCD spectra have not previously been published.

The instruments reported by Holzwarth et al. (21) and by Nafie et al. (30,33) are identical in principle to the system discussed above and differ only in their practical realization. Holzwarth's instrument is based on a Ge PEM, a glower source, a grid polarizer and an InSb detector. Nafie's set-up differs from the 1976 USC instrument (22) only in using a large-diameter InSb detector, thereby enabling the focussing optics to be modified and the "scrambler" PEM to be eliminated. As yet insufficient data have been reported from either instrument to permit very detailed cross comparison and this will not be attemped here.

The experience in our laboratory, which we have detailed at some length, illustrates the importance of the simultaneous optimization of all factors affecting CD sensitivity to the observation of VCD. VCD signals are at best small and must be studied at vibrational resolution ($\sim$10 cm^{-1} in liquids). Artefact signals are very easily generated, optically and electronically. The maximization of the lamp ouput, spectrometer throughput, detector sensitivity, polarizer efficiency, modulator performance and optical quality in all transmitting optical components between the polarizer and detector are each and all of critical significance. In all of these categories many improvements can still be made - and must be in order to further extend the wavelength range and sensitivity of current apparatus. Fortunately, infrared technology is currently developing at a rapid pace, and it is safe to predict that VCD measurement across the conventional IR range will be routine in the not-too-distant future. In particular, the development of the next generation of VCD instrumentation will likely be based on and tied to widely tunable IR lasers which can provide vastly higher photon flux levels than blackbody sources.

RESULTS

The most important result to date is that VCD is finally an observable phenomenon. Prior to 1973, it was by no means obvious that current technology would permit the observation of VCD. While theoretical predictions of the magnitude of VCD and of the CD sensitivity attainable together produced optimism, they involved assumptions which could be challenged. The observation of the C-H stretch VCD of $C_6H_5.CH(OH).CF_3$ by Holzwarth's group (17) finally laid doubt to rest.

Despite the importance of this breakthrough, the generality of the result remained unclear until it was shown that a significant fraction of a large number and variety of unexceptional chiral molecules yielded VCD signals with good signal-to-noise ratio. In the work leading to our 1976 publication (22) some two thirds of the molecules examined gave observable VCD in the hydrogenic stretching region. Subsequent improvements in instrumentation have further raised this percentage, such that the probability of observing VCD in this region of the IR spectrum is now very good for any chiral molecule. Conversely, at this time, the observation of a spectrum in itself is not particularly noteworthy.

The number of VCD spectra that have been observed is by now too large for each to be presented and we will limit our discussion to the illustration of general features of the results. A list of molecules for which VCD has been published is collected in Table 1.

First, to date the largest effects observed correspond to anisotropy ratios ($\Delta A/A$) around 10^{-4}. Since the absorbances at which CD is measured are generally in the range 0.1 - 1.0, the corresponding peak ΔA values are 10^{-5} - 10^{-4}, and these set a minimum requirement for the ΔA sensitivity of any instrumentation which is to be used to measure VCD.

The largest effects observed in our first experiments (22) were in rigid molecules such as camphor (Figure 6) and β-pinene (Figure 7). By contrast, molecules with a large degree of conformational flexibility, such as n-octanol, did not yield definitive VCD signals. In the latter case, since conformational isomers will generate VCD of varying sign and magnitude, their presence leads to a lower observed VCD. It must be pointed out immediately, however, that rigidity is not a guarantee of large VCD, as the comparison of α- and β-pinene (Figure 7) clearly demonstrates. In general, little correlation exists between vibrational and electronic optical activity, if the latter is gauged by $[\alpha]_D$ values. For example, α- and β-pinene have $[\alpha]_D$=48° and 21°, respectively, while the VCD of β-pinene is much larger than that of α-pinene. Likewise, 3-methylcyclohexanone with $[\alpha]_D$=14 has appreciably larger VCD than 3-methylcyclopentanone, whose $[\alpha]_D$=148° (Figure 8).
The large majority of VCD experiments have been carried out with either neat liquids or solutions in CCl_4, the latter being an optimal solvent in the hydrogenic stretching region. Systematic studies of effects due to variation of solvent or to intermolecular interaction have not been carried out so far. However, it is clear from several specific cases that interaction effects do occur. The first experiments on $C_6H_5.CH(OH).CF_3$ involved the neat liquid (17,19). Subsequently (22), it was found on dilution in CCl_4 not only that the O-H stretch VCD changed substantially, an unsurprising result for an alcohol, but that the C-H stretch VCD changed even more dramatically (Figure 9). This result strikingly emphasizes the care that must be taken to eliminate the presence of interaction effects on a spectrum. In the case

TABLE 1.

PUBLISHED VCD SPECTRA

α-$NiSO_4.6H_2O$ α-$NiSO_4.6D_2O$ α-$ZnSeO_4.6H_2O$	Hsu and Holzwarth, 1973
$C_6H_5.CH(OH).CF_3$	Holzwarth et al., 1974 Nafie, Cheng and Stephens, 1975 Nafie, Keiderling and Stephens, 1976
$C_6H_5.CH(NH_2).CH_3$ $C_6H_5.CH[N(CH_3)_2].CH_3$ 3-methylcyclohexanone 3-methylcyclopentanone menthol α-pinene β-pinene 3-bromocamphor borneol poly(4-methyl-1-hexene) tris[3-trifluoromethylhydroxymethylene-d-camphorato] Pr(Pra-Opt) Eu-Opt poly(1-methylpropylvinyl ether)	Nafie, Keiderling and Stephens, 1976
Camphor	Nafie, Keiderling and Stephens, 1976 Keiderling and Stephens, 1976
$[CH(OD)COOD]_2$	Sugeta et al., 1976
$[CH(OH)COOCH_3]_2$	Keiderling and Stephens, 1977
Camphoric anhydride	Clark and Stephens, 1977
alanine	Clark and Stephens, 1977 Diem et al., 1977, 1978
alanylalanine alanylalanylalanine glycylalanine analylglycine	Diem et al., 1978

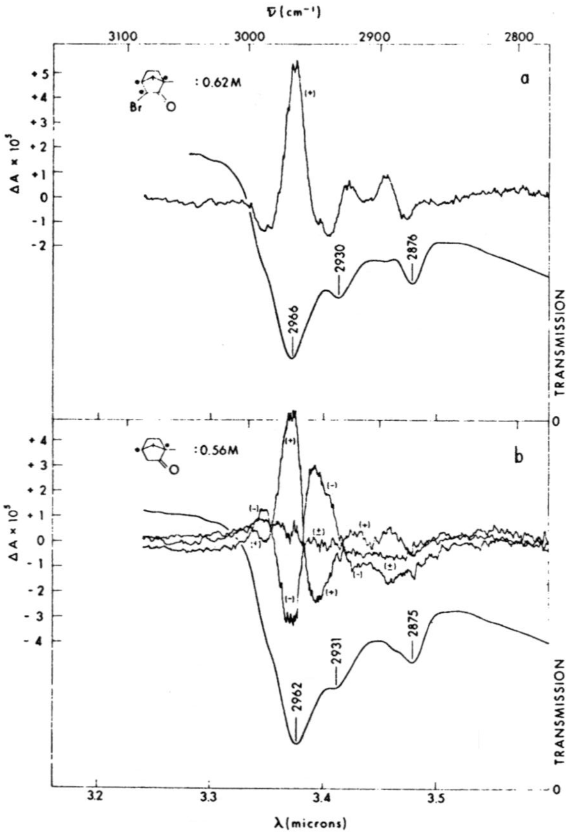

← FIGURE 6

VCD and single-beam transmission spectra of (a) 3-bromocamphor 0.62 M in CCl_4 (pathlength 80 μm, VCD time constant 3 s); (b) camphor 0.56 M in CCl_4 (pathlength 80 μm, VCD time constant 3 s). The VCD of *dl*-camphor may be used as a VCD baseline for 3-bromocamphor.

FIGURE 7 →

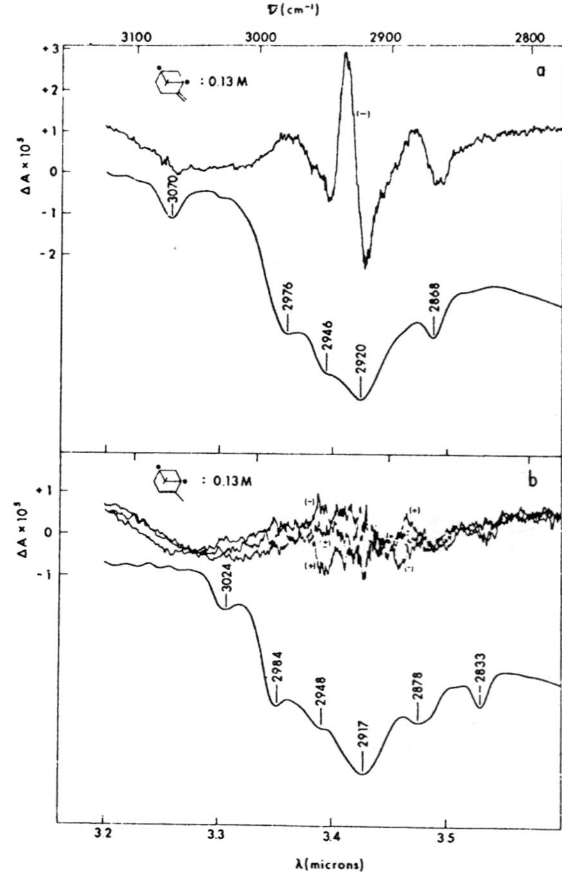

VCD and single-beam transmission spectra of (a) β-pinene, 0.13 M (2% by volume) in CCl_4 (pathlength 200 μm, VCD time constant 10 s); (b) α-pinene, 0.13 M (2% by volume) in CCl_4 (pathlength 200 μm, VCD time constant 10 s). The VCD of *dl*-α-pinene may be used as a VCD baseline for β-pinene.

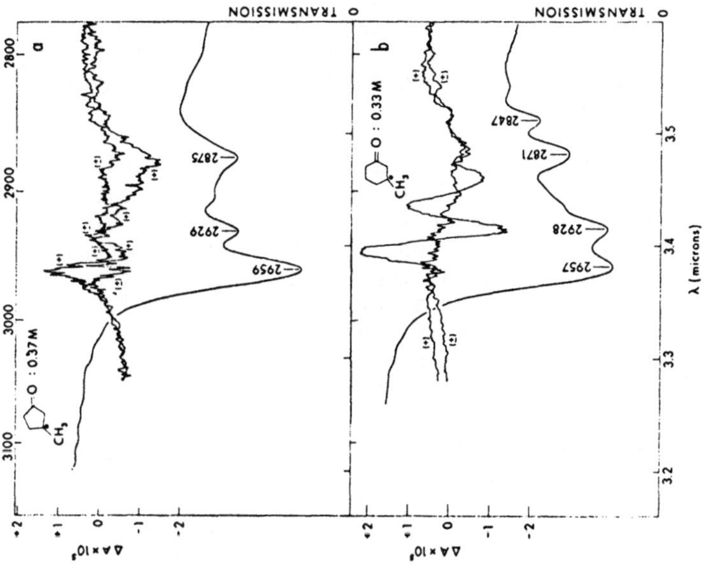

FIGURE 8

VCD and single-beam transmission spectra of (a) 3-methylcyclopentanone (MCP), 0.37 M (4% by volume) in CCl_4 (pathlength 150 μm, VCD time constant 10 s); (b) 3-methylcyclohexanone (MCH), 0.33 M (4% by volume) in CCl_4 (pathlength 150 μm, VCD time constant 10 s).

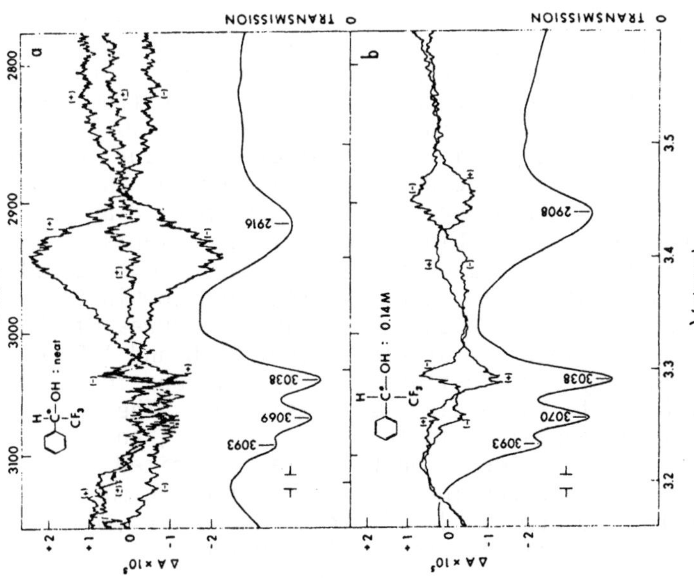

FIGURE 9

VCD and single-beam transmission spectra of 2,2,2-trifluoro-1-phenylethanol (TPE): (a) neat liquid (pathlength 80 μm, VCD time constant 10 s); (b) 0.14 M (2% by volume) in CCl_4 (pathlength 2 mm, VCD time constant 10 s).

of the dialkyltartrate esters (26) where both internal and external hydrogen bonding can occur, the O-H stretch VCD was similarly found to be concentration dependent and only at high dilution is the spectrum of the monomeric, internally hydrogen bonded species obtained.

Once the existence of VCD in fundamentals became established it was of interest to look for effects in overtone and combination bands. Studies of several molecules (23) showed that VCD can be observed comparable in magnitude to that in fundamentals. In camphor, first overtone bands of the C-H stretch modes and of the C=O stretch (Figure 10) and a variety of combination bands yielded measurable VCD. There is, therefore, no reason to believe that transitions owing their existence to anharmonicity are in general significantly more or less chiral than fundamentals.

In the majority of chiral organic molecules the C-H stretch region yields a complex spectrum due to the number of closely spaced transitions. Camphor, for example, possesses 16 C-H stretch modes. With regard to theoretical interpretation, it is clearly advantageous to study modes which are present in smaller number. Outside the hydrogenic stretching region attempts have so far been made to study the C-D stretch in selectively deuterated molecules, the C≡O stretch in metal carbonyls, and the C=O stretch and C=C stretch in various organic molecules.

VCD was reported in the C-D stretch of $C(CH_3)_3.CHDCl$ by Holzwarth's group in 1974. At USC we have also examined the mono-deuterated compounds exo-3-deuterocamphor, exo-3-deuteroisoborneol and α-deuteropropylbenzene (Figures 11-13). In all three cases, the spectrum is more complex than expected for a single vibrational transition. In the deuterated camphor, the VCD spectrum consists of a doublet, with anisotropy ratios of 4.0 and 4.2×10^{-4} at the peaks at 2210 and 2180 cm^{-1} respectively, together with four smaller peaks to lower energy. In the deuterated isoborneol and propylbenzene molecules, the main peak is again accompanied by satellite structure. A possible cause of these effects is suggested by the VCD spectrum of camphor in the same spectral range (Figure 14), which shows that overtone and combination bands are dense throughout this region. The weaker features of the exo-3-deuterocamphor spectrum at 2160, 2140, 2120 and 1090 cm^{-1} coincide in frequency and amplitude with transitions in camphor and can be assigned to the analogous transitions. The doublet structure cannot be explained in this way, however, and must arise from interaction between the C-D stretch excited state and one or more overtone and combination levels (i.e. Fermi resonance) leading to comparable VCD in the resulting, mixed transitions. Similar effects presumably contribute to the multiplicity of the VCD in exo-3-deuteroisoborneol and α-deutero-

FIGURE 10

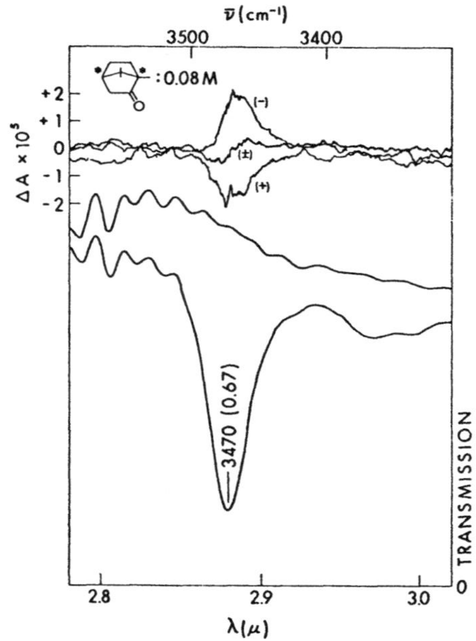

VCD and transmission spectrum of camphor (0.08 M in CCl_4) in the first overtone CO stretching band. The cell has quartz windows and 1 cm pathlength. The peak absorption frequency, in cm^{-1}, and corresponding absorbance is indicated. The spectral bandpass is 0.012 μ (14 cm^{-1} at 2.9 μ) and the VCD time constant is 10 s.

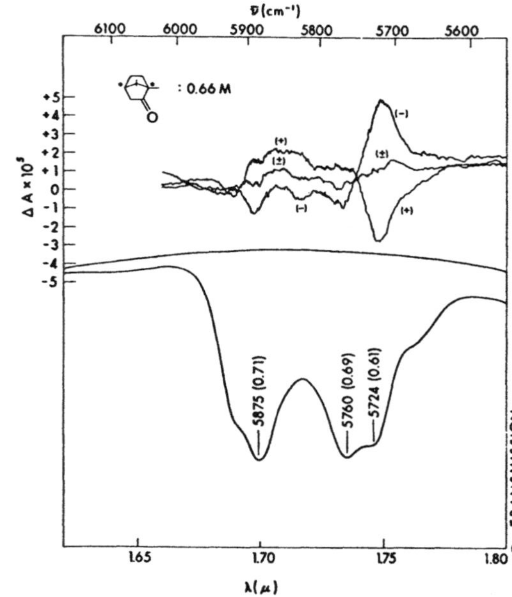

VCD and transmission spectrum of camphor (0.66 M in CCl_4) in the first overtone C–H stretching region. The cell has quartz windows and 1 cm pathlength. Peak absorption frequencies, in cm^{-1}, and corresponding absorbances are indicated. The spectral bandpass is 0.006 μ (21 cm^{-1} at 1.7 μ) and the VCD time constant is 10 s.

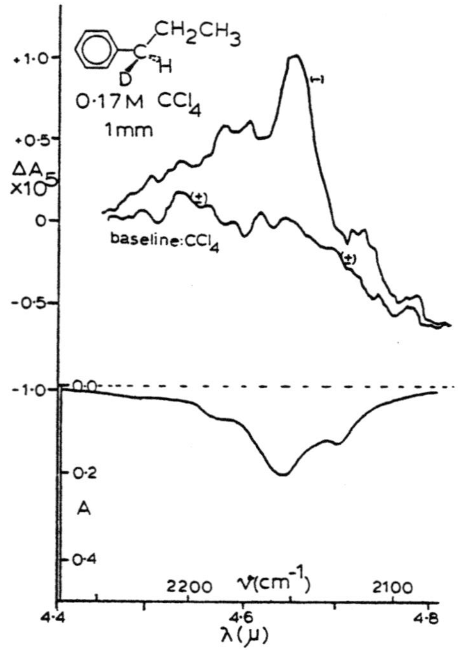

FIGURE 11: VCD of (R)-(-)-C_6H_5.CHD.C_2H_5 in the C-D stretch region. The baseline is provided by the solvent.

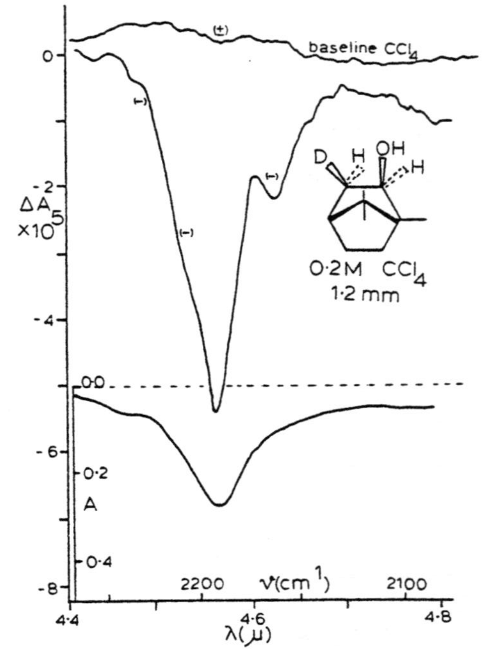

FIGURE 12: VCD of (-)-exo-3-deuteroisoborneol in the C-D stretch region. The baseline is provided by the solvent.

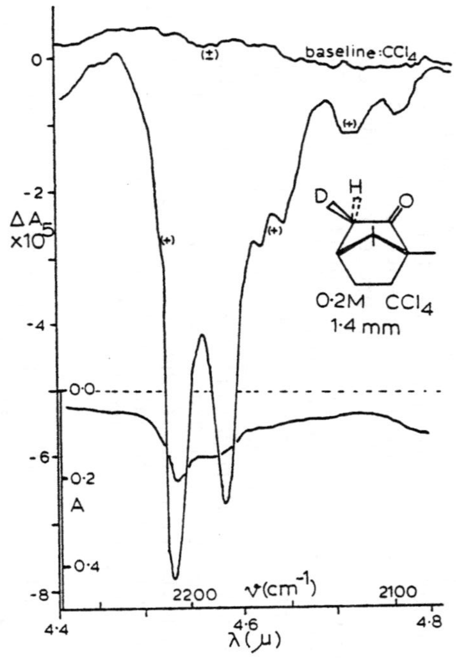

FIGURE 13: VCD of (+)-exo--3-deuterocamphor in the C-D stretch region. The baseline is provided by the solvent.

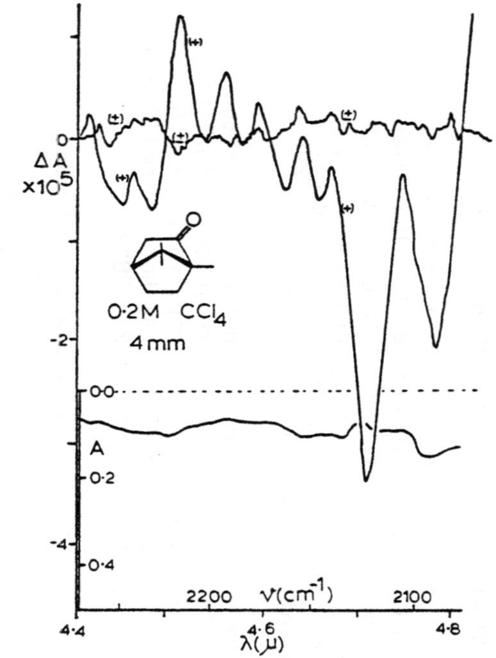

FIGURE 14: VCD of (+)-camphor in the C-D stretch region. The baseline is provided by the racemic compound.

propylbenzene; they may also be present in $C(CH_3)_3.CHDCl$, but the VCD spectrum of this molecule was not published.

The first VCD spectrum obtained for a fundamental carbonyl stretch transition is shown in Figure 15. The molecule is one member of a series of chiral iron carbonyl complexes of the type $Fe(C_5H_5)$ $P(C_6H_5)_3$ (CO)R prepared by Professor T. C. Flood at USC, whose VCD we are currently studying. In the $R=C_2H_5$ complex, the C≡O stretch anisotropy ratio is 2.5 x 10^{-5}.

Subsequently, we have observed VCD in the C=O stretch of organic carbonyl-containing molecules. Figure 16 shows the spectrum of 3-bromocamphor which gives an anisotropy ratio at 1760 cm^{-1} of 7 x 10^{-5}. The corresponding VCD in camphor is smaller, having an anisotropy ratio at 1740 cm^{-1} of 2 x 10^{-5}.

The C=C stretch transition also lies within the accessible spectral range and we have recently observed VCD from this mode for the first time in α- and β-pinene (Figure 17). The anisotropy ratios are 3 x 10^{-5} at 1655 cm^{-1} for α-pinene, and 9 x 10^{-5} at 1640 cm^{-1} for β-pinene.

THEORY

Theoretical aspects of VCD are the subject of the contribution by Professor Moscowitz and will be considered only briefly here.

Prior to the first experimental measurements of VCD a number of calculations of vibrational optical activity had been carried out to assess the magnitudes of effects to be expected. Cohan and Hameka discussed the simple molecules CHDClBr, $C^{35}Cl^{37}ClBrF$ and CHDTBr, whose chirality is a consequence of isotopic substitution (9). Deutsche and Moscowitz presented model calculations for helical polymers (10,12). Schellman reported calculations for 3-, 4- and 5-methylpyrrolidones (16a). Holzwarth and Chabay (14) discussed the predictions of the coupled oscillator model in two specific molecules. These studies yielded results of magnitudes which could reasonably be expected to be measurable and provided moral support in the search for VCD. Unfortunately, none of them has yet been subjected to direct comparison with experiment, as is also true of the later calculations of Snir, Frankel and Schellman for polypeptides (16b,20).

Direct comparison of calculation and experiment has been attempted to date only for a small number of molecules. Faulkner et al. reported (18) results for the C-H stretch of $C_6H_5.CH(OH).CF_3$: for the conformation assumed the calculated rotational strength was of the correct sign but considerably smaller in magnitude than that observed experimentally for the pure liquid.

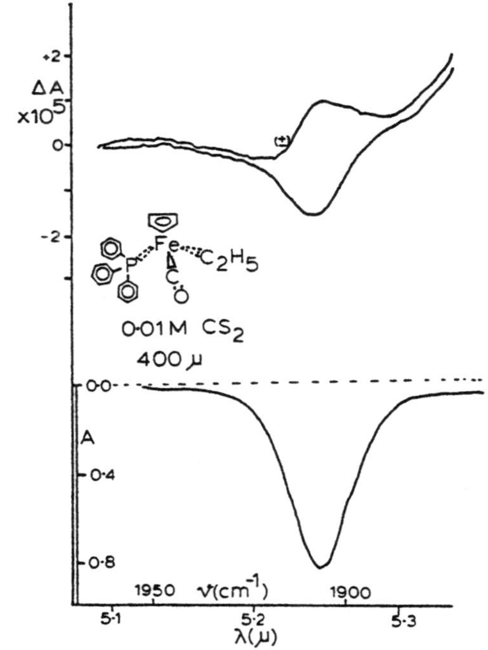

FIGURE 15: VCD of (S)-Fe(C_5H_5) P(C_6H_5)$_3$ CO(C_2H_5) in the C≡O stretch region. The baseline is provided by the racemic compound.

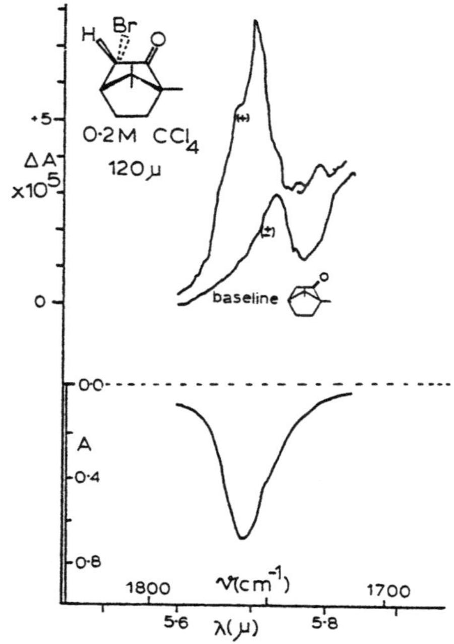

FIGURE 16: VCD of 3-bromocamphor in the C=O stretch region. The baseline is provided by (±)-camphor.

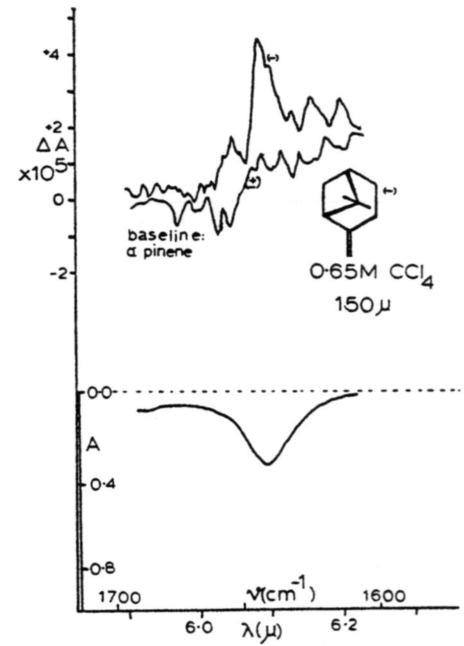

FIGURE 17: VCD of (+)-, (-)- and (±)-α-pinene in the C=C stretch region.

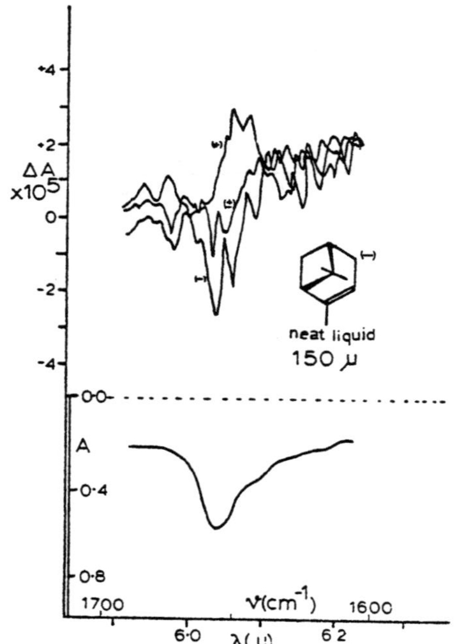

FIGURE 18: VCD of (-)-β-pinene in the C=C stretch region. The baseline is provided by (±)-α-pinene.

Unfortunately, the VCD was subsequently observed to be markedly concentration dependent (22) and in dilute solution to be of opposite sign to that observed in the neat liquid and to that calculated. Faulkner et al. have also reported calculations on the C-H stretch VCD of tartaric acid (31), observed by Sugeta et al. (24). Again, theory and experiment are in substantial disagreement.

The C-D stretch VCD of $C(CH_3)_3.CHD.Cl$ was calculated by Faulkner et al. (18) and in this case reasonable agreement was claimed with the experimental rotational strength reported by Holzwarth's group. The experimental spectrum for this molecule has not been published to date, however, and the precision of this comparison between theory and experiment is difficult to assess.

Not only is the comparison of theory and experiment limited, therefore, but it is also inconclusive due to the structural ambiguities and complexities of the molecules studied. Since all currently available theoretical approaches to VCD involve substantial approximations of unknown accuracy, it is of paramount importance at this time to focus experimental measurements on molecules of definite, known, rigid structure for which vibrational calculations are practicable. Only by the study of such molecules can the validity and the future usefulness of VCD calculations become definitely established.

CONCLUSION

Vibrational optical activity provides an alternative source of information regarding chiral molecules to the electronic optical activity phenomena so well-known and so long studied. The principal potential advantages of VCD over electronic CD lie in the much greater range of transitions accessible in the infrared compared to the near-ultraviolet and the fact that VCD is a ground state property whose analysis, in principle, does not require a knowledge of excited electronic states.

The work of the last five years has shown that VCD is indeed an observable phenomenon, thereby opening the door finally to its actual usage. However, the observation of VCD remains limited to the section of the infrared above $\sim 1600\ cm^{-1}$ and many types of vibrational transition are still out of range of available instrumentation. In addition, it has yet to be shown convincingly that theory can account for observed VCD spectra in molecules of unambiguous configuration and structure. Considerably more development of both instrumentation and theory are thus required before the exploitation of VCD can be contemplated routinely.

ACKNOWLEDGMENTS

We are grateful to Professor T. C. Flood for providing the $Fe(C_5H_5)\ P(C_6H_5)_3\ CO(C_2H_5)$ complex and to Professor L. M. Stephenson for providing samples of exo-3-deutero-camphor and -isoborneol and α-deuteropropylbenzene whose VCD is reported in Figures 11-13 and 15. We also acknowledge financial support by the National Science Foundation and the National Institutes of Health.

REFERENCES

Vibrational Optical Activity*

1. T. M. Lowry, Optical Rotatory Power, Longmans, Green & Co., London, 1935.
2. H. S. Gutowsky, J. Chem. Phys. 19, 438 (1951).
3. W. Fickett, J. Am. Chem. Soc. 74, 4204 (1952).
4. C. D. West, J. Chem. Phys. 22, 749 (1954).
5. H. J. Hediger and H. H. Gunthard, Helv. Chim. Acta 37, 1125 (1954).
6. L. I. Katzin, J. Phys. Chem. 68, 2367 (1964).
7. H. R. Wyss and H. H. Gunthard, Helv. Chim. Acta, 49, 660 (1966).
8. H. R. Wyss and H. H. Gunthard, J. Opt. Soc. Am. 56, 888 (1966).
9. N. V. Cohan and H. F. Hameka, J. Am. Chem. Soc. 88, 2136 (1966).
10. C. W. Deutsche and A. Moscowitz, J. Chem. Phys. 49, 3257 (1968).
11. C. W. Deutsche, D. A. Lightner, R. W. Woody, and A. Moscowitz, Ann. Rev. Phys. Chem. 20, 407 (1968).
12. C. W. Deutsche and A. Moscowitz, J. Chem. Phys. 53, 2630 (1970).
13. Y. N. Chirgadze, S. Y. Venyaminov and V. M. Lobachev, Biopolymers 10, 809 (1971).
14. G. Holzwarth and I. Chabay, J. Chem. Phys. 57, 1632 (1972).
15. E. C. Hsu and G. Holzwarth, J. Chem. Phys. 59, 4678 (1973).
16. J. A. Schellman a) J. Chem. Phys. 59, 4678 (1973); 60, 343 (1974); b) Peptides, Polypeptides and Proteins, Ed. E. R. Blou· F. A. Bovey, M. Goodman, and N. Lotan, Wiley, 1974, p. 320.
17. G. Holzwarth, E. C. Hsu, H. S. Mosher, T. R. Faulkner and A. Moscowitz, J. Am. Chem. Soc. 96, 251 (1974).
18. T. R. Faulkner, A. Moscowitz, G. Holzwarth, E. C. Hsu and H. S. Mosher, J. Am. Chem. Soc. 96, 252 (1974).
19. L. A. Nafie, J. C. Cheng and P. J. Stephens, J. Am. Chem. Soc. 97, 3842 (1975).
20. J. Snir, R. A. Frankel and J. A. Schellman, Biopolymers 14, 173 (1975).
21. I. Chabay and G. Holzwarth, Appl. Opt. 14, 454 (1975).
22. L. A. Nafie, T. A. Keiderling and P. J. Stephens, J. Am. Chem. Soc. 98, 2715 (1976).
23. T. A. Keiderling and P. J. Stephens, Chem. Phys. L. 41, 46 (1976).
24. H. Sugeta, C. Marcott, T. R. Faulkner, J. Overend and A. Moscowitz, Chem. Phys. L., 40, 397 (1976).

25. T. R. Faulkner, Ph.D. Thesis, U. Minnesota, 1976.
26. T. A. Keiderling and P. J. Stephens, J. Am. Chem. Soc. 99, 8061 (1977).
27. R. Clark and P. J. Stephens, Proc. Soc. Photo Inst. Eng. 112, 127 (1977).
28. L. A. Nafie and T. H. Walnut, Chem. Phys. L, 49, 441 (1977).
29. T. H. Walnut and L. A. Nafie, J. Chem. Phys. 67, 1501 (1977).
30. M. Diem, P. J. Gotkin, J. M. Kupfer, A. P. Tindall and L. A. Nafie, J. Am. Chem. Soc. 99, 8103 (1977).
31. T. R. Faulkner, C. Marcott, A. Moscowitz and J. Overend, J. Am. Chem. Soc. 99, 8160 (1977).
32. C. Marcott, T. R. Faulkner, A. Moscowitz, J. Overend, J. Am. Chem. Soc. 99, 8169 (1977).
33. M. Diem, P. J. Gotkin, J. M. Kupfer and L. A. Nafie, J. Am. Chem. Soc. 100, 0000 (1978).

Other

34. G. A. Osborne, J. C. Cheng and P. J. Stephens, Rev. Sci. Inst. 44, 10 (1973).
35. J. C. Cheng, L. A. Nafie and P. J. Stephens, J. Opt. Soc. Am. 65, 1031 (1975).
36. J. C. Cheng, L. A. Nafie, S. D. Allen and A. Braunstein, Appl. Opt. 15, 1960 (1976).
37. M. Grosjean and M. Legrand, Comptes Rendus 251, 2150 (1960); L. Velluz, M. Legrand and M. Grosjean, Optical Circular Dichroism, Academic Press, 1965.
38. M. Billardon and J. Badoz, Comptes Rendus, 262B, 1672 (1966).
39. J. C. Kemp, J. Opt. Soc. Am. 59, 950 (1969).
40. S. N. Jasperson and S. E. Schnatterly, Rev. Sci. Inst. 40, 761 (1969).
41. L. F. Mollenauer, D. Downie, H. Engstrom and W. B. Grant, Appl. Opt. 8, 661 (1969).

* This section references all publications to date relating directly to vibrational optical activity, with the exception of those involving Raman spectral and liquid crystal effects.

A VIBRATIONAL ROTATIONAL STRENGTH OF EXTRAORDINARY INTENSITY. AZIDOMETHEMOGLOBIN A*

Curtis Marcott, Henry A. Havel, Bo Hedlund, John Overend, and Albert Moscowitz
Department of Chemistry and Department of Laboratory Medicine and Pathology, University of Minnesota, Minneapolis, Minnesota 55455, USA

The asymmetric stretching vibration of the N_3^- group in azidomethemoglobin A (azidometHb A) exhibits a vibrational circular dichroism (VCD) band (Fig. 1) of extraordinary and unprecedented intensity. Its rotational strength is three orders of magnitude greater than any heretofore noted. Integration of the VCD band gives a vibrational rotational strength (per tetramer) of 3×10^{-40} (esu cm)^2, a value typical for the electronic rotational strengths associated with the $n \rightarrow \pi^*$ transition in saturated ketones.

The infrared transmission and VCD spectra of an aqueous solution of azidometHb A between 2075 and 1975 cm^{-1} are shown in Fig. 1. The peak height of the ir transmission spectrum is about 2-3%, and the large sloping background is due to the H_2O solvent absorption. The ir band maximum and the VCD band maximum both occur at about 2025 cm^{-1}. The spectra were recorded on the modified Holzwarth-Chabay spectrometer [1] in our laboratory [2]. The VCD trace shown is a single scan taken directly from the instrument without baseline subtraction. Solutions of azide-free methemoglobin A at a concentration comparable to that used for azidometHb A showed no observable VCD in the region of the azide asymmetric stretch, nor did a solution of the azide ion alone. The azidometHb A was prepared by adding 1.0 M aqueous sodium azide to a concentrated metHb A phosphate buffer solution (ionic strength = 0.2, pH = 6.0) such that the ratio azide:heme was 1:1. The concentration of metHb in our

* This work has been supported in part by NSF Grant No. CHE 77-06752 and NIH Grant No. HL 16833.

Stephen F. Mason (ed.), Optical Activity and Chiral Discrimination. 289–292.

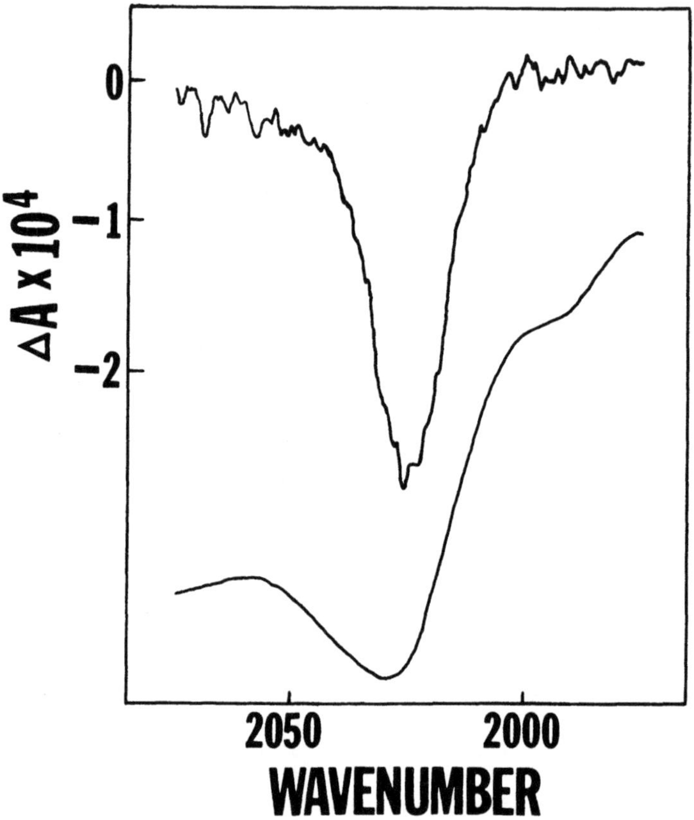

Fig. 1. VCD spectrum (top) and ir transmission spectrum (bottom) of azidomethemoglobin A. The concentration is 4.77 mM (in tetramer) in H_2O. The path length is 0.058 mm, the resolution $\approx$10 cm^{-1}, and the time constant 3 seconds.

stock solution was determined spectroscopically from the cyanide derivative [4,5]. The concentration of azidometHb A in the sample cell was 4.77 mM in tetramer. Solutions this concentrated are required because the strong H_2O solvent absorption necessitates short path lengths (0.058 mm in the case of Fig. 1).

The addition of excess azide ion resulted in no observable change in the VCD intensity, although another peak did appear at 2050 cm^{-1} in the infrared transmission spectrum in addition to the peak at 2025 cm^{-1}. The band at 2050 cm^{-1} is due to ionic azide [6] and shows no VCD. In contrast, the band at 2025 cm^{-1} is the covalently bound azide band [6] and shows the extremely large negatively signed VCD peak whose rotational strength is invariant to different excesses of azide ion. This indicates that nearly all of the azide ion is covalently bound to the iron atom when the azide:heme ratio is 1:1.

In addition to the azide ion, we studied the ligands CO, CN^-, and isopropylisocyanide, which bind to hemoglobin and have ir absorptions in the "water window" between 2200 and 1900 cm^{-1}. None of the metHb A complexes of these ligands showed any VCD detectable with our present instrumental sensitivity. The azide group differs from the other ligands just mentioned in one notable way, namely, in the potentially chiral arrangement of its lone pairs upon binding with the iron and interaction with the chiral hemoglobin environment.

Of late, we have been attempting to extend our fixed-partial-charge (fpc) calculations by explicit modeling of the contributions due to lone pair electrons. We find that specific inclusion of lone pair contributions can eliminate the order-of-magnitude discrepancies frequently observed between fpc calculated and experimental vibrational rotational strengths. Hence, we are inclined to attribute the extraordinarily large rotational strength found for azidometHb A to the presence of the chirally arranged lone pairs noted above.

REFERENCES AND NOTES

1. I. Chabay and G. Holzwarth, Appl. Opt., 14, 454 (1975).
2. We have replaced the homemade germanium modulator described in Ref. [1] with a commercial 57 kHz CaF_2 photoelastic modulator (Hinds International). The signal processing electronics have also been modified and are now similar to that described by Nafie, et al. [3]. The PAR model HR-8 lock-in amplifier has been replaced by two lock-ins: an ORTEC model 9503 followed by a PAR model 186A. Unlike the electronics described in Ref. [3], we do not have a normalization circuit, but rather divide the output of the

PAR 186A by the output of the PAR model 220 lock-in [1] with an ORTEC model 5047 ratiometer. These modifications have resulted in a considerable flattening of the baseline and a significant improvement in signal-to-noise ratio over earlier spectra. See for example, H. Sugeta, C. Marcott, T. R. Faulkner, J. Overend, and A. Moscowitz, Chem. Phys. Lett., 40, 397 (1976).
3. L. A. Nafie, T. A. Keiderling, and P. J. Stephens, J. Am. Chem. Soc., 98, 2715 (1976).
4. O. W. van Assendelft and W. G. Zijlstra, Anal. Biochem., 69, 43 (1975).
5. D. L. Drabkin and J. H. Austin, J. Biol. Chem., 112, 51 (1935).
6. S. McCoy and W. S. Caughy, Biochemistry, 9, 2387 (1970).

CHIRAL DISCRIMINATION

D.P. Craig

Research School of Chemistry, The Australian National University, Canberra, A.C.T., 2600, Australia.

Stephen F. Mason (ed.), Optical Activity and Chiral Discrimination. 293–318.

I. INTRODUCTION

I.1 Brief review of experimental background

Chiral discrimination is a new term for an old idea, going back at least to Pasteur and perhaps before. Where the pair-interaction energy of two molecules or ions is different for the two enantiomers of one acting on the same enantiomer of the other there is chiral discrimination. In its application to small molecules,

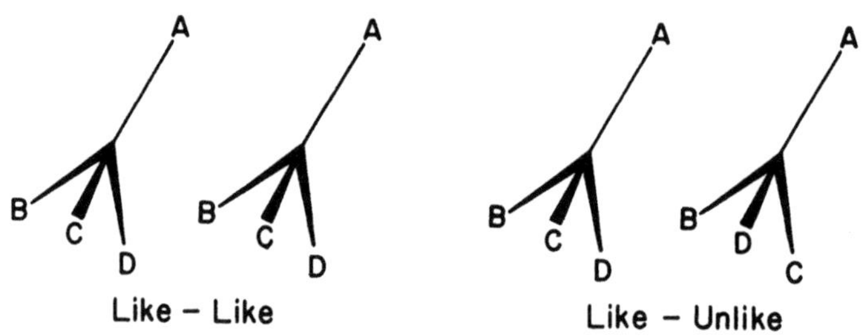

Fig. 1. Interacting molecule pairs. The left hand pair (same chirality) have a different interaction energy from the right hand pair (opposite chirality). The difference is the discrimination energy, defined to take account of relative orientation, or of orientational averaging.

at distances greater than closest approach, where the molecules had some freedom of rotation, the current revival of interest stems from the work of F.P. Dwyer and collaborators in the 1950's [1]. Since then theoretical papers have appeared and many new and precise experimental measurements have been made enabling magnitudes to be estimated. In the discussion to follow the main results and problems are surveyed. Much of the theory as well as the experimental basis has been reviewed quite recently [2,3]. These sources may be used for fuller accounts. In the following a wider viewpoint is attempted, with some emphasis on recent development.

We shall see that the theoretical mechanisms known to give discriminations cover four orders of magnitude in the energies, and a realistic view must be taken of the phenomena in which evidence might be sought. The larger values are compatible with the evidence of gross discrimination already familiar in, for example, the separation of diastereomers by crystallisation. Values at the smaller end of the range are of the same order as, but still smaller than, the molar heats of mixing of enantiomers, but it may well be that new types of experiment will have to be devised to test them, in which, in particular, contributions by closely coupled molecule pairs are avoided. Under such circumstances, the predicted discrimination energies, although very small, are easily within ranges measured in other areas of chemistry and physics.

In experimental measurements, the discriminating energy differences are typically a small fraction of the total intermolecular (or interionic) energy, but in some cases are easily identified. For example this is so in the observation that the (-) form of tris-phenanthroline ruthenium perchlorate is more soluble (4.70×10^{-5}M at 25°C) in (-) 2-methyl-1-butane than the (+) form (4.0×10^{-5}M at 25°C) [4]. Chiral discrimination has been long known, and it is not surprising that the early examples were some in which it was strong, and which remain arresting. Pasteur in 1858 discovered that the mould *penicillium glaucum* in dilute ammonium racemate (i.e. the salt of racemic tartaric acid) grew at the expense of the *dextro* form, without effect upon the *laevo* form. Pasteur also identified diastereomeric pairs, dAdB and dAℓB, which had different solubilities and showed differences in other properties indicating discriminating interactions of magnitudes measured in kJ mol^{-1}. He knew also that enantiomers could excite different physiological responses, as in the different tastes of (+)- and (-)- asparagine, the former sweet and the latter tasteless. Discriminations in such biomolecular interactions are not discussed here, though much is known about them experimentally. Many prime examples are listed by Albert (1973) [5]. The explanation in molecular detail is known in a few cases, such as that of the differential inhibition by d- and ℓ-phenylalanine on the activity of manganese-carboxypeptidase A [6].

Interacting pairs of small molecules are more easily discussed from a theoretical point of view, which is the purpose of this article. Some experiments bearing upon the small molecule limit, and which point to types of discrimination, and magnitudes, which call for theoretical analysis will be given first. It has been mentioned that the different solubilities of diastereomeric pairs are often striking, and came early to notice. Pasteur's separation of racemic acid ((+) tartaric acid) by addition of (-) cinchonine and exploitation of the different solubilities of the diastereomeric pairs (-) cinchonine (+) tartrate and (-) cinchonine (-) tartrate to effect a separation is typical. The diastereomers pack together in their crystals differently, with different binding energies. However, other phenomena yield quantitative energy data more easily. Thus Greenstein and Winitz [7] quote 1240 J mol^{-1} for the difference in heats of solution of the dℓ and ℓ forms of leucine.

Where, as here, the differences appear in the solid phase, and arise from packing differences for bodies of different 'shapes', the question of whether discrimination can appear at longer distances, beyond the contact separation, does not arise. The long range parts must then be swamped by much larger contact terms. Where no crystal phase is present, the magnitudes are often smaller. Dwyer and Davies [8] followed the racemization of the optically labile (+)- and (-)-tris-o-phenanthroline nickel chloride in aqueous (+)-bromcamphorsulphonate anion (0.012M) and in aqueous cinchoninium cation (0.054 M). These processes ended in equilibria with one enantiomer in excess over the other, and the equilibrium constants and their temperature dependences were measured. From the results the heat content differences were $\Delta H = 1.6$ and 1.8 kJ mol^{-1} respectively and the Gibbs standard free energies $\Delta G° = 0.30$ and 0.34 kJ mol^{-1}. One interprets these to be the heats and free energies required to convert the more stable enantiomer into the less stable in the presence of the chiral solvent. Absence of a solid phase does not necessarily mean that the differences arise from ions other than those in close contact: even where solute and chiral solvent ion bear the same ionic charge there may, for large ions, be contacts maintained by dispersion forces.

Heats of mixing may well be a valuable source of data on the discriminating interactions of chiral molecules. In the example of tartaric acid we know that the racemic mixture crystallises from an aqueous solution of the mixed enantiomers. In apparent contrast it is found [9] that in mixing aqueous solutions of (+) and (-) forms of tartaric acid heat is absorbed, the enthalpy of mixing being +1.9 J mol^{-1} at 25.6°C. Thus it would appear that the interaction of unlike species is unfavourable compared with like species. Interpretation of this result without knowledge of the entropy change is not straightforward, and probably not simply a matter of short-range contacts (in the crystal) giving a different result from long-range interactions between randomly oriented species (in solution), however when there are more results enabling such comparisons to be made progress should be possible. A second example, threonine [9], has a molar heat of mixing of opposite enantiomers of -5.4 J mol^{-1} at 25.6°C; this exothermic behaviour indicates greater stability in unlike contacts, again contrasting with the crystal packing, in that racemic threonine in aqueous solution forms crystals which individually consist of all (+) or all (-) molecules. We note in passing that this behaviour in crystals formed from racemic aqueous solution is not common, some 200 cases only being so far known [10].

In yet another type of experiment, Bosnich and Watts [11]

measured the rates and equilibrium constants for the process in which racemic $[Ni(phen)_3]C\ell_2$, dissolved in the pure optically active solvent (-) 2,3-butanediol, develops an excess of one enantiomer over the other. The standard enthalpies and entropy changes are $\Delta H° = -498$ J mol^{-1} and $\Delta S° = -1.17$ JK^{-1} mol^{-1}, the (-) form of the nickel complex cation being the more stable in (-) solvent. The discrimination in the effect of the same solvent on $[Co(en)_2C\ell_2]^+$ had earlier been shown by Bosnich and Watts to be much greater, $\Delta G° = 3.8$ kJ mol^{-1}, which they supposed might be due to the differential formation of hydrogen bonds from ethylene diamine to the solvent. Such specific chemical factors may enter in many cases, giving discriminations even where the bonds are extremely weak in the chemical sense. Another potential source of discrimination, also chemical in origin, is that to which Gillard has drawn attention in the chiral coordination complexes of α,α'-bipyridyl, o-phenanthroline, and others [12]. This is that an initial chemical reaction with the solvent, namely covalent aquation, gives in small amounts a new complex capable of bonding to another chiral species. This bonding is different for the two enantiomers, and leads to discrimination. Many of these examples are typical in the important respect that they do not distinguish long-range from short-range (contact) interactions. The latter, which are expected to be large, may well dominate over the smaller, though more numerous, long-range terms. Experiments in the vapour phase, such as molecular beam scattering with chiral molecules, are a potential source of finer details of discriminating forces. It is helpful to distinguish the chemical mechanisms of discrimination from those not involving chemical bonding to a chiral solvent, or bonds to another chiral species. Only the latter are treated in this article, in the spirit of discussing discrimination as an aspect of non-bonded intermolecular interactions.

Where discriminating interactions depend, among other terms, on differences in ion or molecular packing, the distances between centres of mass are not the same for dA - dB and dA - ℓB pairs. The difference in the heats of sublimation must include a term reflecting this, as well as others contributed by specific contact interactions. An example will illustrate. The nitroxide of 2,2,5,5 - tetramethyl-3-hydroxypyrrolidine crystallises in pure enantiomeric crystals, and also as a pseudo-racemate. The pure (+)-form crystallises with unit cell volume of 0.926×10^{-27} m^3, compared with 0.938×10^{-27} m^3 for the pseudo-racemate [13]. The difference, expressed as a change in molecule-molecule separations, is 0.4%. If the crystal binding energy is dispersive in origin, going with molecular separation as R^{-6}, and if we assign to the pairwise binding the representative value of 12 kJ mol^{-1}, the contribution to the discrimination is about 0.3 - 0.4 kJ mol^{-1}, a value which, by itself, could account for observed discrimination in many cases. As will be elaborated later, it is not possible to make any general distinction between simple distance dependence and effects arising from differential dA - dB and dA - ℓB interactions at the same distance for the reason that the meaning of 'same distance' is not well defined in systems, like these, which have no centre or symmetry, and thus no natural origin from which to measure distance.

The types of interaction, in which discrimination is to be discussed in Section III and later sections, are briefly sketched in the following paragraphs.

I.2 Molecular interactions. Interaction between permanent electric moments

The dominant electrostatic energy term between ions A and B is the charge-charge term, $V_{AB} = Z_A Z_B / R_{AB}$, where Z_A and Z_B are the charges (including sign) and R_{AB} the separation distance. For molecules carrying no net charge the leading term is the dipole-dipole interaction

$$V_{AB} = \mu_i^A \, \mu_j^B \, R_{AB}^{-3} \, (\delta_{ij} - 3 \, \hat{R}_i \hat{R}_j) \qquad (I.1)$$

in which μ_i^A is a component of the dipole moment operator for A, δ_{ij} is the Kronecker delta, and $\hat{R}_i$ a component of a unit vector along $\tilde{R}_{AB}$. In general, the range of the electrostatic interactions, i.e. the dependence on separation distance, is $R_{AB}^{-n-n'-1}$, n and n' being the multipole order for the centres. n takes the values 0,1,2 ... for charge, dipole, quadrupole ... sources. Evidently the discrimination, which is the difference between interactions of the same order, has the same range as the substantive interaction; for example a discrimination arising from the quadrupole-quadrupole interaction goes as R_{AB}^{-5}.

I.3 The dispersion interaction

Two molecules in ground states attract at all distances and at all relative orientations with approximate distance dependence R_{AB}^{-6}. The interaction arises from dipole coupling in the second order, according to expression (I.2),

$$\Delta E = - R_{AB}^{6} \, \beta_{ij} \beta_{k\ell} \sum_{n^a, n^b}{}' \frac{\langle 0|\mu_i|n^a\rangle\langle n^a|\mu_k|0\rangle\langle 0|\mu_j|n^b\rangle\langle n^b|\mu_\ell|0\rangle}{E(n^a) + E(n^b)} \qquad (I.2)$$

$\beta_{ij} = \delta_{ij} - 3\hat{R}_i \hat{R}_j$ and n^a is an index for the excited states of molecule a. Summation over the repeated indices i, j, k, and ℓ is implied. (I.2) is the leading term; there are other terms in higher inverse powers of R_{AB}. Also there is a magnetic analogue in which the electric moment operators μ_i are replaced by the magnetic moment operators $m_i = -(e/mc)(\underset{\sim}{r} \times \underset{\sim}{p})_i$, m being the electron mass, c the velocity of light and $\underset{\sim}{r}$ and $\underset{\sim}{p}$ respectively the position and momentum of the electron. The magnitude of the magnetic analogue of (I.2) is much smaller, by about α^4 times, where $\alpha = 1/137$ is the fine structure constant. We shall refer to magnetic effects in dispersive interactions later in connection with discrimination. The interaction (I.2) is responsible for the attractive forces in many molecular crystals, notably those of the aromatic hydrocarbons, where the heats of sublimation are often 60-100 kJ mol^{-1}.

I.4 Repulsive short-range interactions

The interpenetration of closed electron shells, as in the close approach of two ground state helium atoms, is opposed by an exchange force varying rapidly with distance. The dependence of repulsive energy with distance is exponential; over short intervals near the touching separation the variation can be represented as approximately R^{-12}. This variation with distance is so steep that the view of atomic contacts as that of contacts of hard spheres is often quite adequate, and gives an acceptable physical picture for many purposes. These short-range forces, in equilibrium with dispersive interactions, determine the packing of neutral molecules in molecular crystals, and even in ionic crystals also do so, on account of their very strong distance dependence.

Where a chiral ion has a strongly-held sheath of solvent molecules, the shape of the ion, carrying the chiral information, may show discrimination of the short-range or contact type at larger distances. It has not yet been possible to identify separately this effect, arising from the propagation of chiral surface structure by solvent, in experimental measurements, though it is likely to play some part where ions carry large charges. Not all molecules and ions have surface conformations which contain chiral information; for example in a large molecule a single asymmetric centre may be deeply sequestered, and the 'surface' configuration may be unaffected. In such cases, when they can be identified, the discrimination must be by long-range forces.

I.5 The resonance coupling

If one molecule of an identical pair A and B is in an

electronically excited state, the excitation can be exchanged, leading to an interaction between the two molecules varying as the inverse cube of distance, and proportional to the oscillator strength of the transition. The energy, for dipole allowed transitions, is given by expression (I.3) in which the operator is the dipole coupling (I.1).

$$\begin{aligned}\Delta E &= \langle A^*B|\mu_i^A\mu_j^B\, R^{-3}\,(\delta_{ij} - 3\hat{R}_i\hat{R}_j)|AB^*\rangle \\ &= R^{-3}\,\langle A^*|\mu_i^A|A\rangle\langle B|\mu_j^B|B^*\rangle\, \beta_{ij}\end{aligned} \qquad (I.3)$$

If we know the transition moments for the transition between the ground state A and the excited state A* we may evaluate expression (I.3). For example a pair of molecules at 0.5 nm separation, oriented with transition moments of 0.1 enm parallel to one another and orthogonal to the separation vector $\underset{\sim}{R}$, repel, their potential energy being $\sim$11.2 kJ mol^{-1} or 929 cm^{-1}. Resonance interaction appears as a spectral splitting in systems of like molecules, such as molecular crystals (the Davydov splitting). There is again a pure magnetic analogue, with order of magnitude less by α^2 or $\sim 5 \times 10^{-5}$.

II. THE CALCULATION OF DISCRIMINATING INTERACTIONS

The discriminating interactions appear as small changes to the intermolecular coupling energies just listed, and may be considered according to the same scheme of classification into electrostatic, resonance, dispersive and short-range repulsive. The calculation of discrimination brings up in an acute form the old problem of getting a small energy term as the difference of very large quantities. This has made very difficult *ab initio* calculations of molecular binding energies, which are differences between large atomic quantities only slightly modified by bonds. Discrimination energies are much smaller than the energies of most chemical bonds, and the problem of *ab initio* calculations so much more difficult for that reason. Empirical calculations open the way to treating large parts of the total energies as common to the dA-dB and dA-ℓB pairs, so that these common parts do not need to be calculated at all. The effort can be confined to dealing with the energy difference; however there are many pitfalls, especially on how the ℓB molecule is to be related to its dB counterpart, as will emerge later. The empirical calculations to be described give the order of magnitude of discrimination to be expected from the various interaction mechanisms, the dependence on angles of orientation and the dependence on distance. They apply to pairwise interactions; in the bulk the total discrimination is expected to be close to the appropriate sum of pairwise intermolecular parts, but three-body terms, for example, are thought to come into interactions in the liquid state up to a few percent of the total energies, and in principle may discriminate also.

Two molecules at long distances in the vapour or liquid phases (separated by tens of molecular diameters or more) have relative orientations that are uncorrelated. A discriminating interaction, like other intermolecular terms, is then found after random averaging over all relative orientations. At short distances, where the forces between the molecules may be strong enough to cause partial orientation, non-random averages of different kinds are required. For example, if an axis in one molecule is strongly correlated with an axis in another, a rotational average about the common axis is needed instead of a randon average. In the limit of molecules in intimate contact there is no relative motion and the interaction energy is required in the fully locked configuration, specified by relative angles and separation distance.

An approximation in much of what follows (as in I.1) is to represent the electrical properties of a molecule by its electric and magnetic moments as *point* moments. This is convenient, and realistic so long as the molecule is separated from those with which it interacts by distances large compared with molecular dimensions. This condition is rarely satisfied, and the approximation has to be applied cautiously in optically active systems for the additional reason that there is often no natural choice of multipole origins as there is, for example, in molecules possessing a centre of symmetry. The choice of origin is arbitrary, and usually unspecified: the centre of mass is often used. Since this choice affects the separation distance R between coupled molecules, the interpretation of results depends on it. As is well known, the total interaction energy found by using moments of *all* orders in the summed result is independent of origin choice at any distance beyond that of charge overlap. In such complete work the variation of moment magnitudes between origins exactly compensates variations of origin separations, but where only the leading moment(s) appears there can be errors. As we have noted, it is not a trivial problem to prescribe how the pair dA-ℓB is to be generated from the pair dA-dB in such a way that the interaction energy difference includes only discriminating terms: no generally applicable procedure is known. The method used here for convenience is to generate ℓB from dB by reflection in a plane containing the origins in A and B. Finally, the first non-vanishing electric moment of a distribution of charges is independent of origin choice: in an uncharged molecule the dipole moment is the same for any origin. However obviously the *dipole-dipole interaction* between two such molecules fixed in space is still not origin independent, insofar as it depends on the separation of the origins arbitrarily chosen.

The calculation of molecular interactions may be done to take account of the detailed distribution of charges in each molecule, instead of using only the moments of the distribution at an arbitrary origin as just described. This is the better method, when practicable, for calculating the interaction of molecules that are close together, separated by distances not large compared with molecular dimensions. Where molecules are close the point multipole moment expansion is an acceptable approximation only when the static charge distribution, or the transition charge distribution, is well concentrated near the point chosen as origin. In many chiral molecules this does not apply. In hexahelicene, and metal complex ions such as $Ni^{II}(o\text{-phen})_3$, to take two examples, the electrons affected by certain optical transitions may be remote from the centre of mass. Thus one can see that two molecules near to one another will couple strongly through their 'surface' charges in zones of closest approach and much less elsewhere. The differences will be especially marked where the distance dependence is strong, being greater for the R^{-6} law for dispersion interaction than for the R^{-3} law for static dipolar interaction.

This approach through charge distributions has been used in working out intermolecular forces in molecular crystals by the so-called transition monopole method [14], and its performance has been compared, again in molecular crystals, with that of coupled point multipoles [15]. Knowledge of the static or transition moment distribution, which is the starting point for the method, is available when we have the molecular wave functions or may simply assume them to learn about orders of magnitude and to make comparisons with other methods. An example for the electrostatic interaction is given in (IV.3) and for the dispersion interaction in (V.3).

In the important class of chiral systems of symmetry D_n, including especially D_3, the problem of origin choice does not arise. Typical examples of D_3 are the tris-phenanthroline and tris-bipyridyl complexes of transition metals. The origin is taken at the intersection of the threefold and twofold axes, which coincides with the metal atom. In symmetries C_n, including C_1 (lack of any symmetry except the trivial identity operation) there is no natural origin.

III. FORMS OF INTERACTION HAMILTONIAN

It is usual to adopt a perturbation viewpoint, separating the total energy into parts for the free molecules, assumed unaltered throughout, and parts for the coupling of the free molecules.

All the interactions to be used in the following text are included in expression (III.1), with $\hbar = m = c = 1$

$$H = H_a + H_b + H_E - \sum_i \underset{\sim}{p}_i \cdot \underset{\sim}{A}_i + \frac{1}{2} \sum_i \underset{\sim}{A}_i^2 + H_{rad} \qquad \text{(III.1)}$$

where

$$H_E = - \sum_{jp} \frac{Z_p}{r_{jp}} - \sum_{rq} \frac{Z_p}{r_{rq}} + \sum_{pq} \frac{Z_p Z_q}{r_{pq}} + \sum_{ij} \frac{1}{r_{ij}} \qquad \text{(III.2)}$$

and

$$H_{rad} = \frac{1}{8\pi} \int (\underset{\sim}{E}^{\perp 2} + \underset{\sim}{B}^2) dV \qquad \text{(III.3)}$$

H_a and H_b are free molecule Hamiltonians, H_E is the electrostatic energy of interaction between the electrons and nuclei of A and those of B, the quantities Z_p and Z_q being nuclear charges and r_{iq} separations between particles. The two terms in bold type in eq. III.1 are coupling terms to the radiation field with vector potential $\underset{\sim}{A}_i$ at particle i; $\underset{\sim}{p}_i$ is the momentum operator for the particle. $\underset{\sim}{E}^{\perp}$ is the transverse electric field and $\underset{\sim}{B}$ the magnetic induction field; H_{rad} is thus the Hamiltonian for the radiation field, as an integral over the space of the system. The Hamiltonian H includes the electrostatic terms H_E and the radiative correction terms. In the first order of perturbation theory, only the electrostatic terms appear, giving the changes in the energy due to permanent electric dipole moments, and higher moments, of molecule A interacting with moments on B. In the second order of perturbation theory the electrostatic terms are the source of dispersion interaction. The magnetic coupling, even in a static field, is from the $\underset{\sim}{p}.\underset{\sim}{A}$ term, which also gives a dispersive energy. The radiative corrections (from the fourth and fifth terms of (III.1)) contribute to the dispersive interactions only at long distances, and are not important in discrimination. However, the radiative coupling itself gives a discriminating interaction between identical systems one of which is in an excited state as in Section I.5 (resonance discrimination). For that term, as in many other cases involving radiation, the Hamiltonian (III.1) is more conveniently written in its so-called multipolar form [16], obtained from (III.1) by a canonical transformation. It is here written with the dipole terms only

$$H = H_a + H_b + H_{rad} - \sum_{i=a,b} \underset{\sim}{\mu}^{(i)} . \underset{\sim}{E}^{\perp(i)} - \sum_i \underset{\sim}{m}^{(i)} . \underset{\sim}{B}^{(i)} \qquad \text{(III.4)}$$

where $\underset{\sim}{\mu}^{(i)}$ is the electric dipole moment operator for molecule i, $\underset{\sim}{m}^{(i)}$ the magnetic dipole moment operator, and $\underset{\sim}{E}^{\perp(i)}$ the transverse field at molecule i. The dipole terms shown in (III.4) are in many cases sufficient. The electrostatic interactions (III.2) are in (III.4) contained in the coupling through the transverse field $\tilde{E}^{\perp}$, and calculations using (III.4) are generally simpler than with (III.1).

IV. ELECTROSTATIC INTERACTIONS

IV.1 Form of coupling operator

Following conventional ideas, H_E in (III.2) is expanded into a multipole series the form of which is quoted elsewhere, in the present notation [17]. Each term of the expansion gives the operator for the coupling of a multipole moment in one molecule to a multipole moment in the other, the result being displayed symbolically as

$$\begin{aligned} H_E &= C_aC_b + (C_aD_b + C_aQ_b + \ldots + \text{terms with a and b interchanged}) \\ &\quad + D_aD_b + (D_aQ_b + D_aO_b + \ldots + \text{terms with a and b interchanged}) \\ &\quad + Q_aQ_b + \text{etc.} \end{aligned} \qquad \text{(IV.1)}$$

and consisting of the charge-charge interaction, the interaction of net charge with dipole, quadrupole, octupole, etc. moments, the dipole-dipole, dipole-quadrupole, dipole-octupole interactions, and higher terms. A first step in finding the discrimination in electrostatic interactions is to identify the electric moments which produce an electric field possessing chirality. Such a field, acting on another chiral molecule, will exert different forces according to the relative handedness of the pair.

IV.2 The elementary symmetry arguments

The symmetry of a molecule is the set of transformations which leave its Hamiltonian operator invariant. In the absence of external fields the transformations are simply those which map the (rigid) molecule on itself; for a molecule to be chiral there can be no such transformation which is an improper rotation, namely a rotation combined with inversion. We thus arrive at the symmetry groups compatible with chirality: C_n, D_n, T, O and I. There are no known examples of the last three in chiral molecules: the common case of a molecule lacking all symmetry is, in a formal way, that of C_1, and many chiral inorganic systems are known in D_3.

We may begin with the interactions of static electric charge distributions. Here the external field of one molecule is the sum of partial fields each having as its source a multipole component. The combination of cources is chiral if its symmetry is that of one of the chiral groups. A dipole moment is achiral, having planes of reflection symmetry (twofold improper axis); a quadrupole moment component is also achiral, having a centre of symmetry (improper axis of unit order); it is readily shown that any single multipole component possesses elements of symmetry that make it achiral. One must use combinations of multipoles; the simplest combination that is chiral is a dipole and a quadrupole with individual choice of axis to remove any common symmetry plane. If we choose a quadrupole moment component symmetric to reflection in the (xy) plane and anti-symmetric to (xz) and (yz) reflections, we may then have a dipole moment in the (xz) plane not lying along x or z. This gives a chiral moment-pair which can be described as a skew dipole-quadrupole (Fig. 2). It is not difficult to get a set of rules to find the chiral combinations in other allowed symmetry groups [2].

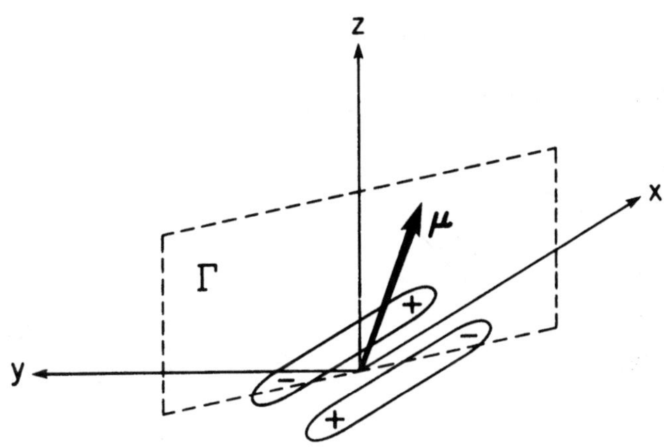

Fig. 2. The skew dipole-quadrupole. The dipole $\underset{\sim}{\mu}$ lies in a plane which is not a symmetry plane of the quadrupole component. This is the simplest chiral combination of moments.

It is assumed in this statement that the point multipoles are centred at the same point in the molecule; if that were not so a chiral field could be produced by two point *dipoles* located at different points. However, discussion will be continued in the single-centre framework. Discrimination will exist if the pairwise A-B coupling of electric moments is different for ℓA-ℓB and ℓA-dB, dB being generated from ℓA, as before, by some admissible procedure, such as reflection in a plane containing the multipole centres in A and B. The magnitude of the discrimination depends on the particular plane chosen. We must next analyse the effect of averaging over the relative molecular configurations accessible to the complex molecules. If the molecules are far apart, with a total interaction energy very small compared with the thermal energy, all relative orientations are equally probable. It is well known that when any purely electrostatic interaction is randomly averaged in this way (with the separation R kept fixed) the result vanishes. There is no net interaction and, *a fortiori*, no discrimination. As the total interaction increases, approaching the thermal energy, there is a weighting of relative orientations in favour of those giving strong coupling, and the averaged electrostatic energy is given by

$$\langle H_E \rangle = \frac{\int\int d\omega d\omega' \langle H_E, \omega\omega' \rangle \exp\{-\langle H_E, \omega\omega' \rangle / kT\}}{\int\int d\omega d\omega' \exp\{-\langle H_E, \omega\omega' \rangle / kT\}} \qquad \text{(IV.1)}$$

ω and ω' are orientations of the molecules, k is Boltzmann's

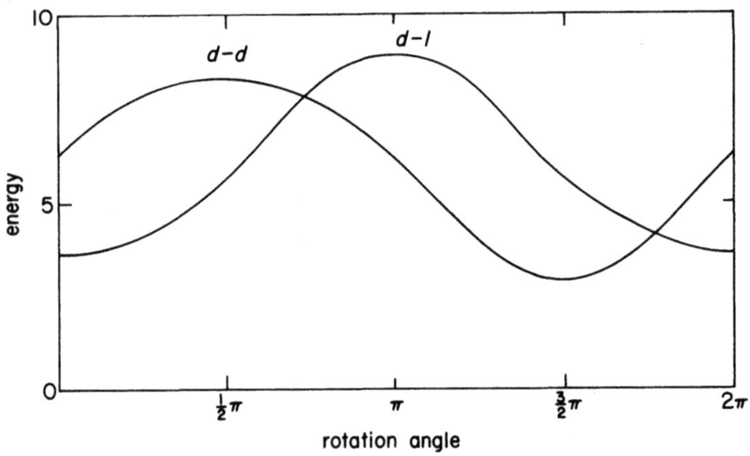

Fig. 3. Quadrupole-quadrupole energy for chiral molecules represented by skew dipole-quadrupoles. The dipole moments are aligned head-to-tail, and one system rotated relative to the other about the common (z) axis. For moments $Q_{xx} = -Q_{yy} = Q_{xz} = 1$ em^{-20}, $Q_{xy} = 0$, and separation 0.5 nm the discrimination energy is 297 J mol^{-1}. Taken, with permission, from D.P. Craig and P.E. Schipper, Proc. Roy. Soc., (1975), A342, 19.

constant and T the absolute temperature. The discriminating term of lowest order is that in $(kT)^{-2}$ and its distance dependence is so strong (R^{-17}) that it can be altogether ignored as a contributor to the discrimination of nearly free molecules or ions [18]. At shorter distances still, where the tumbling motion of the two molecules is replaced by strongly correlated relative motion, larger energies are found. For example, if the dipole-dipole energy is big enough to impose a head-to-tail alignment of the molecules, the discrimination depends solely on the quadrupole-quadrupole term, in the case where each molecule is represented by a skew dipole-quadrupole. The discrimination falls off as R^{-5}. If the quadrupole component magnitudes after projection on the plane normal to the dipole axis are 1 em^{-20} and the separation 0.5 nm the dA-dA interaction at the energy maximum is more stable than the dA-ℓA at its most stable by 300 J mol^{-1}, as shown in Fig. 3. Thus in such locked configurations the electrostatic term is of an order of magnitude that could lead to measurable effects. It is of course also the case that rotational averaging about the common dipole axis wipes out all quadrupole-quadrupole contributions.

IV.3 Electrostatic discrimination in extended systems

The analysis has dealt with interactions between molecules acting on one another through their electrostatic fields simulated by the fields of point multipoles. The molecules are taken to be small compared with their separation distance. It is quite another case (cf Section II) if molecules are close enough to make it necessary to include the structure of their charge distributions. For example, a long polymer may have $-NH_2$ and -OH groups attached regularly along a hydrocarbon chain. These groups have dipole moments, and if the chain imposes a helical structure upon the groups we can model them by a set of helically arranged dipoles transverse to the chain axis as in Fig. 4. One can construct a model crystal by parallel packing of helices of the type of Fig. 3. In a cubic lattice of cell dimension 0.5 nm in which each dipole moment is 0.1 enm, calculations of the discrimination in interactions with a skewed

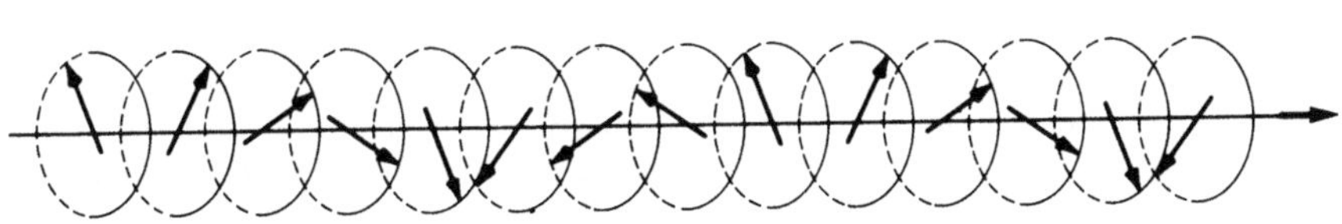

Fig. 4. Transverse dipoles packed chirally on a right-handed helix.

dipole placed at one lattice site (dipole moment 0.1 enm, quadrupole components 1 em^{-20}) give values 1-3 kJ mol^{-1} for various values of helicity [18]. Similarly large values are found in interactions between smaller molecules in which the moments are represented as based at 2 or more different molecular sites: it is however difficult to be sure that the magnitudes chosen as starting data for such calculations are realistic. Again, in such cases, the electrostatic term vanishes in freely rotating systems, while in the crystal the field sources are fixed and only the test 'molecule' is free to rotate.

V. DISPERSIVE DISCRIMINATION

V.1 The electric-magnetic term

The dispersion interaction in expression (I.2), unlike the interactions of permanent electric (and magnetic) multipoles, persists after random rotational averaging. The same is true of the discriminating parts of the dispersion interaction as was first found by Mavroyannis and Stephen [18]. Although small these interactions may play a part in liquids and, conceivably, in scattering and other phenomena in the vapour phase. The results in this section are derived and more fully described elsewhere [19].

Returning to the Hamiltonian (III.1), we need to take account of terms arising from the magnetic dipole transition moments of molecular transitions. It may be shown that the appropriate Hamiltonian for molecules at distances much smaller than the wavelengths of the low molecular electronic transitions is given by (V.1).

$$H = H_a + H_b + H_E + H_M \qquad (V.1)$$

where $H_E = R^{-3}\ \beta_{ij}\mu_i^{(a)}\mu_j^{(b)}$; $H_M = R^{-3}\ \beta_{ij}m_i^{(a)}m_j^{(b)}$

and $\beta_{ij} = (\delta_{ij} - 3\hat{R}_i\hat{R}_j)$

$m_i^{(a)}$ is the i-th component of the magnetic moment operator for molecule A, and summation over repeated indices is assumed. H_E and H_M are written in dipole approximation, without taking account of higher multipole interactions. The higher multipoles also give rise to dispersion interactions, usually of much less importance. The dispersion interaction now gives an energy depression ΔE,

$$\Delta E = -\sum_{n^a,n^b}{}' \frac{\langle 0^a0^b|H_E + H_M|n^an^b\rangle\langle n^an^b|H_E + H_M|0^a0^b\rangle}{E(n^a) + E(n^b)} \tag{V.2}$$

in which $E(n^a)$ and $E(n^b)$ are transition energies and the sum is over the excited states n^a and n^b of the molecules. If only the H_E term of the operator is used we have the usual dispersion interaction given before in (I.2). If only the H_M term is used we have the magnetic analogue, smaller by 6-8 orders of magnitude. In both of these cases, the magnitudes go as R^{-6}, and depend upon four transition moments. If one molecule is replaced by its enantiomer either *two* or none of the moments enter with changed signs and the overall sign is unaltered. These terms are non-discriminating. The electric-magnetic cross-term from (V.2) is given in (V.3):

$$\Delta E_{E-M} = -2\mathrm{Re}\sum_{n^an^b}{}' \frac{\langle 0^a0^b|H_E|n^an^b\rangle\langle n^an^b|H_M|0^a0^b\rangle}{E(n^a) + E(n^b)} \tag{V.3}$$

This is a discriminating term, because each numerator in the sum is the product of two electric moments and two magnetic. If one molecule is replaced by its enantiomer, generated by inversion, *one* electric moment only changes sign so that there is an overall change of sign in Δ_{E-M}. The form (V.4) is found by substitution for H_E and H_M,

$$\Delta E_{E-M} = -2R^{-6}\beta_{ij}\beta_{k\ell}\,\mathrm{Re}\sum_{n^a,n^b}{}' \frac{\langle 0|\mu_i|n^a\rangle\langle n^a|m_k|0\rangle\langle 0|\mu_j|n^b\rangle\langle n^b|m_\ell|0\rangle}{E(n^a) + E(n^b)} \tag{V.4}$$

$$= -2R^{-6}\beta_{ij}\beta_k\,\mathrm{Re}\sum_{n^a,n^b}{}' \frac{(R_{ik})_{0n^a}(R_{j\ell})_{0n^b}}{E(n^a) + E(n^b)} \tag{V.5}$$

where the matrix elements are components of the optical rotatory tensor $\underset{\sim}{R}$

$$(R_{ik})_{0n^a} = \langle 0|\mu_i|n^a\rangle\langle n^a|m_k|0\rangle \tag{V.6}$$

The transformation properties of this tensor depend upon those of the electric and magnetic dipole moments, respectively a polar and an axial vector. Thus $\underset{\sim}{R}$ transforms under rotations as a second rank tensor, but under inversion is antisymmetric.

Its contraction $\sum_i \mu_i m_i = \underset{\sim}{\mu}.\underset{\sim}{m}$ transforms as a psuedo-scalar; that is, it is invariant to rotations but changes sign under inversion. The imaginary part determines the molecular optical rotatory power.

The average value of the discrimination (V.5) over all relative orientations, each assumed equally probable, is

$$\Delta E^{Av}_{E-M} = -\ (4/3)R^{-6}\ \frac{\mathrm{Re}\{R^{\|}(n^a)R^{\|}(n^b)\}}{E(n^a) + E(n^b)} \tag{V.7}$$

where only one level in each molecule is supposed to contribute, and $R^{\|} = \underset{\sim}{\mu}.\underset{\sim}{m}$. The discrimination (V.7) favours unlike (*d*-*ℓ*) interactions over like. The magnitude is about 4 orders of magnitude ($\sim\alpha^2$) less than the electric-electric dispersive attraction, and is thus very small, in the order 1 J mol^{-1}, and varies with distance in the same way ($\sim R^{-6}$) as the dispersion interaction itself. The discrimination is pairwise additive, so that one enantiomer of species A dissolved in an optically pure solvent B might be favoured by up to about 10 J mol^{-1} over the other enantiomer.

With partial locking of orientation, free rotation being confined to one axis, the discrimination is six times that of (V.7). The discriminating term (V.7) has to be doubled to give the discrimination energy, namely the difference in interaction between *d*-*d* and *d*-*ℓ* pairs.

V.2 The R^{-9} pure electric discrimination

Expression (V.2) is the second-order perturbation correction to the ground state energy, under the perturbation by electric and magnetic dipole coupling. The pure electric part in expression (I.2) is non-discriminating, and the mixed electric-magnetic part (V.7) is discriminating. We now consider the third order perturbation correction which, as first pointed out by Schipper [20] can at short range be comparable in magnitude to the term (V.7) although, on account of its R^{-9} dependence on distance, it quickly gets much smaller. The reason is that the third order term has a purely electric part which is discriminating, namely

$$\Delta E^{(3)} = R^{-9}\ \beta_{ij}\beta_{k\ell}\beta_{mn} \sum_{stuv}$$

$$\frac{\langle 00|\mu_i^{(a)}\mu_j^{(b)}|su\rangle\langle su|\mu_k^{(a)}\mu_\ell^{(b)}|tv\rangle\langle tv|\mu_m^{(a)}\mu_n^{(b)}|00\rangle}{(E_s + E_u)(E_t + E_v)} \tag{V.8}$$

where for simplicity the interacting molecules a and b are supposed each to have two active excited states only, labelled as in Fig. 5

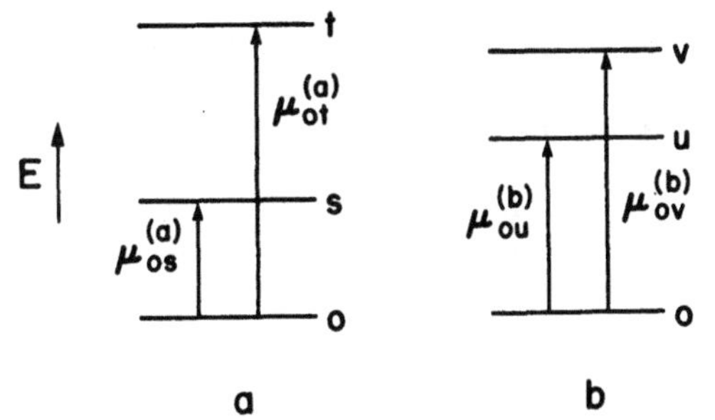

Fig. 5. Energy diagrams of dipole-accessible states of molecules a and b coupled by third-order purely electric dispersion forces.

The sum in (V.8) is taken over the two states s and t in molecule a, and u and v in b. Averaging over all angles gives

$$\Delta E^{(3)} = \frac{2}{3} R^{-9} \frac{(\underset{\sim}{\mu}^{(a)}_{os} \times \underset{\sim}{\mu}^{(a)}_{st} \cdot \underset{\sim}{\mu}^{(a)}_{to})(\underset{\sim}{\mu}^{(b)}_{ou} \times \underset{\sim}{\mu}^{(b)}_{uv} \cdot \underset{\sim}{\mu}^{(b)}_{vo})}{(E_s + E_u)(E_t + E_v)} \qquad (V.9)$$

the discrimination being twice the energy shift (V.9). The term owes its existence to the fact that in a chiral molecule each of the transitions $s \leftarrow 0$, $t \leftarrow s$, $0 \leftarrow t$ is dipole-allowed. The vector products $\underset{\sim}{\mu} \times \underset{\sim}{\mu}$ transform like axial vectors, and the bracketed scalar triple products transform like pseudo-scalars. The symmetry is the same as that of the $\underset{\sim}{\mu}.\underset{\sim}{m}$ term. It is possible to show that for electric moments of 0.1 enm, and magnetic moments of 1 Bohr magneton and for the most favourable directions of these moments, the third order energy $\Delta E^{(3)}$ is of the same order of magnitude as $\Delta E_{E\text{-}M}$ in (V.7) at a separation of 1 nm, though at longer distances it drops off much more quickly. The requirement that the bracketed quantities in the numerator of (V.9) should be non-zero is satisfied if the three transition moments are non-co-planar.

V.3 Dispersion discrimination at short distances: the importance of molecular electronic structure.

When molecules interact at distances long compared with the

molecular dimensions, the treatment of electronic properties in terms of the leading, or leading two, point moments is valid. At short distances, as already seen (Section II) the leading point moment approximation is not satisfactory. The R^{-6} distance law for dispersion energy has the feature, when molecular size is comparable to R, that contributions to the interaction energy by the molecular regions that are close outweigh remote regions. It is then better, in principle, to base calculations on the wave functions, which describe the distribution of the electrons through the volume of the molecules, than to use merely the matrix elements, such as the transition moments, from which the electronic structural detail has been lost. The interaction between *moments* is thus to be replaced by that of *densities*.

A method based on two earlier treatments of dispersion energy at short distances [21,22] has recently been used for the calculation of dispersive discrimination [23]. The method will be illustrated by a simple example of the calculation of the dispersion energy itself, rather than the discrimination, between two identical diatomic molecules, and taking account of only one transition in each. One molecule has available atomic orbitals $a(\nu)$ and $b(\nu)$, and the other $c(\nu)$ and $d(\nu)$ (Fig. 6) and the transitions entering into the calculation of the dispersion interaction are between the molecular orbitals:

$$\left.\begin{aligned} &(2-2S)^{-1}\ (a-b) \leftarrow (2+2S)^{-1}\ (a+b) \\ &(2-2S)^{-1}\ (c-d) \leftarrow (2+2S)^{-1}\ (c+d) \end{aligned}\right\} \qquad (V.10)$$

S is the overlap integral between the atomic orbitals at a and b,

$$S = \langle a|b\rangle = \langle c|d\rangle$$

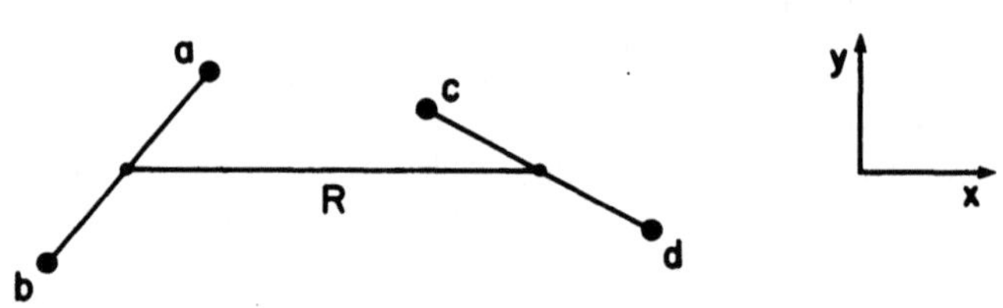

Fig. 6. Coupled diatomic molecules ab and cd with centres separated by R

The expression for the dispersion energy in this simple case, analogous to expression (V.2), is

$$\Delta E = -\frac{<(a+b)(1)(a-b)(1)|H'(1,2)|(c+d)(2)(c-d)(2)>^2}{2\varepsilon}(4-4S^2)^{-4} \quad (V.11)$$

ε being the common excitation energy for the transitions shown in (V.10). Electrons 1 and 2 are the active electrons in the two molecules. The product

$$(4-4S^2)^{-1}(a+b)(1)(a-b)(1) = (4-4S^2)^{-1}\{a(1)a(1) - b(1)b(1)\} \quad (V.12)$$

is the transition charge density in the left-hand molecule of Fig. 6. We may visualize this density as made up of a charge of $1/(4-4S^2)$ electrons distributed in the atomic orbital at a and the same charge at b. The quantity

$$(4-4S^2)^{-2}<a(1)a(1) - b(1)b(1)|H'|c(2)c(2) - d(2)d(2)> \quad (V.13)$$

is the interaction of two such charge densities on the two molecules under the interaction H', which has not yet been specified. H' is the total intermolecular interaction:

$$H' = \frac{1}{r_{ac}} + \frac{1}{r_{ad}} + \frac{1}{r_{bc}} + \frac{1}{r_{bd}} - \frac{1}{r_{a2}} - \frac{1}{r_{b2}} - \frac{1}{r_{c1}} - \frac{1}{r_{d1}} + \frac{1}{r_{12}} \quad (V.14)$$

However, when the integrations over electron coordinates are done contributions by the first four terms vanish, and those from the next four cancel. Only the last term in (V.14) gives a result, so that (V.13) consists of four terms, each of the form,

$$(4-4S^2)^{-2} \quad <a(1)a(1) \; |\frac{1}{r_{12}}| \; c(2)c(2)> \quad (V.15)$$

namely the repulsion of charges of $(4-4S^2)^{-1}$ electrons in atomic orbitals on atoms a and c. To a good approximation these charges can be treated as concentrated at the nuclei instead of distributed in the orbitals, and

$$(4-4S^2)^{-2} <a(1)a(1) \; |\frac{1}{r_{12}}| \; c(2)c(2)> \approx \frac{1}{r_{ac}} (4-4S^2)^{-2} \quad (V.16)$$

This is the repulsion energy of point charges each of magnitude $1/(4-4S^2)$, or 'monopoles', the approximation (V.16) being the monopole approximation. The dispersion energy (V.11) is now given in terms of the monopole interactions

$$\Delta E \approx \frac{1}{(4-4S^2)^4} \cdot \frac{\left(\frac{1}{r_{ac}} - \frac{1}{r_{bc}} - \frac{1}{r_{ad}} + \frac{1}{r_{bd}}\right)^2}{2\varepsilon} \quad (V.17)$$

This can be evaluated for any geometry, the value of S being found from the basis atomic orbitals. Applications to real systems are more complicated, though not more difficult in principle. It is not essential to make the approximation in (V.16), and the work of Coulson and Davies [23] on the dispersion interactions of polyenes at short range was done without it.

The relationship of expression (V.17) to the usual dipole approximation can be readily recovered. Each of the four inverse distances r_{ac}^{-1}, is expanded about R.

We find the result, correct to terms in R^{-3},

$$\Delta E \sim \frac{1}{R^3(4-4S^2)^4} \cdot \frac{4x_a x_c - 8y_a y_c}{2\varepsilon} \qquad (V.18)$$

where x_a is the coordinate of nucleus a in Fig. 5. We next obtain (V.18) by applying the dipole approximation in place of the full expression (V.11). The dipole approximation to the operator (V.14) is

$$\frac{1}{R^3}(x_1x_2 - 2y_1y_2) \qquad (V.19)$$

x_1 being the displacement operator for electron 1. Then, noting the moments for first of the transitions (V.10),

$$(4-4S^2)^{-1} \langle(a+c)(1)|2x_1|(a-c)(1)\rangle = 2x_a/(4-4S^2) \qquad (V.20)$$

and evaluating (V.11) with the help of the operator (V.19), we find an expression for the dispersion energy identical with (V.18). Thus (V.18) is the result when only the *first moment* of the charge distribution (*i.e.* the dipole moment) is allowed for, whereas (V.17) takes account in an approximate way of the detailed character of the charge distribution throughout the molecular space.

Kuroda, Mason, Roger and Seal [23] have used these methods, including the monopole approximation, to calculate the dispersion energy between two D_3 tris-(butadienyl) metal complexes, both for an active pair and a *meso* pair. With the help of molecular wave functions, the accuracy of which is not finally established, but which can be listed, they find the calculated discrimination energy to be greater by nearly five orders of magnitude, even at 10 nm separation, than given in the dipole approximation. Moreover, the values at short distances are found to be large enough to explain observed discriminations in D_3 complexes such as $[Ni(o\text{-}phen)_3]C\ell_2$ and $[Co(acac)_3]$.

VI. THE RESONANCE DISCRIMINATION

We now refer to the resonance interaction already described in Section I. It is not associated with the coupling of molecules in their ground states, but with a pair of identical molecules one of which is in an excited state. The excitation can be exchanged between molecules, and there is a resonance splitting of the excited level. The resonance discrimination is connected with the excitation exchange process. Detailed examination shows that this discrimination is a retarded interaction, depending on the coupling by photons emitted by one molecule and absorbed by the other. One of the emission and absorption events depends on the electric dipole transition moment of the molecule, the other on the magnetic moment. At short distances the energy shift attributable to this coupling is [19].

$$E_{E-M} \cong \varepsilon_{ijk} \hat{R}_k \frac{1}{\lambda\!\!\!^{-} R^2} \{\langle n^a|\mu_i|0\rangle\langle 0|im_j|n^b\rangle + \langle n^b|\mu_i|0\rangle\langle 0|im_j|n^a\rangle\} \qquad \text{(VI.1)}$$

ε_{ijk} is the Levi-Civita symbol, $\varepsilon_{ijk} = 1$ for cyclic order of three different indices, -1 for anticyclic order, and zero for any two indices the same. $\hat{R}_k$ is the k component of a unit vector along $\underset{\sim}{R}$, and $\lambda\!\!\!^{-}$ is the reduced characteristic wavelength of the transition $n^a \leftarrow 0$. We can relate its magnitude to the electric-electric resonance coupling by noting that the magnetic moment has replaced the electric moment in one factor of each term inside curly brackets, and also that the reduced wavelength has replaced one power of R in the denominator. The combined effect is to reduce the electric-electric term by between 4 and 5 orders of magnitude, giving a result that could hardly exceed 1 J mol^{-1} under any conditions so far envisaged for pairwise interactions. In crystals there are more interesting possibilities.

VII. SHORT-RANGE TERMS

The approach to these discriminating terms must be different. Two molecules or ions in close contact will either be locked in a single relative configuration or conceivably may have partial or complete freedom about a single rotational axis. The attractive forces maintaining such contacts will under most conditions be strongly directional, and relative motion is improbable. The second point of difference from long range coupling is the difficulty of defining what is meant by discrimination energy. If we take a d-d pair of like ions or molecules at touching distance, or a d-D pair of unlike, it is not easy to relate their interaction energy with the d-ℓ or d-L pairs at *their* touching distance. The contacts are different, and there are different centre of mass separations. Also the orientations of the diastereomeric pairs are unrelated, or at least not related in any general sense. Thus the concept of discriminating pairwise interactions is not very useful. The useful quantity is the summed interaction expressed in the lattice energy or heat of sublimation in the case of a crystal, and in solution thermodynamic quantities such as heats of solution. So far as pairwise terms are concerned it is of course possible to compare d-d and d-ℓ energies, each pair being in contact in its most stable configuration. The energy difference in this case is not the discrimination in the earlier sense, inasmuch as the centre of mass separations are different, and the relative molecular orientations not related by any symmetry operation.

Such energy differences are measured in kJ mol^{-1}, some thousands of times greater than any available from long range terms between freely or partially rotating molecules. As an indication one can refer to calculations on a model system [24] on the short-range interactions of two chirally substituted methyls joined either to form a *meso* or optically active ethane molecule. The case is that in which the ethane is 2,3-dicyanobutane, Fig. 7.

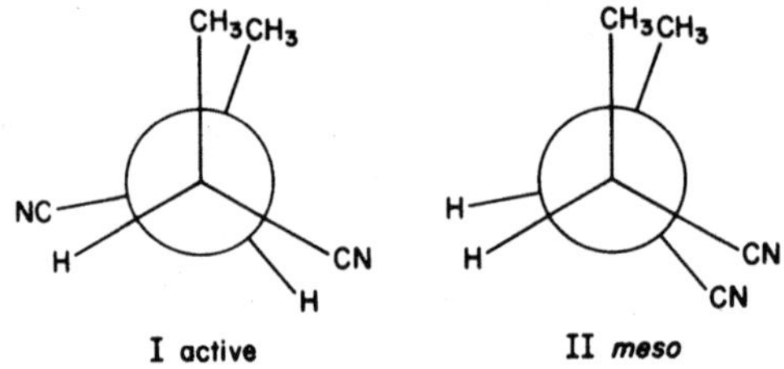

Fig. 7. Projections of active and *meso* forms of 2,3-dicyanobutane

The figure illustrates the two ways of joining the moieties. *Ab initio* calculations were made as a function of dihedral angle (rotation about the C-C bond). There are energy maxima at the eclipsed configurations, and minima between. The more stable form at its minimum is 3.0 kJ mol^{-1} favoured over the less stable. Such energy differences are likely in real situations, and must dominate over discriminations from non-contact interactions, where the two types occur together in solutions. The contacts between such groups approximate to hard surface contacts; one might say that the d-d or d-ℓ pair which is preferred is the one able to fill the available space more efficiently, so increasing the binding energy by shortening inter-group distances.

An interesting specific proposal to account for discrimination in a chiral solvent, arising from short-range packing forces, was made by Bosnich and Watts [12] to account for discrimination between the forms of tris (o-phen)Ni(II) in (-)-2,3-butanediol as described in Section I. In such complex ions the planar o-phenanthrolines are attached to form a three-bladed propeller, the screw sense being opposite in the two chiral antipodes. The proposal is that solvent molecules pack between the o-phenanthrolines, the methyl group being oriented inward toward the metal, with the 2,3-hydroxyl groups pointing outward, capable of hydrogen bonding to other solvent molecules. The disposition of these -OH groups is not the same under the constraints imposed by the two chiral forms of the complex ion, and their hydrogen bonding ability need not be the same, account being taken of the hydrogen bonded structure of the solvent itself. Thus discrimination may be a consequence, reflecting different degrees of solvent-breaking in the two situations. This illustrates that the conformations of interacting molecule pairs may have to be included in packing considerations, as well as the space-filling properties of atoms or small groupings.

A natural extension to the concept of efficient space filling in solution is the statistical treatment of a classical fluid of nonspherical molecules. A first attempt by Sawford (unpublished) is to treat the chiral system as an aggregate of spherical atoms, with force system based on atom-atom potentials. The calculations for two dimensions showed significant differences for like-like and like-unlike mixtures, the excluded volume being smaller, and the energetics more favourable, for the like-like case. Further details are given elsewhere [2]. Such statistical approaches are a promising line on the problem of short-range interactions.

I thank Dr T. Thirunamachandran for his comments on the manuscript.

REFERENCES

1. F.P. Dwyer, N. Gyarfas and M.F. O'Dwyer, Nature, 168, 29, (1951), and later papers.
2. D.P. Craig and D.P. Mellor, Topics in Current Chemistry, 63, 1, (1976).
3. S.F. Mason, Ann. Rep., 73, 53, (1976).
4. K. Mizumachi, J. Coord. Chem., 3, 191, (1973).
5. A. Albert, Selective Toxivity, 5th Ed., Chapman, 1973.
6. F.A. Quoicho, et al, Symp. Quantit. Biol., 36, 561, (1971).
7. J.P. Greenstein and M. Winitz, Chemistry of the Amino Acids, Vol. I, p.645, John Wiley, New York, 1961.
8. F.P. Dwyer and N.R. Davies, Trans. Faraday Soc., 50, 24, (1954).
9. S. Takagi, R. Fujishoro and K. Amaya, Chem. Commun., 480, (1968).
10. M. Leclercq, A. Collet and J. Jacques, Tetrahedron, 9, 1, (1976).
11. B. Bosnich and D.W. Watts, Inorg. Chem., 14, 47, (1975).
12. R.D. Gillard, in press.
13. B. Chion, A. Lajzerowicz, A. Collet and J. Jacques, Acta. Cryst., B32, 339, (1976).
14. R. Silbey, J. Jortner, M. Vala and S.A. Rice, J. Chem. Phys., 42, 2948, (1965).
15. D.P. Craig and T. Thirunamachandran, Proc. Phys. Soc., 84, 781, (1964).
16. E.A. Power and S. Zienau, Phil. Trans. Roy. Soc. (Lond.), A251, 427, (1959).
17. D.P. Craig and P.E. Schipper, Proc. Roy. Soc. (Lond.), A342, 19, (1975).
18. C. Mavroyannis and M.J. Stephen, Mol. Phys., 5, 629, (1962).
19. D.P. Craig, E.A. Power and T. Thirunamachandran, Proc. Roy. Soc., A322, 165 (1971).
20. P.E. Schipper, in course of publication.
21. F. London, J. Phys. Chem., 46, 305, (1942).
22. C.A. Coulson and P.L. Davies, Trans. Faraday Soc., 48, 777, (1952).
23. R. Kuroda, S.F. Mason, C.D. Roger and R.H. Seal, Chem. Physics Letters, 57, 1, (1978).
24. D.P. Craig, L. Radom and P.J. Stiles, Proc. Roy. Soc., A343, 11, (1975).

THE DISPERSION AND OTHER CHIRAL DISCRIMINATIONS

S. F. Mason

Chemistry Department, King's College,
London WC2R 2LS, U.K.

1. INTRODUCTION

Before the time of Pasteur, the problem of optical isomerism, as Thomas Graham put it in 1842, 'defeated every attempt at explanation' [1]. The sodium-ammonium salts of tartaric and racemic acid, Graham continued, 'not only coincide in the proportion of their water and other constituents, and in the composition of their acids, but also in their external form, having been observed by Mitscherlich to be isomorphous' [1]. Aware of Herschel's correlation of the sign of the optical activity with the particular hemihedral morphology in quartz crystals [2], it appeared to Pasteur that Mitscherlich's observation might be incomplete, and by 1848 he established that this was indeed the case. By hand-sorting, the crystals of sodium ammonium racemate were separable into two enantiomorphous types with an equal and opposite specific rotation in solution [3]. From the doctrine, prominent in French stereochemistry from the time of Haüy to that of Le Bel [4] that, 'the molecule and the crystal are images of each other', Pasteur deduced that the individual molecules of (+)- and (-)-tartaric acid have mirror-image morphologies, like the corresponding crystal forms.

In his review lectures on molecular dissymmetry of 1860, Pasteur identified three types of chiral discrimination [5]. Firstly, enantiomeric discrimination, shown by the majority of racemates, which crystallize as such, only a small proportion forming active enantiomorphous crystals separable by hand-picking. Secondly, diastereomeric discrimination, manifest in the different physical and chemical properties of (+)-A(+)-B

Stephen F. Mason (ed.), Optical Activity and Chiral Discrimination. 319–337.

and (-)-A(+)-B, notably, unequal solubilities, which affords a general procedure for the optical resolution of racemates. The third discrimination, reflecting Pasteur's growing preoccupation with microbiology, appeared in the biochemical selectivity of micro-organisms between the enantiomers in a racemic mixture.

In energy terms, data relevant to the chiral discriminations are relatively sparse. The energy-differential between the *active* pair (+)-A(+)-B and the *racemic* or *quasi-racemic* pair (-)-A(+)-B is generally smaller for enantiomeric molecules than their diastereomeric counterparts, and the chiral discrimination energy between a given set of molecule-pairs is normally larger in the solid than in the fluid phase [6,7]. For enantiomeric substances, the *racemic* interactions are generally larger than the corresponding *active* interactions. As yet, only some 250 cases are known of spontaneous optical resolution by the crystallization of a racemate either from solution or from the melt [8]. A calorimetric study of some 50 chiral organic compounds has shown [9] that the free energy change in the formation of racemic crystals from an equimolecular mixture of the crystals of the corresponding (+)- and (-)-enantiomer lies in the range 0 to -8 k J mol^{-1}.

Enantiomeric discriminations in solution are substantially weaker ($\sim$ J mol^{-1}) than in the solid phase, and in some cases exhibit the converse preference. Measurements of the heat of mixing of aqueous (+)- and (-)-tartaric acid solutions give a mixing-enthalpy at 25.6°C of +1.9 J mol^{-1}, showing that the *racemic* d-ℓ contacts are energetically less stable than the *active* d-d or ℓ-ℓ contacts in solution [10]. Measurements of the heats of solution show, however, that the *racemic* d-ℓ contacts are preferred in the solid state, the lattice of the racemic crystal being more stable by 9.52 k J mol^{-1} than that of (+)-tartaric acid [11]. In contrast, the mixing of aqueous solutions of (+)- and (-)-threonine is exothermic, with a mixing-enthalpy of -5.4 J mol^{-1} [10], but again the preferred contacts in solution (d-ℓ) are the converse of those in the solid state (d-d and ℓ-ℓ), the lattice of (-)-threonine being more stable than that of the corresponding racemate by 170 J mol^{-1} [11]. Racemic threonine is a representative of the minor class giving a conglomerate of individual (+)- and (-)-crystals on separating from solution [8].

Diastereomeric discrimination energies in solution are commonly of the same order ($\sim$ k J mol^{-1}) as those differentiating *active* from *racemic* enantiomeric contacts in the solid phase. The Λ-(+)-isomer of the tris-ethylenediamine complex $[Co(en)_3]Cl_3$ is 4.2 times as soluble in (+)-diethyltartrate at 25°C as the Δ-(-)-isomer [12], corresponding to a diastereomeric discrimination of 3.6 k J mol^{-1} in favour of the Λ-(+)-

form. A similar discrimination energy, again in favour of the Λ-(+)-form, is derived from the equilibrium isomer-ratio, (4:1) of the optically-labile tris-oxalate complex, $K_3[Cr(ox)_3]$, in (+)-diethyltartrate solution at 30°C [13]. The diastereomeric discrimination in these cases is dependent upon hydrogen-bonding between the complex ion and the hydroxyl-groups of (+)-diethyltartrate, since the discrimination is drastically reduced in solutions of the isopropylidene derivative of (+)-diethyltartrate, where the hydroxyl groups are blocked. The significance of hydrogen-bonding for diastereomeric discrimination in the solid state is demonstrated by a recent series of X-ray crystal structure analyses of the less-soluble diastereomers formed by racemic ethylenediamine transition-metal complexes with (+)-tartaric acid [14].

Chiral discriminations due primarily to the dispersion forces are expected to be rather weaker than those arising from hydrogen-bonding. The former are universal, however, and, where they are characterised, a comparison of the expectation-value with observation is feasible. A recent study of the diastereomeric discrimination in aqueous solution between two stereochemically-rigid D_3 metal complexes, one neutral and both devoid of accessible hydrogen-bonding groups, provides an order of magnitude measure of the probable dispersion energy-differential between an *active* and a quasi-*racemic* molecule-pair. A solution (0.12M) of Δ-(-)-$[Ni(phen)_3]Cl_2$, (phen = 1,10-phenanthroline) preferentially extracts the Δ-(-)-isomer of $[Cr(pd)_3]$, where pd = 2,4-pentandionate(acetylacetonate), from a carbon tetrachloride solution of the corresponding racemate [15]. The enantiomeric excess of the latter complex in the aqueous solution is 1.5%, corresponding to a discriminating free energy of 75 J mol^{-1} at ambient temperature in favour of the *active* pair Δ-A Δ-B.

2. THE DISPERSION CHIRAL DISCRIMINATION ENERGIES

Theoretical treatments of chiral discrimination are based generally upon a point-multipole expansion of the charge distribution in each molecule of an interacting pair in order to evaluate the intermolecular potential. In a study of the dispersion energy between two randomly-orientated molecules, Mavroyannis and Stephen [16] showed, by taking into account the magnetic as well as the electrostatic interaction, that the energy is invariably larger for the *racemic* pair (-)-A(+)-B than for the *active* pair (+)-A(+)-B. For a separation R_{AB} between the two molecules, the chiral discrimination in the dispersion energy ($\Delta E_D = act\text{-}E_D - rac\text{-}E_D$) follows from the relation [16],

$$\Delta E_D = (8/3) \sum_{n(A)} \sum_{m(B)} R^A_{on} R^B_{om} [R^6_{AB} (E^A_{on} + E^B_{om})]^{-1} \quad (1)$$

where R^A_{on} refers to the rotational strength of the electronic transition with the energy E^A_{on} of molecule A, and R^B_{om} and E^B_{om} have the corresponding significance for molecule B.

More recently Craig and coworkers [6,17,18] have extended theoretical studies of chiral discrimination to the energies arising from the permanent electric and magnetic moments, as well as the corresponding moments mutually induced in enantiomeric molecules, for fixed and semi-locked as well as random mutual orientations, including the retardation effects and resonance discrimination in the photoexcited state. The dispersion energy treatment has been carried to the third order of perturbation theory by Schipper [19] who demonstrates a discrimination for randomly orientated enantiomers, proportional to $(R_{AB})^{-9}$, arising from the signed triple scalar product of the electric dipole transition moments of each of the molecules A and B, e.g.,

$$T^A_{o,s,t} = [\vec{\mu}_{ot} \times \vec{\mu}_{ts} \cdot \vec{\mu}_{so}]_A \quad (2)$$

The experimentally-determined rotational strengths of enantiomers, and the corresponding transition energies, when substituted into equation (1) afford chiral discrimination energies which are generally too small and are not always correct in sign [20,21]. The problem is common to the recent developments of chiral discrimination theory based upon the point-multipole expansion of the intermolecular potential. The areas of the CD bands of Δ-(-)-$[Ni(phen)_3]^{2+}$ [22] and of Δ-(-)-$[Cr(pd)_3]$ [23] give the rotational strengths of these enantiomers over the range 185 - 1000 nm (Figure 1). The substitution of these rotational strengths and the measured transition energies into equation (1) gives a discrimination for the randomly orientated molecules of $0.5(R_{AB})^{-6}$ J mol^{-1}(Å)$^{-6}$ in favour of the quasi-racemic Δ-Λ pair, compared with the 75 J mol^{-1} contrary discrimination observed [12] at a concentration where the mean separation of the two species is some 24 Å. The dependency on $(R_{AB})^{-6}$ implies that the chiral discrimination derives mainly from the molecule pairs with a separation smaller than the mean value, but if it is assumed that the two molecules are locked at the distance of closest approach ($\sim$ 10 Å), where the two C_3 axes are collinear, the electromagnetic chiral discrimination amounts only to 3×10^{-6} J mol^{-1}.

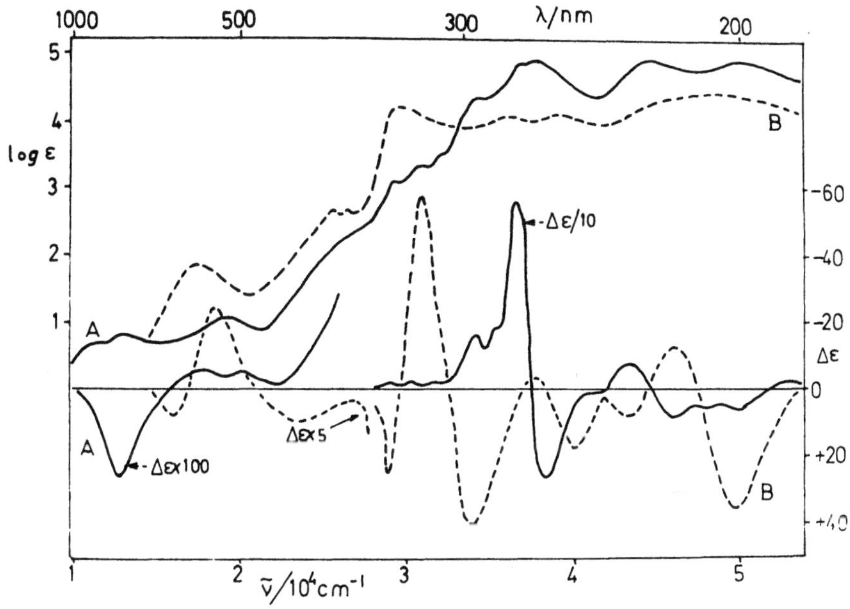

Figure 1. Absorption spectra (upper curves) and CD spectra (lower curves) of (A), Δ-(-)-$[Ni(phen)_3]^{2+}$, and (B), Δ-(-)-$[Cr(pd)_3]$.

The electromagnetic chiral discrimination of equation (1) and its analogues for fixed or semi-locked molecular configurations [17] are limited principally by the sum rule for the rotational strengths of each of the chiral species, $\sum_n R_{on} = 0$. Each positive rotational strength has a negative counterpart of generally comparable magnitude, often close on the frequency scale in the CD spectrum (Figure 1). Each positive term in the sum of equation (1) is accompanied by a comparable negative term, whether the *active*-pair or the *racemic*-pair be considered, and while exact mutual cancellation is precluded by the inequality of the energy-denominators for corresponding positive and negative terms, the overall sum is nonetheless small. An analogous zero-sum rule governs the triple scalar product of electric dipole transition moments for a given chiral species employed for the third-order chiral discrimination in the dispersion energy [19]. The adoption of the closure procedure over the complete set of **state** functions, following the proof of Condon [24] for the sum rule governing the rotational strengths, shows that the terms of equation (2) sum to zero over all states,

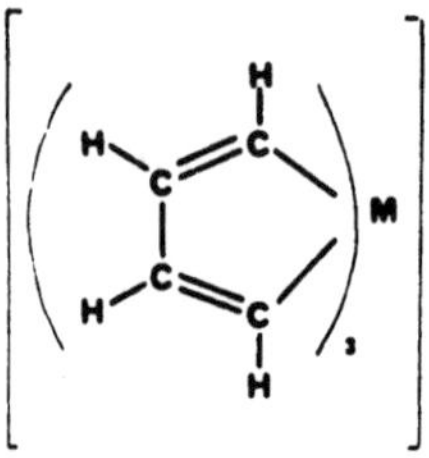

1

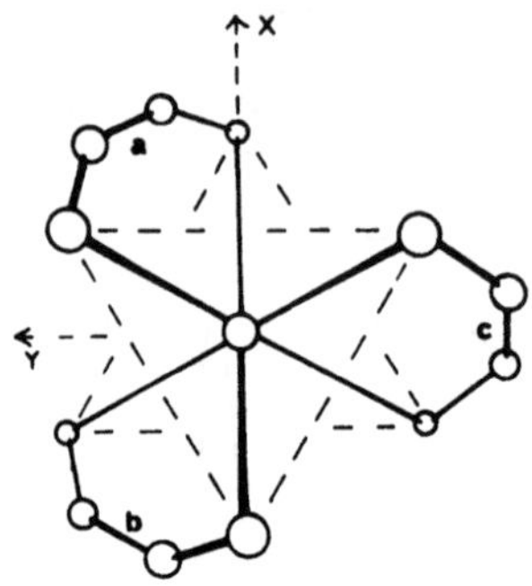

2

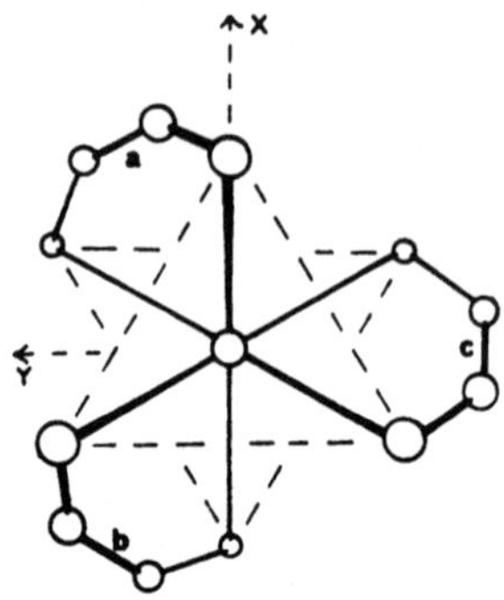

3

$$\sum_s \sum_t [\vec{\mu}_{ot} \times \vec{\mu}_{ts} \cdot \vec{\mu}_{so}] = 0 \qquad (3)$$

The third-order chiral discrimination does not vanish, again on account of the inequality of the energy-denominators, but the resultant discrimination energy is no larger than that deriving from the rotational strengths (equation 1) even at small intermolecular separations.

It has long been recognised, e.g. Boltzmann (1874) [25], that optical activity is an effect in which the finite extension of the chiral molecule cannot be neglected in relation to the generally much larger wavelength of light. Analogy suggests that the finite extension of dissymmetric molecules in relation to their mean separation cannot be ignored in the energy terms of the chiral discrimination. The latter principle has been investigated by calculating the dispersion energy between two tris(butadienyl) chelate model complexes (1), of the same or of opposite chiralities in which each transitional charge distribution has an extension dictated by the stereochemistry of the D_3 molecular system [21]. For molecules belonging to the dihedral groups D_p, the intermolecular distance R_{AB} becomes well-defined by the separation between the intersection point of the p-fold and the orthogonal two-fold rotational axes in the individual molecules.

3. THE EXTENDED TRANSITION MONOPOLE MODEL

Each of the model compounds considered (1) consists of three mutually perpendicular butadienyl chelate rings bonded to a central metal atom with octahedral coordination. The M-C and the C-C bonds have lengths of 2.00 and 1.40 Å, respectively, and the $\angle$CMC and $\angle$CCC bond angles have the respective angles of 90° and 120°. The C-H bonds have a length of 1.05 Å and the van der Waals radii are taken as 1.20 and 1.85 Å for the hydrogen atom and the carbon atom, respectively. A right-handed molecule (2) with the Δ-configuration is related to the corresponding left-handed molecule (3) with the Λ-configuration by a reflection of the atoms through the XY molecular plane, retaining the right-handed Cartesian frame.

The calculation of the dispersion energy between a *racemic*-pair, (2) and (3), or an *active*-pair, both of type (2), followed the method of Coulson and Davies [26] based upon the transition monopole treatment of London [27]. The electronic transitions contributing to the dispersion energy are confined to the $\pi \rightarrow \pi^*$ manifold of the butadienyl chelate rings. Taking into account only the Coulombic contribution to the intermolecular potential, V_E, the dispersion energy between a pair of D_3 systems (1) is given by,

$$E_D = -2 \sum_{n(A)} \sum_{m(B)} [(A_o B_o | V_E | A_n B_m)]^2 [E_{on}^A + E_{om}^B]^{-1} \tag{4}$$

where the molecular ground state, A_o or B_o, is a simple product of the three individual ligand functions, and each excited state is a symmetry-adapted linear combination of simple products of ligand excited π-configurations. The Coulomb potential V_E operates between the transition charge density q_μ^{on} on atom μ of the molecule A in the transition $A_o \rightarrow A_n$ and that of atom ν in the molecule B, q_ν^{om}, in the transition $B_o \rightarrow B_m$,

$$V_E = \sum_{\mu(A)} \sum_{\nu(B)} q_\mu^{on} q_\nu^{om} (1/R_{\mu\nu}) \tag{5}$$

For a given separation R between the central metal atoms of two D_3 systems (1), there is a discrimination between the *racemic*-pair, (2) and (3), and an *active*-pair, both of the type (2), on account of the difference between corresponding $R_{\mu\nu}$ values in the two cases.

The resonance term of Förster [28] and Dexter [29] for energy-transfer in the lowest photoexcited state was computed for two D_3 systems (1) by a similar procedure. The resonance quantity, E_R^2, representing the square of the Coulomb energy between the two transitional charge distributions, has the AO-expanded form,

$$E_R^2 = \left\{ \sum_{\mu(A)} \sum_{\nu(B)} [2e^2 (C_{2\mu} C_{3\mu})_A (1/R_{\mu\nu}) (C_{2\nu} C_{3\nu})_B] \right\}^2 \qquad (6)$$

where the AO coefficients $C_{2\mu}$ and $C_{3\mu}$ of atom μ in the molecule A, and $C_{2\nu}$ and $C_{3\nu}$ of the atom ν in the molecule B, refer in each case to the E combination in D_3 of the three $\pi_2 \text{-}^1\pi_3^*$ butadiene configurations for the energy-transfer process, $A(E) \rightarrow B(E)$, or to the corresponding A_2 combinations for the transfer process, $A(A_2) \rightarrow B(A_2)$. Theoretically the E excited state of (1) deriving from the lowest butadiene π-excitation ($\pi_2 \rightarrow \pi_3^*$) lies at a lower wavenumber, by 1280 cm^{-1}, than the corresponding A_2 state in the point-dipole exciton approximation [30].

The computations of the dispersion energy E_D from equation (4) and the resonance terms E_R^2 from equation (6) for both the *racemic*-pair and the *active*-pair refer to three primary initial mutual orientations, each affording a semi-locked average E_D and discrimination ΔE_D. The two D_3 molecules (1) have collinear $C_3(Z)$ axes, R = (001), in the first orientation, and collinear $C_2(Y)$ axes, R = (010), in the second, while the two molecules are separated along the body diagonal of a cube bounded by the molecular frames, R = (111), in the third. For each initial orientation, E_D and E_R^2 were calculated with a fixed separation R between the central metal atoms of the two molecules for rotational increments of 15° of one or both of the molecules about the individual $C_3(Z)$ axes over the range 0 to $2\pi/3$ and the individual $C_2(Y)$ axes over the 0 to π range.

Values of E_D and E_R^2 for a set of 400 spatially homogeneous or equivalent mutual orientations were averaged to approximate or simulate the corresponding values for random mutual orientations of the two molecules. The vectors from the metal atom of molecule (1) to the centroid of each triangular face of a circumscribed regular icosahedron provide a set of 20 spatially homogeneous directions. The rotation of each of the two molecules to bring a given pair of vectors into collinearity affords, when sequentially repeated over the range of all possible pairwise combinations of the two vector sets, the 400 values of E_D, E_R^2, and their *racemic-active* differentials, averaged as the icosahedrally-based mean values listed (Tables 1 and 2). The averages over rotations for semi-locked molecule-pair configurations are recorded for the dispersion energy, E_D,

Table 1. The dispersion energy E_D (J mol^{-1}) of two D_3 molecules (1) for the *active*-pair (ΛΛ) and the differential in that energy between the *active* and the *racemic* molecule-pairs, $\Delta E_D = E_D(\Lambda\Lambda) - E_D(\Lambda\Delta)$. Averages over rotations for semi-locked molecule-pair configurations are recorded, together with the icosahedrally-based average, and the power n of the $\Delta E_D/R^n$ relation.

Orientation	R = (001)		R = (010)		R = (111)		I_h
Rotation axis:	Z	Y	Z	Y	Z	Y	av.
R = 10A°							
$-E_D$	214	228	286	217	218	217	233
$-\Delta E_D/10^{-1}$	+58	+57	+185	+9.5	+25	-0.21	+21
R = 20A°							
$-E_D/10^{-2}$	322	281	258	244	260	262	267
$-\Delta E_D/10^{-4}$	+195	+151	+580	+43	+34	+3.5	+57
R = 100A°							
$-E_D/10^{-6}$	202	167	143	140	154	156	157
$-\Delta E_D/10^{-9}$	+46	+35	+139	+12	+5.3	+1.3	+14
n(20-50A°)	8.05	8.06	8.04	7.97	8.62	7.55	8.07
n(100-250A°)	8.00	8.00	8.00	8.00	8.00	8.00	8.00

and the chiral discrimination energy, ΔE_D (Table 1), and the resonance term, $E_R{}^2$, and its chiral differential, $\Delta E_R{}^2$ (Table 2), together with the power n of the R^{-n} dependency of the discrimination.

4. CONTACT DISCRIMINATION

When the two molecules (1) are in van der Waals contact the chiral discrimination in the dispersion energy is found to lie in the range ± 1 k J mol^{-1} with an *active* or a *racemic* preference dependent upon the particular mutual orientation of the two molecules (Figure 2). For two molecules with collinear C_3(Z) axes the minimum separation is the smaller for the *active*-pair, but with collinear C_2(Y) axes a minimum

Table 2. The resonance term E_R^2 (cm^{-2}) for energy-transfer between two D_3 molecules (1) in the lowest E excited state for an *active* molecule-pair (ΛΛ) and the differential in the term between *active* and *racemic* molecule-pairs, $\Delta E_R^2 = E_R^2(\Lambda\Delta)$. Averages over rotations for semi-locked molecule-pair configurations are listed, together with the homogeneous icosahedrally-based average, and the power *n* of the $\Delta E_R^2/R^n$ relation.

Orientation:	R = (001)		R = (010)		R = (111)		I_h
Rotation axis:	Z	Y	Z	Y	Z	Y	av.
R = 10A°							
$E_R^2/10^2$	236	243	278	225	240	218	240
$\Delta E_R^2/10^2$	-40	-30	-98	-6.5	-12	+0.68	-10
R = 20A°							
E_R^2	231	256	479	419	238	266	294
ΔE_R^2	-16	-11	-44	-3.6	-2.2	-0.26	-4.6
R = 100A°							
$E_R^2/10^{-4}$	128	144	318	281	128	160	174
$\Delta E_R^2/10^{-6}$	-39	-29	-118	-9.8	-4.4	-1.1	-12
n(20-50A°)	8.04	7.99	7.96	7.94	8.32	7.65	7.96
n(100-250A°)	8.00	8.00	8.00	8.00	8.00	8.00	8.00

separation, retaining parallel or antiparallel C_3 axes, is found for the *racemic*-pair, the corresponding minimum for the *active*-pair requiring the C_3 axes to be mutually orientated at the tetrahedral angle (Figure 2). These steric features are evident in the packing modes of *active* and *racemic* D_3 molecules observed in crystal structures.

The D_3 homochiral diastereomer, K{Λ-(+)-[Ni(phen)$_3$]Λ-(-)-[Co(ox)$_3$]}.2H O, crystallizes in the space group $P2_1 3$ with Z = 4 [31]. Each of the four formula units in a unit cell is a member of a homochiral column of alternating complex cations and anions with parallel C_3 axes along the column direction.

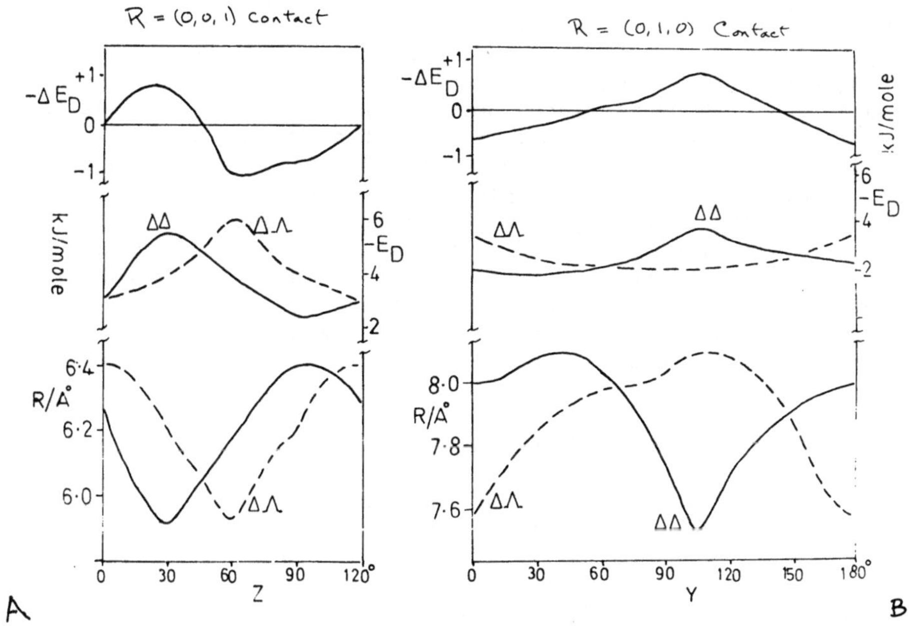

Figure 2. The relations between the distance R/Å between the metal atoms of two D_3 molecules (1) (bottom curves), the dispersion energy, E_D/K J mol^{-1} (middle curves), and the dispersion energy differential, $\Delta E_D = E_D(\Delta\Delta) - E_D(\Delta\Lambda)$ (top curves) and (A), the angle of rotation about the Z-axis for the collinear C_3(Z) axis mutual molecular orientation and (B), the angle of rotation about the Y-axis for the collinear C_2(Y) axis mutual molecular orientation, at van der Waals contact between the two molecules for the *active* pair (full curves) and the *racemic* pair (broken curves).

The four columns are made up of two parallel pairs, and the column direction of one pair lies at the tetrahedral angle to that of the other pair.

In the crystal lattice of *rac*-[Cr(pd)$_3$] and isomorphous [M(pd)$_3$] complexes crystallizing in the space group $P2_1/c$ with Z = 4, the D_3 complexes pack in parallel pairs of homochiral columns directed along the molecular C_3 axes, all molecules of one column having the Δ-configuration while those of the parallel adjacent column have the Λ-configuration [32]. The C_3 axes of the four molecules in the unit cell of the $P2_1$ crystal of Δ-(−)-[Cr(pd)$_3$], in contrast, are so orientated that only one pair are parallel, or nearly so, while each of the other three pairs mutually lie close to the tetrahedral angle [33]. The steric requirements for the efficient packing of D_3 structures along the direction of the molecular C_3 axes, and in the orthogonal plane, notably along the direction of the molecular C_2 axes, are competitive in the crystal of Δ-(-)-[Cr(pd)$_3$], whereas in the *rac*-[Cr(pd)$_3$] crystal structure these requirements have a more complementary character. The *racemic* crystal is the more compact with a unit cell volume of 1714 Å^3 compared with the corresponding volume of 1806 Å^3 for the *active* crystal. The latter is the more soluble in a range of common solvents. At ambient temperature Δ-(-)-[Cr(pd)$_3$] crystals are 9 times more soluble in *n*-hexane than the corresponding racemate, representing a free energy discrimination of 7 k J mol^{-1} in favour of the racemic lattice.

5. NON-CONTACT DISCRIMINATION

At separations larger than the contact distance for two molecules (1) the number of pair-configurations which give rise to a dispersion energy differential in favour of the *active*-pair is reduced, and the *racemic* preference becomes general at separations ⩾ 15 Å for all of the semi-locked rotational averages (Table 1). At a separation of 10 Å, no orientation of the two molecules (1) with collinear C_3(Z) axes affords an *active* preference (Figure 3). The contour diagram for the (111) semi-locked configuration shows a prominent peak of *active* preference for a C_2(Y) rotation of one molecule by 60° and the other by 150° at a 10 Å intermolecular separation (Figure 4) and, with attenuation, at larger separations. The peak dominates in the average over the C_2(Y) rotations in the (111) mutual molecular orientation, which remains the sole semi-locked configuration with a mean *active* preference at 10 Å (Table 1).

At separations ⩾ 15 Å the general *racemic* preference exhibits a R^{-n} proportionality which converges to the power $n = 8$ in the far zone, ⩾ 100 Å, from higher or lower values in the near zone, 20 - 50 Å (Table 1). A similar pattern is observed for the resonance term governing energy-transfer between two D_3 molecules (1) in the lowest photoexcited state of E symmetry, $E_R{}^2$(E) (Table 2). The differential in the resonance term, $\Delta E_R{}^2$, similarly shows a *racemic* preference at separations ⩾ 15 Å, but the ratio of the differential, $\Delta E_R{}^2$, to the resonance term, $E_R{}^2$, is generally larger, approximately by an order of magnitude, than the corresponding dispersion-energy ratio at a given separation, either for a particular semi-locked configuration or for the icosahedrally-based average (Tables 1 and 2).

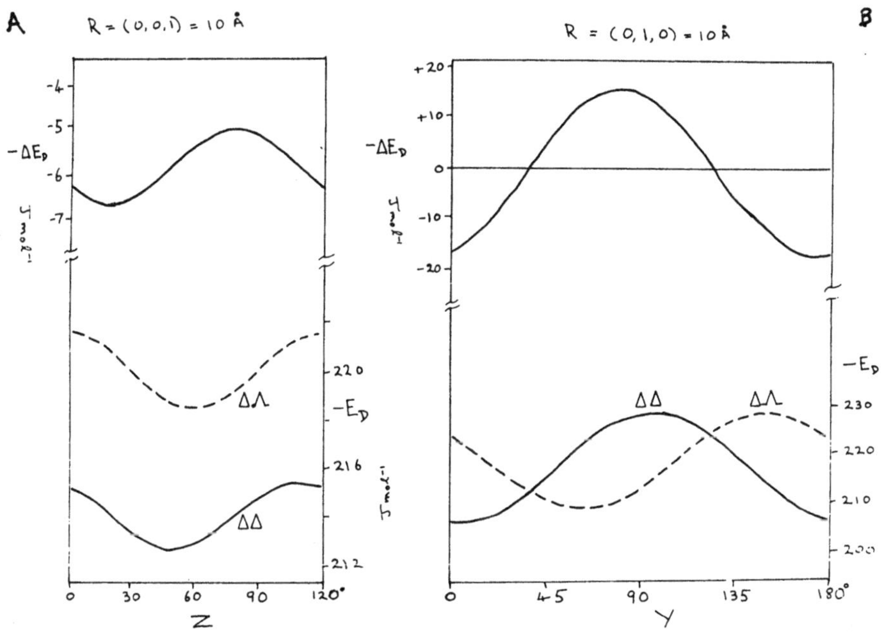

Figure 3. The relations between the dispersion energy E_D/J mol^{-1} (lower curves) for the *active* pair (ΔΔ) (full curves) and the *racemic* pair (ΔΛ) (broken curves) of molecules (1) at a distance of 10Å between the metal atoms, and the energy differential, $E_D = E_D(\Delta\Delta) - E_D(\Delta\Lambda)$, (upper curves) and the angle of rotation about the line of centres for (A), collinear C_3(Z) axes, and (B), collinear C_2(Y) axes.

In terms of a multipole expansion of the transitional charge distributions of two D_3 molecules (1), the R^{-8} long-range behaviour of the chiral discrimination in the dispersion energy and in the resonance term suggests a leading contribution to the differentiations from interactions between the dipole and quadrupole moments of molecule A with the corresponding moments of molecule B [34]. However, a transition of the symmetry type, $A_1 \rightarrow A_2$ in a D_3 molecule is necessarily devoid of a quadrupole moment yet the resonance term for energy-transfer between two molecules (1) through the lowest A_2 photoexcited has a chiral differential $\Delta E_R^2(A_2)$ which for the icosahedrally-based average is proportional to R^{-8}. Although each of the two components of the lowest energy $A_1 \rightarrow E$ transition in the molecule (1) has an electric dipole and quadrupole moment, and these two moments are phase locked, e.g.

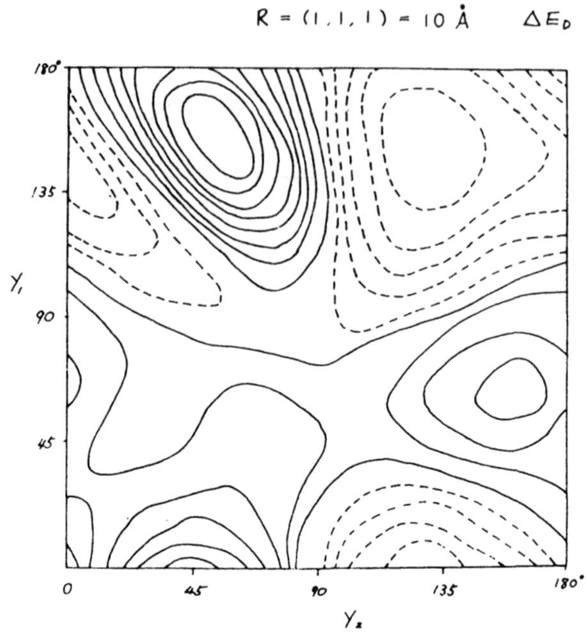

Figure 4. The relations between the dispersion energy differential, $\Delta E_D = E_D(\Delta\Delta) - E_D(\Delta\Lambda)$, for two D_3 molecules (1) and the rotation of each molecule about the $C_2(Y)$ axis for the (1,1,1) mutual molecular orientation at a separation of 10Å between the two metal atoms. The full contours represent an *active* preference and the broken contours a *racemic* preference with a contour spacing of 3.2 over the range +18 to -25 J mol^{-1}.

for the E_Y component, into the positive combination $+\mu_y\theta_{xz}$ for the Δ-configuration (2) and into the corresponding negative combination $-\mu_y\theta_{xz}$ for the Λ-configuration (3), these combinations afford a net zero chiral discrimination on rotational averaging, e.g. over the $C_2(Y)$ rotations for the semi-locked collinear $C_2(Y)$ mutual molecular orientation.

In the extended transition monopole model, chiral discriminations in the dispersion energy, or in the resonance term for energy-transfer, originate from the differing interatomic distance $R_{\mu\nu}$ (equation 5) for the *active* molecule-pair and the corresponding *racemic* pair at a given intermolecular separation R and particular mutual orientation. For two dihedral molecules, each with space-fixed rotational axes, the difference between the interatomic distances $R_{\mu\nu}$ may be

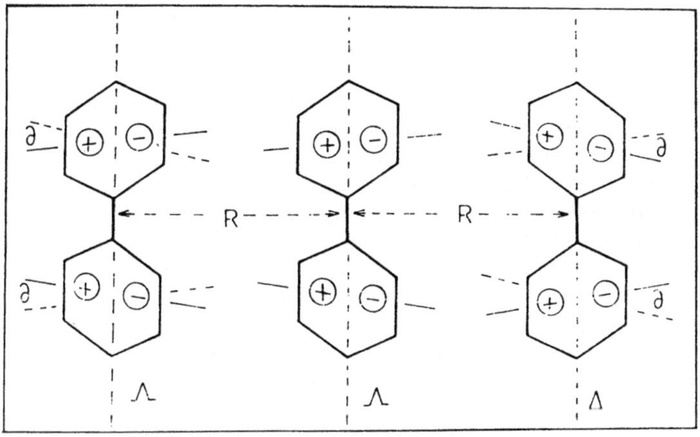

Figure 5. The intramolecular torsional displacement δ of the *ortho* and the *meta* CH groups in chiral biphenyl, with spaced-fixed D_2 rotational axes, required to interconvert the Λ and the Δ stereochemical configuration, affording corresponding mutual molecular orientations in the *active* pair (ΛΛ) and in the *racemic* pair (ΛΔ) for a given separation R.

expressed as a function of the distance R separating the intersection points of the molecular axes and an intramolecular atomic displacement, δ, or set of such displacements, required to convert one of the enantiomers into the antipodal configuration, i.e. to convert an *active* pair into a *racemic* pair, or *vice versa*. In the case of D_2 biphenyl, δ is the torsional displacement of the *ortho* and *meta* C-H groups required to transform the Λ-isomer into its Δ-antipode (Figure 5). The ratio of the chiral discrimination in a quantity dependent upon the square of the Coulomb potential between individual atoms of the two molecules to the quantity in question, notably, the resonance term (equation 6) and the numerators of the dispersion-energy expression (equation 4), gives a power-series relation in (δ/R) on expansion of the squared-potential, e.g. for E_R^2,

$$\Delta E_R^2/E_R^2 = C_1(\delta/R)^2 + C_2(\delta/R)^4 + \cdots \qquad (7)$$

where the signed numerical coefficients are dependent upon the particular transitional charge distribution and the molecular stereochemistry. From the leading term of equation (7), the

R^{-8} proportionality of the chiral discrimination (Tables 1 and 2) follows from the long-range R^{-6} dependency of the resonance term, E_R^2 (equation 6) and its analogues in the numerator of the dispersion-energy relation (equation 4).

The chiral discrimination in the dispersion energy at non-contact distances in the near zone is of the correct order ($\sim$ J mol^{-1}) for the less distant separations (Table 1). Chiral discrimination in the intermolecular transfer of photoexcitation between molecules with dihedral symmetry has not been observed as yet. For D_3 systems with a lowest excited state of E symmetry the discrimination is expected to be readily detectable (Table 2). Recently it has been shown that the Stern-Volmer constant for the quenching of the fluorescence of R-(-)-1,1'-binaphthyl in n-hexane solution is 1.90 times larger for S-(-)-N,N-dimethyl-α-phenylethylamine than for the R-(+)-enantiomer [35].

6. THE CHIRAL DISCRIMINATION OF THE WEAK NUCLEAR INTERACTION

The discovery that parity is not conserved in the weak interactions connected with nuclear β-decay stimulated a search for the chiral discrimination implied by the preferred anti-parallel linear and angular momentum of the electrons in high-energy β-rays [36-40]. Racemates and achiral substances have been irradiated with β- or γ-rays from radioisotopes [37,38], or with longitudinally-polarized electrons from particle accelerators [39,40], with the expectation of an enantiomeric selectivity either in the radiolysis or in the photolysis due to the concomitant circularly-polarized bremsstrahlung. In some cases significant selectivities are reported [38,39].

Following proposals that the singular chirality of the majority of natural products has arisen from the cumulative discrimination of the weak nuclear interaction over the course of time [36,41-43], recent attention has been directed to the energy-difference between an enantiomer and its antipode expected from the parity non-conserving interaction of the nuclei with the core and valency electrons [44,45]. The weak interaction couples the momentum and the spin of electrons with a non-zero charge density at the nucleus, in the s-orbitals, and mixes them into levels of opposite parity, notably, the p-orbitals. The parity non-conserving potential V_{PNC} is proportional to the spin of the electron and, for closed-shell systems, it is necessary to take the spin-orbit coupling potential V_{SO} into account [44,45]. The PNC energy-shift of a singlet state $\langle S|$ has the form,

$$\Delta E_{PNC} = \sum_T \langle S|V_{PNC}|T\rangle\langle T|V_{SO}|S\rangle [E_S - E_T]^{-1}$$

where the $|T\rangle$ refer to corresponding triplet states. The energy-shift ΔE_{PNC} is a pseudoscalar quantity, increasing the energy of one enantiomer and reducing that of the other. The shift is strongly-dependent upon the nuclear charge, being proportional to Z^5 [44,45]. Calculations of ΔE_{PNC} based upon the available wavefunctions of chiral dialkyl sulphides [46] and twisted ethylene [47] give values of the order of 10^{-23} and 10^{-18} eV [48], respectively. Although these values are below the present limits of detection, they support the proposed origin of the molecular chirality of natural products in the cumulative discrimination of the parity non-conserving weak interaction.

REFERENCES

[1] T. Graham, Elements of Chemistry, H. Bailliere, London, 157 (1842).

[2] J. Herschel, Trans. Camb. Phil. Soc., 1, 43 (1822).

[3] L. Pasteur, Compt. Rend., 28, 477 (1849).

[4] S.F. Mason, Topics Stereochem., 9, 1 (1976).

[5] L. Pasteur, Alembic Club Reprint, No.14, Edinburgh (1948).

[6] D.P. Craig and D.P. Mellor, Topics Current Chem., 63, 1 (1976).

[7] S.F. Mason, Ann. Rep. Chem. Soc., A73, 53 (1976).

[8] A. Collet, M.J. Brienne and J. Jacques, Bull. Soc. Chim. France, 127 (1972); 494 (1977).

[9] M. Leclercq, A. Collet and J. Jacques, Tetrahedron, 32, 821 (1976).

[10] S. Takagi, R. Fujishiro, and K. Amaya, J.C.S. Chem. Comm., 480 (1968).

[11] M. Matsumoto and K. Amaya, Chem. Letters, 87 (1978).

[12] M. Yamamoto and Y. Yamamoto, Inorg. Nuclear Chem. Letters, 11, 833 (1975).

[13] A.F. Drake, J. Levey and S.F. Mason, work in progress.

[14] Y. Kushi, M. Kuramoto and H. Yoneda, Chem. Letters, 135, 339, 663 and 1133 (1976).

[15] E. Iwamoto, M. Yamamoto and Y. Yamamoto, Inorg. Nuclear Chem. Letters, 13, 399 (1977).

[16] C. Mavroyannis and M.J. Stephen, Molec. Phys., 5, 629 (1962).

[17] D.P. Craig, E.A. Power and T. Thirunamachandran, Proc. Roy. Soc., A322, 165 (1971).

[18] D.P. Craig and P.E. Schipper, Proc. Roy. Soc., A342, 19 (1975).

[19] P.E. Schipper, Chem. Phys., 26, 29 (1977).

[20] S.F. Mason, Proc. Roy. Soc., A297, 3 (1967).

[21] R. Kuroda, S.F. Mason, C.D. Rodger and R.H. Seal, Chem. Phys. Letters, 57, 1 (1978); Molec. Phys., (1979) in press.

[22] S.F. Mason and B.P. Peart, J. Chem. Soc. Dalton, 949 (1973).

[23] S.F. Mason, R.D. Peacock and T. Prosperi, J. Chem. Soc. Dalton, 702 (1977).

[24] E.U. Condon, Rev. Mod. Phys., 9, 432 (1937).

[25] L. Boltzmann, Pogg. Ann. Phys. Chem., Jubelband, 134 (1874).

[26] C.A. Coulson and P.L. Davies, Trans. Faraday Soc., 48, 777 (1952).

[27] F. London, J. Phys. Chem., 46, 305 (1942).

[28] T. Förster, Modern Quantum Chemistry, ed. O. Sinanoglŭ, Academic Press, N.Y. and London, (III), 93 (1965).

[29] D.L. Dexter, J. Chem. Phys., 21, 836 (1953).

[30] S.F. Mason, Inorg. Chim. Acta Rev., 2, 89 (1968).

[31] K.R. Butler and M.R. Snow, J. Chem. Soc.(A), 565 (1971).

[32] B. Morosin, Acta Cryst., 19, 131 (1965).

[33] R. Kuroda and S.F. Mason, J. Chem. Soc. Dalton, in press (1979).

[34] A.D. Buckingham, Adv. Chem. Phys., 12, 107 (1967).

[35] M. Irie, T. Yorozu and K. Hayashi, J. Amer. Chem. Soc., 100, 2236 (1978).

[36] T.L.V. Ulbricht, Quart. Rev. Chem. Soc., 13, 48 (1959).

[37] T.L.V. Ulbricht and F. Vester, Tetrahedron, 18, 629 (1962).

[38] K.L. Kovacs and A.S. Garay, Nature, 254, 538 (1975).

[39] W.A. Bonner, M.A. Van Dort and M.R. Yearian, Nature, 258, 419 (1975).

[40] M.M. Ulrich and D.C. Walker, Nature, 258, 418 (1975).

[41] Y. Yamagata, J. Theor. Biol., 11, 495 (1966).

[42] D.W. Rein, J. Mol. Evolution, 4, 15 (1974).

[43] V.S. Letokhov, Phys. Letters, 53A, 275 (1975).

[44] B.Y. Zel'dovich, D.B. Saakyan and I.I. Sobel'man, JETP Letters, 25, 94 (1977).

[45] D.W. Rein and P.G.H. Sandars, 'Handed Interactions in Handed Molecules', internal report, Clarendon Laboratory, Oxford, (1978).

[46] J.S. Rosenfield and A. Moscowitz, J. Amer. Chem. Soc., 94, 4797 (1972).

[47] T.D. Bouman and A.E. Hansen, J. Chem. Phys., 66, 3460 (1977).

[48] D.W. Rein, R.A. Hegstrom and P.G.H. Sandars, to be published.

DIASTEREOISOMERIC DISCRIMINATIONS IN CHIRAL METAL COMPLEXES

R.D. Gillard

Department of Chemistry, University College Cardiff, Wales, U.K.

ABSTRACT. The differing properties (both in the solid state and in solution) of diastereoisomers of the type $(DA)(DB)_n$ and $(DA)(LB)_n$ (A and B being differing sources of chirality) are rather arbitrarily discussed as intra- or inter-molecular effects.

a) Intramolecular interactions are exemplified by contrasting isomers as follows: $[Pt(L\text{-pro-}\bar{O})(S\text{-sar-}\bar{O})]$ with $[Pt(L\text{-pro-}\bar{O})(R\text{-sar-}\bar{O})]$

: $[Cu(L\text{-ala-}\bar{O})_2]$ with $[Cu(L\text{-ala-}\bar{O})(D\text{-ala-}\bar{O})]$.

: D- with L-$[M(L\text{-ala-}\bar{O})_3]$.

: S- with R-$[M(\text{dipeptide})_2]^-$.

: D- with L-$[M(\text{Chel})_2(S\text{-Lig})]^{n+}$,

where chel is a bidentate chelator, and Lig represents an amino-acid or hydroxy-acid like L-glutamic acid, which is potentially more than bidentate.

b) Intermolecular:

Selective ion-pairing and classical resolution are illustrated for 1:1 interactions, generally of equally but oppositely charged species, as follows:

Triple, e.g. $[M(en)_3]^{2+}$ with $[M'(chel)_3]^{3-}$.

Double charges, e.g. $[Ru(phen)_3]^{2+}$ with $[Pt(S_5)_3]^{2-}$.

Single charges, e.g. $[M(en)_2X_2]^+$ with $[Co(EDTA)]^-$, where en is 1,2-diamino-ethane, phen is 1,10-phenanthroline, X is a uni-

Stephen F. Mason (ed.), Optical Activity and Chiral Discrimination. 339–351.

dentate anion, and EDTA is the 1,2-diamino-ethane-N,N,N',N'-tetra-acetate anion.

Finally, the importance of kinetic effects, such as nucleation, as against thermodynamic effects (solubility) is described.

INTRODUCTION. We know that shaking hands (a right-right combination) is different from holding hands (right-left). There is clearly a difference of free-energy, which is often related (for a pure substance) to the entropy of mixing, RTln2. Many manifestations of this exist, such as the observable drop in temperature on mixing equal amounts of the enantiomers of coniine, or the differing solubilities of racemic and chiral forms of the same compound in achiral solvents. For example, the solubilities at 21°C in water (g/l) of DL - mer-$[Co(gly\text{-}\bar{O})_3]$ and of D-(+)-mer $[Co(gly\text{-}\bar{O})_3]$ are 8.8 and 5.8 respectively.

My brief today is to outline the experimental features of discrimination in metal complex compounds, and I will introduce this by means of examples. The type of metal complex which has been resolved by classical means has been extended a good deal recently; one of the more interesting areas concerns iron-transport in bacteria. The bacterial discrimination between enantiomers discovered by Pasteur for simple organic molecules was extended (ten years ago) to a clear distinction between the D- and L-forms (I and II) of $[Co(gly\text{-}\bar{O})_3]$ (the purple mer-isomer shown was used).

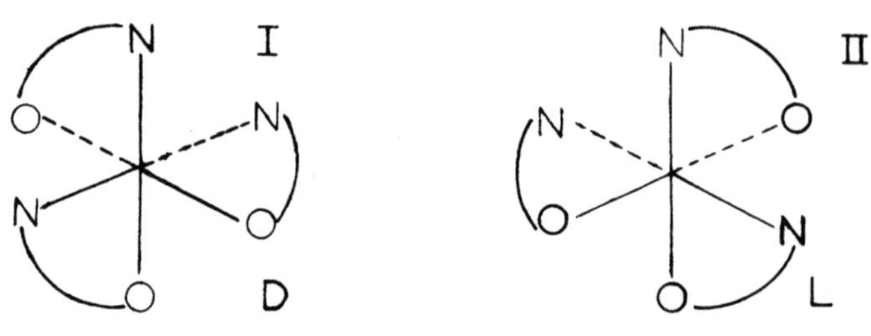

Proteus vulgaris in a medium containing DL-mer-$[Co(gly\text{-}\bar{O})_3]$ (with glucose as energy source) removed [1] stereospecifically the L-isomer: the D-isomer was unaffected (and had $\Delta\varepsilon_{535}$ = +2.74). This is probably the simplest resolution of the substance. (In a similar way, a strain of Pseudomonas stutzeri will [2] bring about stereoselective reduction of $[Co(en)_2L]^{3+}$, (where L = bipy or phen) when either complex ion is the sole source of nitrogen).

The iron-transporting species are, in general, similar octahedral chelated species, relatively specific either for ferric or for ferrous ions. Those low molecular weight compounds which facilitate the mobilization of iron(III) are called siderophores, and two groups have recently studied the optically active complexes of thiohydroxamates in a most instructive way.

Murray and his associates crystallized [3] the tris-complex (IV) of N-methyl formothiohydroxamate(III)

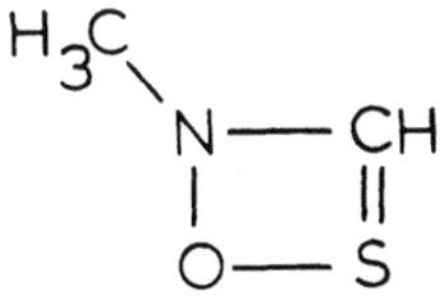

III

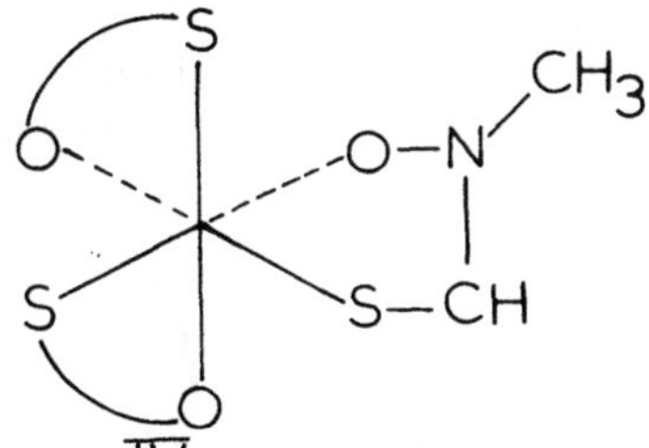

IV

(which complex had been isolated from Pseudomonas fluorescens grown on n-paraffin as the sole carbon source). Remarkably, the compound had spontaneously resolved (Z = 1 in space group R3).

Raymond, continuing his work on siderophores, discovered [4] that it was possible to resolve the related tris-complex (VI) of the di-anion (V) of thiobenzohydroxamate with iron(III).

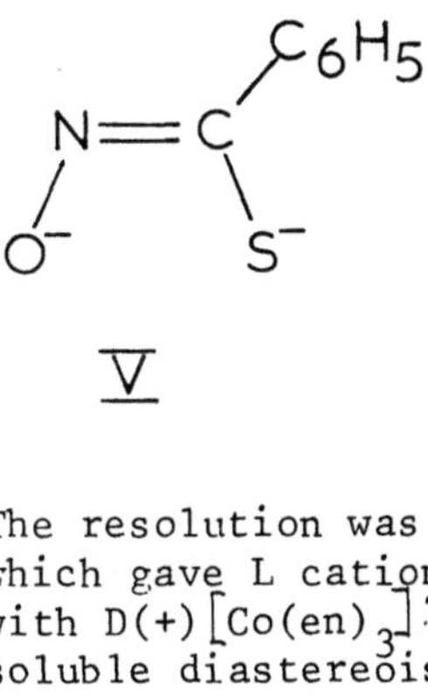

V

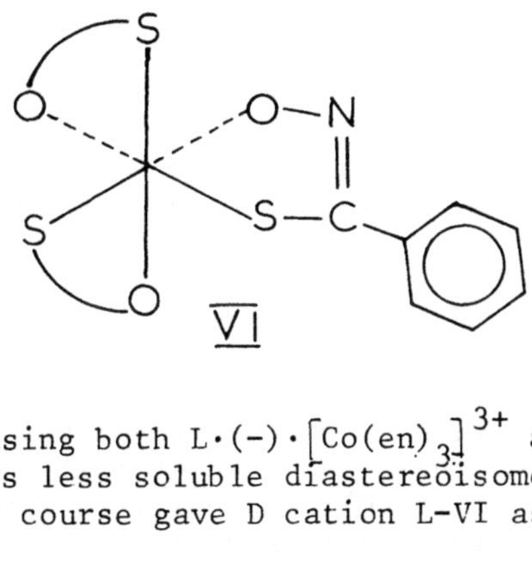

VI

The resolution was achieved using both $L\cdot(-)\cdot[Co(en)_3]^{3+}$ as agent, which gave L cation D anion as less soluble diastereoisomer, and with $D(+)[Co(en)_3]^{3+}$ which of course gave D cation L-VI as less soluble diastereoisomer.

These examples will suffice to establish a major feature which distinguishes most organic systems from coordination compounds (particularly the chiral ones so far studied). This feature is the highly polar nature of the metal derivatives, often obvious in a formal ionic charge and consequent water solubility. This is a major theme which I believe makes a worthwhile, albeit woolly, distinction in the general area of diastereomeric discrimination within metal complex compounds. I shall illustrate this theme by successive consideration of intra molecular effects and then intermolecular effects.

a) Intramolecular effects.

Let us consider the four co-ordinated uncharged complex molecules, bis-amino acidateplatinum(II). If we use L-prolinate as one residue, its nitrogen atom necessarily has the S configuration (because the fusion of the 2 five-membered rings is mutually cis). Now, with sarcosinate (N-methylglycinate) as the second chelator, we have the two possible trans-diastereoisomers shown as VII and VIII.

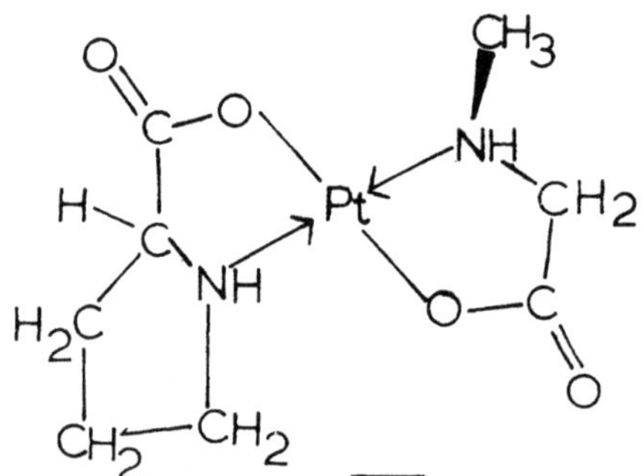

VII

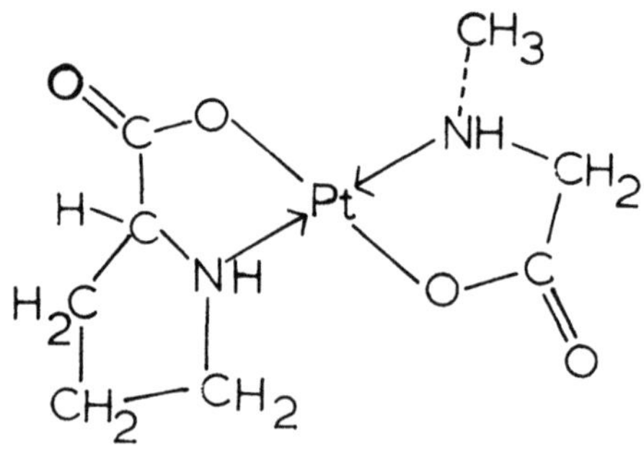

VIII

Because of the kinetic inertia of platinum(II), these isomers can readily be separated (in slightly acid solution) and interconverted (in basic media), so that we can readily measure the equilibrium constant (by using their very different c.d. spectra at 336 nm: for the S-sarcosinate, $\Delta\varepsilon$ = -0.55, for R-sarcosinate, $\Delta\varepsilon \approx 0$, and for the equilibrated mixture, $\Delta\varepsilon$ = -0.32).

$$\text{So,}\quad K = \frac{[\mathrm{Pt(pro\text{-}\bar{O})(S\text{-}sar\text{-}\bar{O})}]}{[\mathrm{Pt(pro\text{-}\bar{O})(R\text{-}sar\text{-}\bar{O})}]} \simeq 1$$

From models, or from conformational prediction of favoured structure, there is no doubt that the more stable diastereoisomer will be [Pt(Spro$\bar{O}$)(S-sar-$\bar{O}$)]. However, in water, there is [5] little difference in stability.

This emphasises a most important point, which lies at the core of chiral discrimination in coordination compounds. The conformational predictions (whether based upon models or on calculations using inter-atomic potential functions) are relevant to isolated molecules (i.e. those in the vapour phase). Differential solvation is particularly likely for aqueous solutions, and we therefore have the difficulty expressed in the Scheme:

"vapour"	(S-pro-$\bar{O}$)(S-sar-$\bar{O}$)	$\rightleftharpoons$	(S-pro-$\bar{O}$)(R-sar$\bar{O}$)	K $\ll$ 1
aqueous solution	(S-pro-$\bar{O}$)(S-sar-O) $\downarrow\uparrow$	$\rightleftharpoons$	(S-pro-$\bar{O}$)(R-sar$\bar{O}$) $\downarrow\uparrow$	K $\simeq$ 1.0
solid	(S-pro-$\bar{O}$)(S-sar-O)		(S-pro-$\bar{O}$)(R-sar$\bar{O}$)	
	(more soluble: <u>ca</u> 150g.dm^{-3})		(less soluble: <u>ca</u> 10g.dm^{-3})	

The interplay of solvation energy and lattice energy often gives such unpredictable results, even for systems as simple as these platinous complexes.

Other difficulties have prevented a full understanding of 'simple' complex compounds. We are interested in equilibrium ratios of isomers, so an obvious approach was to evaluate the equilibrium constants (where A is an amino acidate and S solvent) in a kinetically labile system, containing such metals as copper or nickel.

$$M^{2+} + LA^{-} \rightleftharpoons M(LA)^{+} \begin{cases} \overset{K_{LL}}{\rightleftharpoons} M(LA)(LA) \\ \underset{K_{LD}}{\rightleftharpoons} M(LA)(DA) \end{cases}$$

The question to be answered is this: does the first chiral centre (the asymmetric carbon of A in MAS_n) lead to discrimination between adding a D as against an L? The potentiometric titration curves superposed for M = Cu^{2+}, A = ala-$\bar{O}$, and this could, at first sight, be taken to mean that there is no discrimination. However, the simple comparison in Scheme 2 by no means represents reality, and while the equality of empirical K_{LL} and K_{DL} is a fact, it may be irrelevant in studies of chiral discrimination for the following reasons.

First, it takes no heed of geometrical isomerism. When bis-glycinatocopper(II) crystallizes from water, the product is the solid <u>cis</u>-isomer (IXa) as its monohydrate. However, if this is equilibrated with mother liquor or (a recent discovery due to P. O'Brien) heated dry, the product (presumably the more stable solid) is the <u>trans</u>-isomer. (Xa).

* This value is, of course, identical to K_{1D}, that for the enantiomeric process. The suggestions, which still appear in the literature, that, for any ligand, K_{1L} and K_{1D} differ are wrong.

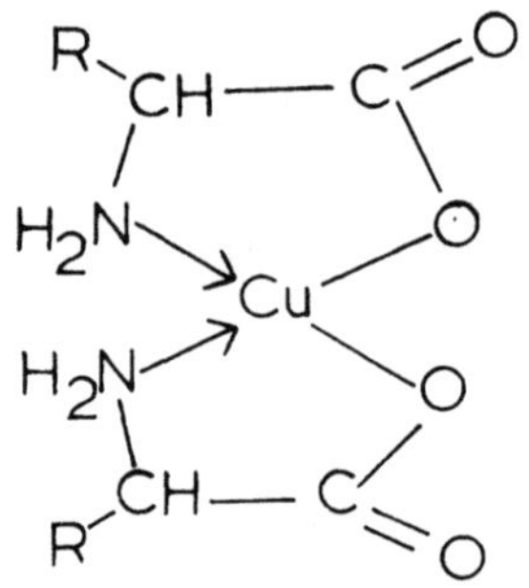

IX a: R = H
b: R = CH_3

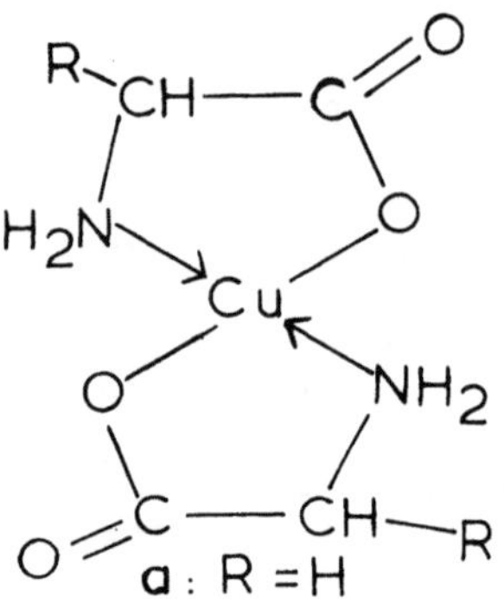

X a: R = H
b: R = CH_3

Conversely, when bis-L-alaninatocopper(II) crystallizes from water, the first product is the Cambridge-blue trans-isomer. When left to equilibrate, this disappears, and the more stable solid, a much more pleasing Oxford-blue forms. So, clearly, we must consider the presence in solution of both cis- and trans-isomers: (indeed, the basis for the elevation of the trans-structure in solution to a position of dogmatic dominance is not at all clear). To our first approach, through K_{LL} and K_{LD}, we must add two further equilibria, that is trans-LL $\rightleftharpoons$ cis-LL and trans-LD $\rightleftharpoons$ cis-LD.

Secondly, although there is no evidence for the binding of a third amino-acidate to cupric ions, there is yet another novel equilibrium which needs to be incorporated. When bis-L-alaninato-copper(II) is taken to pH values higher than 9.5, the c.d. per copper ion decreases. Clearly, a new species is present, and its formation constant from [Cu(L-A_2)] as against that from [Cu(LA)(DA)] may well be different. So, finally, our superimposed titration curves for L-alanine as against D-L-alanine may result from fortuitous cancellation among three discriminatory equilibrium constants.

One area where apparent intramolecular discrimination has been found is [6] for M (in Scheme 2) = Ni^{2+} or Co^{2+} and A = histamine. Here $K_{LD} > K_{LL}$ (with ΔG = 2.1 and 1.3kJmol^{-1} for Ni and Co, respectively). Unfortunately, this is not really a fair comparison, since the ligand is terdentate and the complex is six coordinated. As is clear from the sketches XI and XII, in order for two L-terdentate ligands to occupy all six coordination positions, the complex must have the cis-structure (XI), whereas for the centric D and L-complex to be bis-terdentate, the structure must be trans-(XII), and the difference in geometry of bonding in the horizontal plane alone could give rise to measurable effects in relative stability. In a related way, K_{LD} and K_{LL} for histidine with cobalt(II), nickel(II), and zinc(II) differ [7] by 0.12, 0.26 and

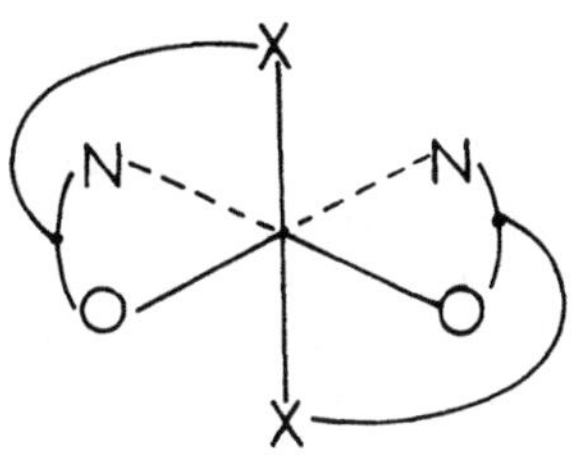

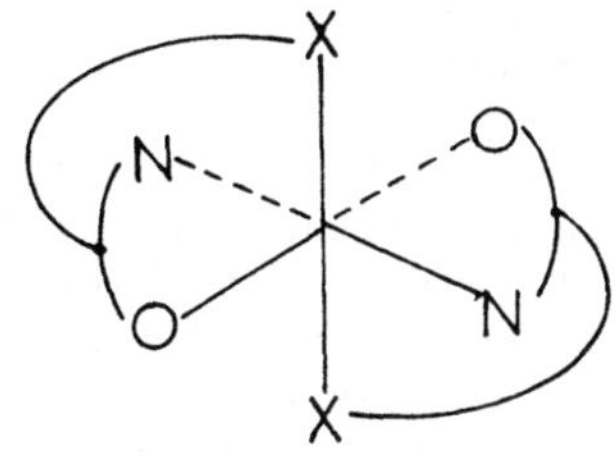

XI LL   LD XII

0.13 respectively, i.e. if there were no structural change, the free energy differences between the mixed and pure complexes would be about 4kJmol^{-1}. It is a most intriguing fact that for both the nickel [8] and zinc [9] bis-complexes made from racemic histidine, spontaneous resolution into crystals of M(L-his)$_2$ and M(D-his)$_2$ occurs. That is, while, in solution, the trans centric structure (XII) is more stable, the the crystal, the cis optically active structure (XI) packs better.

When we add, in our study of intramolecular effects, the chirality of octahedral metals to that of the organic ligand, then the situation becomes easier in some respects (because of kinetic inertia, it is possible to separate metastable solids containing cobalt(III), chromium(III), and so forth) but more complex in others. Thus, although it is feasible to separate the four isomers of tris-L-alaninatocobalt(III), i.e. D-mer, L-mer, D-facial, L-facial, and indeed, the crystal structures of both mer isomers are available (L-(-)-mer.H_2O [10] and D-(+)-mer.H_2O [11]) establishment of the stability relations among four isomers (here and for other amino acids) is frustrated by the gross insolubility of the D facial form. Further, the solubility relations can cause unexpected difficulties. In general, crystallization to constant properties has been used to obtain individual substances. Occasionally, however, the purified substance is not a simple isomer, but an adduct of two isomers. For example, on crystallizing the mixture of the two meridional isomers of tris-L-valinatocobalt(III), the purple-red crystals are actually [12] the quasi-racemate DL-mer-[Co(L-val-O)$_3$] whose saturated aqueous solution is 6.9 x 10^{-2}M, as against 9.5 x 10^{-2}M for pure L-mer-[Co(L-val-O)$_3$].

A similar frustration occurred in attempting to evaluate the properties of the two diastereoisomers of the peptide complex bis-L-alanyl-L-leucinato cobaltate(III) in order to establish equilibrium ratios. Here, in systems of a type which rarely yield crystals, the preparative solution gave some good crystals of $NH_4[Co(L\text{-ala-L-leu-}\bar{O})_2]2H_2O$, but these too turned out to be a quasi-racemate, containing a 1:1 ratio of the two diastereoisomers.

A recent development of interest in connection with such quasi-racemates is the work [13] in the solid state by Whuler and his school. Since one objective of effort on chiral discrimination is to understand the difference in packing forces (and geometries) for racemic as against active substances, their findings are of great value. At room temperature, the 'active racemates' $\{(+)[Co(en)_3](-)[Cren_3]\}\ Cl_6\ 6.1\ H_2O$ and $\{(+)[Co(en)_3](-)[Cr(en)_3]$ $(SCN)_6.H_2O$ show that the complex ions have M-N bond lengths independent of M. However, in structures determined at 133°K for the thiocyanate, and at 123°K for the chloride, the complex ions have recovered their individual characters: the Co-N and Cr-N bond lengths become different.

As a final example of intramolecular discriminatory effects, the diastereomeric cases reprecented by tris-chelated octahedral complexes like $[M(chel)_2(S\text{-ligand})]^{n+}$ are valuable. Where S is a bidentate ligand only, say an L-amino acid, and chel is a ligand like 1,2-diaminoethane or acetylacetonate, the relative stabilities <u>in solution</u>, of $D[M(chel)_2(S\text{-lig})]^{n+}$ and $L[M(chel)_2(S\text{-lig})]^{n+}$ do not differ by much (although occasionally solubilities may be surprisingly divergent). However, when S possesses a third polar site, selective interactions between this third site and the chiral framework $M(chel)_2$ become possible, and, if there is some mechanism whereby such interactions are sizeable (e.g. hydrogen bonding) the diastereoisomers may be very different in stability. For example, the major product of reacting $DL\text{-}[Co(en)_2(CO_3)]^+$ with L-glutamic acid contains the diastereoisomer $D(+)[Co(en)_2(L\text{-glu-}\bar{O})]^+$, in which [14] the γ-carboxylate of the L-glutamate forms a very good hydrogen bond to an N-H group of the chiral D-bisethylenediamine framework. In perhaps a similar way, when L-cysteinate forms a tris-complex with cobalt(III), there is only one major product, $[Co(\underline{N},\underline{S}\text{-cyst})_3]^{3-}$ which is quite remarkable since in this system, we have the possibilities of geometric (meridional vs facial), linkage (<u>N</u>,<u>S</u>- or <u>N</u>,<u>O</u>- or <u>O</u>,<u>S</u>- bidentate attachment of a single cysteine) and optical (D- or L-cobalt) isomerism. Clearly, intramolecular discrimination is at work, but we are still short of detailed knowledge of the relevant kinetics and thermodynamics during this complicated synthesis, and the structure of the product is at best tentative.

b) <u>Intermolecular effects</u>

When oppositely equally charged chiral species interact, we have (crudely) the three concentration ranges outlined in Scheme 3 where only one enantiomer of the cation has been added to the racemic anion.

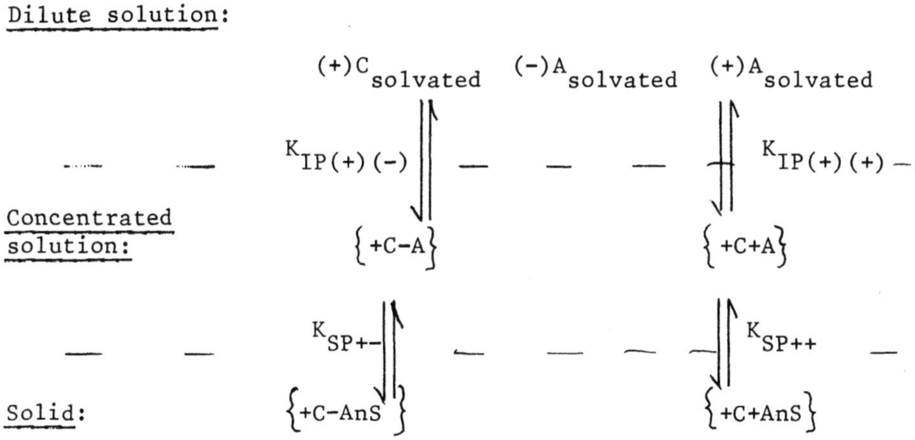

The equilibrium constants which have been dominant in considering diastereoisomeric discrimination are those for selective ion-pairing K_{IP} and for solubility product, in optical resolutions. Added to the factors discussed under the heading of intramolecular discrimination (i.e. polar solvation and hydrogen-bonding) there is now the effect of charge neutralization.

Just as in the case shown (one among many) of insolubility resting on equality of charge (I = insoluble, S = soluble)

	ClO_4^-	SO_4^{2-}	PO_4^{3-}
Cs^+	I	S	S
Ba^{2+}	S	I	S
La^{3+}	S	S	I

so it has proved operationally useful to use the criteria of charge and size in designing resolutions. The reasons why this charge equality is in general useful are not yet clear, but a few examples will show that it has some utility. The discussion is restricted to cases where both chiral gegen-ions are complex ions though it is worth mentioning that Werner and Basyrin [15] used (+)$[Coen_3]Br_3$ to resolve 2,3-dimethylsuccinic acid in 1913!

Triply charged ions:

A fair number of highly satisfactory resolutions via the grossly differing aqueous solubilities of the diastereoisomeric salts $[M(chel)_3]^{3+}[M^*(chel')_3]^{3-}$ has been achieved. For example, D(+)$[Co(en)_3]^{3+}$ gives very clean precipitation of one enantiomer of the anions D$[Co(C_2O_4)_3]^{3-}$, (5-membered rings [16])D$[Co(CO_3)_3]^{3-}$

(4-membered rings [14]), and D[(Co(malonate)$_3$]$^{3-}$ with 6-membered rings [18], and D(+)[Coen)$_3$]$^{3+}$ and D(-)[Rh(en)$_3$]$^{3+}$ have resolved [19] [M(C_2O_4)$_3$]$^{3-}$, with M = V, Cr, Mn, Fe, and Co. The less soluble diastereoisomers are again D cation D anion. (The c.d. spectra for kinetically labile species were measured for the solids as CsCl discs).

In such resolutions, the two ionic components are present in equal numbers, and the high formal charges are neutralized locally (i.e. for the geometries established so far, chiefly by Snow and his school, there are no short cation-cation or anion-anion contacts. There are, however, strikingly rich patterns of short hydrogen bonds involving the oxygen atoms of the coordinated di-carboxylates and the amino groups pf the coordinated amines, and such strong chirally selective hydrogen-bonding has also been involved to explain resolutions of cobalt ammines by the anions of organic acids like tartrate).

In line with this view, it has not been altogether a simple matter to resolve the triply charged cations [ML$_3$]$^{3+}$ where L = 2,2'-bipyridyl or 1,10-phenanthroline and M = Co, Cr, Rh. Formally, these cations have no polar sites like N-H groups, and further, until recently, no convenient resolving anion of triple charge was available. However, the stereospecifically formed green ion (XIII) [Co(L-cys)$_3$]$^{3-}$, mentioned earlier, itself unfortunately rather unstable, may be oxidized to its yellow tris-sulphinate, which seems likely, from its extremely simple carbon-13 magnetic resonance spectrum, to have the _facial_ structure (XIV).

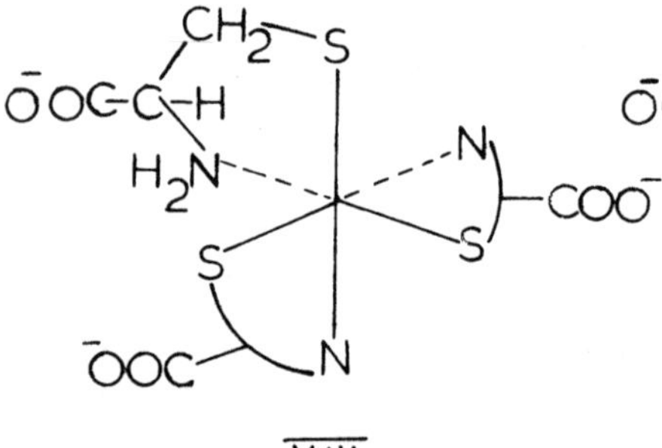

XIII

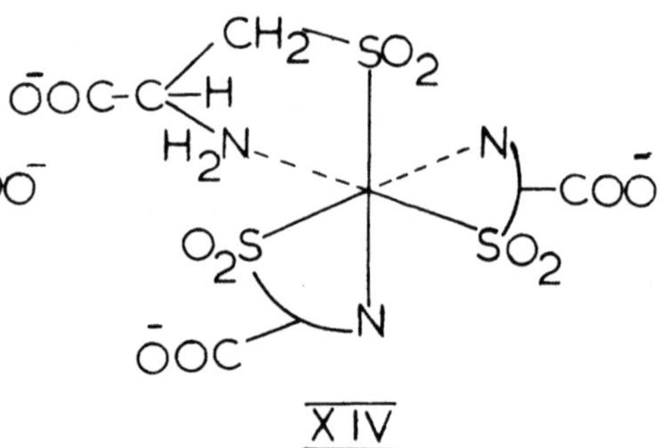

XIV

Optically pure samples of many 3+ cations all, so far, with the D- configuration are readily obtained [20] through this tris-cysteine-sulphinate. They include [Rh(phen)$_3$]$^{3+}$ [Rh(bipy)$_3$]$^{3+}$ [Co(bipy)$_3$]$^{3+}$ [Co(en)(phen)$_2$]$^{3+}$ and [Rh(phen)$_2$(NH_3)$_2$]$^{3+}$.

Doubly charged ions

In connection with this question of the relative sizes of

influences transmitting chiral information in solids, one might argue that, as Basolo put it, "birds of a feather flock together". That is, a class a cation may well interact better (or more selectively) with a class a anion than with a class b, and vice versa. So, in addition to the ideas of charge and size, and of selectivity arising through specific polar interactions, we might expect that more polarizable anions could be resolved by more polarizable cations. This has been supported by a rather novel resolution of the species $[PtS_{15}]^{2-}$, achieved by Dr. Wimmer recently. He used (+) $[Ru(1,10\text{-phen})_3]^{2+}$, and obtained crystals of the (+)(+)diastereoisomer.

This is, of course, only the third purely inorganic molecule to be resolved, the others being Werner's 'hexol' and F.G. Mann's bis-aquobissulphamatorhodate(III). It is the first inorganic binary species to be resolved, the first to contain rings larger than 4-membered, the first to contain a transition element of the third row. It is even more remarkable in that, on crystallising its ammonium salt, $(NH_4)_2PtS_{15}$, under conditions appropriate to rapid racemization in solution, the solid obtained contains only one enantiomer of the tris-pentasulphidoplatinate(IV) ion. The yield is higher than 50% so this constitutes a spontaneous asymmetric crystallization, also the first for an inorganic molecule.

Singly charged ions:

Again, there are several examples of effective chiral discrimination by a -1 resolving agent, such as (+) [Co(edta) $^-$], whereedta signifies the hexadentate tetra-anion of 1,2-diamino-ethane-N,N',N'-tetra-acetic acid. The less soluble diastereomeric salts are [21]:

(+) $[Rh(en)_2(C_2O_4)]$ (+) [Co(edta)] and (-) $[Co(en)_2(C_2O_4)]$ (+) [Coedta] which are isomorphous, and

(+) $[Rh(en)_2(NO_2)_2]$ (+) [Co(edta)] and (-) $[Co(en)_2NO_2)_2]$ (+) [Co(edta)] which are mutually isomorphous.

We may clearly apply the so-called rule of less soluble diastereoisomers as formulated by Jaeger to say that the ions (+) $[Rh(en)_2(NO_2)_2]^+$ and (-) $[Co(en)_2(NO_2)_2]^+$ have the same configurations. However, that is a thermodynamic argument, based on solubility, and I would like to conclude by mentioning two or three examples of clear kinetic influences in chiral discriminations, as yet a further experimental complication.

1) In resolving tris-oxalatochromate(III) using natural strychnine, Mead and Johnson [22] found that, depending in an

erratic way upon the exact conditions, either hand of $[Cr(C_2O_4)_3]^{3-}$ might be present in the diastereoisomeric solid product.

2) With (+) $[Coen_3]^{3+}$, Sargeson [23] obtained only inactive precipitates with (±) $[Co(C_2O_4)_3]^{3-}$, while Vaughn, Magnuson, and Seiler [16] used this as a resolution. Both phenomena have been observed (several times each during ten years) in my own laboratory, with both (+) $[Co(en)_3]^{3+}$ and (-) $[Rhen_3]^{3+}$. Although very subtle temperature factors may be involved, it seems more likely that nucleation and growth kinetics are at issue.

3) In our first experiment upon the diastereoisomers $[Co(en)_3][Co(L\text{-}cysU)_3]$, the cation in the less soluble (i.e. first crystallising) solid was (-) $[Coen_3]^{3+}$. This is the case: we have the original diastereoisomer, and it is indeed the salt of (-) $[Coen_3]^{3+}$. In every subsequent experiment, we obtain, as our precipitate the salt of (+) $[Coen_3]^{3+}$.

Perhaps I may re-echo a remark by Professor Craig, who said that, in the understanding of chiral discrimination, we are at a very early stage. That is most certainly true, and my own caveat would be that we should apply thermodynamic arguments only to systems where we are confident that we know what species are present, and that we are dealing with an equilibrium situation.

In trying to devise models for real systems, where the experimental facts are not subject to such ambiguities as I have mentioned, it is quite clear that the specific solvation effects of water will need to be included. I shall return to this subject tomorrow in treating one well-known case of chiral discrimination in coordination chemistry (the Pfeiffer effect) in detail.

REFERENCES

1. R.D. Gillard, J.R. Lyons and C. Thorpe, J. Chem. Soc. (Dalton), 1584 (1972).
2. L.S. Dollimore, R.D. Gillard and I.H. Mather, J. Chem. Soc. (Dalton), 518 (1974).
3. K.S. Murray, P.J. Newman, B.M. Gatehouse and D. Taylor, Australian J. Chem., 31, 983 (1978).
4. K. Abu-Dari and K.N. Raymond, Inorg. Chem., 16, 807 (1977).
5. R.D. Gillard and O.P. Slyudkin, J. Chem. Soc. (Dalton) 152 (1978).
6. J.H. Ritsma, J.C. van der Grampel and F. Jellinek, Rec. trav. Chim., 88, 411 (1969).
7. P.J. Morris and R.B. Martin, J. Inorg. Nuclear Chem., 32, 2891 (1970).
8. K.A. Fraser and M.M. Harding, J. Chem. Soc. A., 415 (1967).
9. M.M. Harding and S.J. Cole, Acta Cryst. 16, 643 (1963).

10. R. Herak, B. Prelesnik and I. Krstanovic, Acta Cryst. 34B, 91 (1978).
11. M.G.B. Drew, J.H. Dunlop, R.D. Gillard and D. Rogers, J. Chem. Soc., Chem. Comm., 42 (1966).
12. R.D. Gillard, N.C. Payne and D.C. Phillips, J. Chem. Soc. A, 973 (1968).
13. A. Whuler, C. Bronty and P. Spinat, Acta Cryst., 34B, 425 (1978) and references therein.
14. R.D. Gillard, R. Maskill and A. Pasini, J. Chem. Soc. A., 2268 (1971).
15. A. Werner and M. Basyrin, Chem. Ber., 46, 3229 (1913).
16. J.W. Vaughn, V.E. Magnuson and G.J. Seiler, Inorg. Chem., 8, 1201 (1969).
17. R.D. Gillard, P.R. Mitchell and M.G. Price, J. Chem. Soc. (Dalton) 1211 (1972).
18. R.J. Gene and M.R. Snow, J. Chem. Soc. A., 2981 (1971).
19. R.D. Gillard, D.J. Shepherd, and D.A. Tarr, J. Chem. Soc. (Dalton) 594 (1976).
20. L.S. Dollimore and R.D. Gillard, J. Chem. Soc. (Dalton) 933 (1973).
21. R.D. Gillard and L.R.H. Tipping, J. Chem. Soc. (Dalton) 1241 (1977).
22. A. Mead and C.H. Johnson, Trans. Faraday Soc., 35, 681 (1939).
23. A.M. Sargeson in "Chelating Agents and Metal Chelates", eds. F.P. Dwyer and D.P. Mellor, Academic Press, New York, 194 (1964).

THE ORIGIN OF THE PFEIFFER EFFECT

R.D. Gillard

University College, Cardiff, P.O. Box 78, Cardiff,
CF1 1XL, Wales.

ABSTRACT. The intriguing enhancement of optical activity when a racemic (and "kinetically labile") metal complex of the type $(ML_3)^{2+}$ (M = Zn,Cd, Ni,Co etc; L = 2,2'-bipyridyl or 1,10-phenanthroline) is added to a solution of a cationic 'environment' compound (usually an alkaloid) is the best example of the 'Pfeiffer effect'.

Salts of the methyl-1,10-phenanthrolinium cation (which is essentially planar) also serve to enhance environmental rotations. This is explained by the formation of a covalent hydrate, introducing new chiral centres [*C(2)H(OH) and *NH(M)]. The concept of covalent hydration (giving new chiralities in the ligands of $(ML_3)^{2+}$) and selective hydrogen bonding of the covalent hydrate gives a novel explanation of the known properties of Pfeiffer systems.

Chiral discrimination is one of the major themes of modern studies in stereochemistry, as represented by the present Advanced Study Institute. Such discrimination is evident, on a macro-scale, every time we get dressed in shoes or gloves, but it comes to the attention of chemists on a molecular scale through various phenomena. All depend, in some way or another, kinetically or thermodynamically, on the difference between the interactions right-right and right-left*. The attempt to understand this difference is a major pre-occupation of chemists.

*This difference corresponds exactly to that between shaking hands and holding hands.

Stephen F. Mason (ed.), Optical Activity and Chiral Discrimination. 353–367.

If, to a solution of an optically active substance, with $[\alpha]_\lambda$, say, of x^o, we add some other optically inactive solute, and there is no change of homogeneity, we do not, in general, expect x to change significantly. It is our tacit belief in Oudeman's law |1| that underpins most of this expectation. Oudeman stated that the molecular rotation of the salts of optically active acids or bases have a limiting constant value. Clearly, if this is not true, then there must be some interaction between the solutes present, and gross departures indicate gross effects. For example, although malic acid has a very small optical rotation for sodium yellow light, $[\alpha]_D \simeq 1^o$, the addition of salts of the linear, optically inactive* uranyl ion gives a solution with $[\alpha]_D \simeq 500^o$. Here, clearly, formation of complex species gives chiral strength to the electronic transitions of the uranyl ion.

One particular enhancement of optical activity is often called [2] the Pfeiffer effect [3] . I wish in this lecture to offer a new molecular description of the origin of the chiral discrimination involved. The Pfeiffer effect was discovered by Pfeiffer and Quehl [4], who measured values of specific rotation for solutions of optically active natural products such as alkaloids first alone and then with racemic metal complexes added. The values differed, although there was no obvious way in which the natural product ("environmental compound") and the metal complex ("Pfeiffer compound") could interact. As a typical example quoted [5] from the original work, $[\alpha]_D$ for a solution of cinchonine hydrochloride in water was +5.29°, whereas on adding the complex salt tris-1,10-phenanthrolinezinc(II) as its sulphate the final value was -2.46°. The definition [6] of molar Pfeiffer rotation (at T^o and λ_{nm}) is

$$Pm = \frac{P}{[E][C]l}, \text{ where } P = \pm(\alpha_{E+C} - \alpha_E) \quad \alpha_{E+C} \text{ is the rotation}$$

of the solution containing the environment compound E and the complex C, α_E is the rotation of the solution of E alone, [E] and [C] denote molar concentrations, and l is the solution path length in metres. The subject of the Pfeiffer effect has been extensively reviewed from various viewpoints [7] and no attempt is made here to give complete details.

The mystery in Pfeiffer's original observations was perhaps more evident than in some later cases of induced equilibrium shifts in labile racemates. How could a complex of kinetically labile, colourless, zinc ion (admittedly tris-chelated and therefore having, at highest, D_3 symmetry - i.e. a racemic species),

* This phrase refers to solutions of such salts as uranyl nitrate or sodium uranyl acetate. The latter, of course, forms enantiomorphous crystals

which is not resolvable by classical methods,* contribute optical activity.

To restate the observations, when a racemic mixture of a labile metal complex (P) is added to a solution containing some chiral molecules (E) additional optical activity is sometimes developed. For the present, attention will be restricted essentially to the systems involving water, and these divide into two types on a charge basis.

(1) The Pfeiffer compound (P) may have the same electrical charge as the environmental species (E), as in the systems where P is $ML_3^{\ 2+}$ (M = Zn,Cd,Ni,etc; L = 1,10-phenanthroline or 2,2'-bipyridyl), and E is also cationic, e.g. (-)strychninium ion, or (+)cinchoninium ion.≠

(2) The Pfeiffer racemate may have the charge opposite to that of the environment compound, as in (a) P = $[ML_3]^{2+}$: E = (+)-3-bromo-camphor-9-sulphonate ion, or in (b) P = $[M(C_2O_4)_3]^{3-}$ ion, where M = cobalt(III) or chromium(III) and E = an alkaloid cation like strychninium.

In category (2) there are obviously possibilities of specific counter-ionic attractions of an ion-pairing kind, which do not exist in (1). For this reason, since we are used to resolutions (i.e. solubility differences arising from specific right-right as against right-left cation-anion interactions, this second class of Pfeiffer system is both less mysterious** and more complicated than the cases under (1). If we can explain the first group, where like-charged species somehow interact to give more chirality, then, mutatis mutandis, we can explain the second.

I will therefore restrict my detailed discussion to those systems where a cationic (usually, though not always alkaloid) environment compound somehow transmits its chiral coding to alter the zero optical rotation of a racemic metal cationic complex.

* Typical exchange rates for Zn^{2+} systems are high, as they are for Co(II) and Cu(II) which also manifest Pfeiffer effects.

≠ Cinchonine hydrochloride, dissolved in water, exists chiefly as the mono- and di-protonated cations. At pH6, the mono-cation predominates [8].

** Cf. the optical resolution through classical diastereoisomers of tris-oxalatochromate(III) species achieved using strychninium by Johnson and Mead.

This is the strange group*, where despite the inevitable electrostatic repulsion between the cationic Pfeiffer compound and the cationic environment compound, strong Pfeiffer enhancement of rotation occurs even at low concentrations in water.

Let us collect the facts which relate to the cation:cation Pfeiffer effect.

(1) The size of the effect: the magnitude of the effect on rotation is roughly proportional [9] to [E] and to [C]. The degree of resolution obtained, for those few systems where the "labile" metal complex possesses enough optical stability to be studied as an isolated enantiomer (mostly Ni^{2+} complexes), in terms of the excess of one enantiomer over the other is somewhere between 0 and 3%. In free energy terms, this corresponds to a rather small amount, since the distinction between DL and D is RT ln2. Typical values quoted for the experimental free energy of discrimination are from 0.1 to 1 $kJmol^{-1}$.

The predominance of the ions $[ML_3]^{2+}$ as Pfeiffer racemate is pleasing, because these are compounds of high interest and importance. They have been used for an enormous range of purposes, which include, inter alia: antibacterial studies; the photodecomposition of water; analyses of metal ions (particularly iron) in manifold circumstances; fundamental studies of the rates and mechanisms of electron transfer and of substitution at metal centres, and work on the exciton c.d. arising from the helical arrangement of the three blades of the propeller.

(2) The effect is manifested <u>only</u> for metal complexes containing N-heterocyclic ligands, such as 1,10-phenanthroline, 2,2'-bipyridyl, (where the effects seem to be considerably smaller than with phenanthroline), and 2,-(2-pyridyl)-imidazoline.

(3) On the basis of a few observations, the effect is much larger for protonated alkaloidium salts than for their methylated analogues. The alkaloids which have most commonly been employed to generate Pfeiffer effects in $[ML_3]^{2+}$ are: strychnine, brucine, quinine, cinchonine and nicotine.

(4) Water seems to be the best solvent by far for generating cation:cation Pfeiffer effects. It seems clear [6,10] that the effect does not occur in alcohol solvents and the few other solvents studied (DMF, glacial acetic acid) have been applied <u>only</u> to cation:anion cases. Cation:cation effects have been reported only for water, except for the single case of $[ZnP_3]^{2+}$:

* Clearly, examples of anion:anion Pfeiffer effect would be equally strange: so far, none has been reported.

cinchonine in acetic acid. mentioned by Kirschner and Magnell [11], without detail.

(5) All those classes of substance which are known to modify water structure and activity cause marked changes in the Pfeiffer effect. The alcohols and urea depress it, and indeed this has been the basis for a view of its origin based upon 'hydrophobic bonding' between the Pfeiffer cation and the environmental cation [12].

From 3,4, and 5, the Pfeiffer effect may well be related to hydrogen bonding, and from 5, the ability of water to form two hydrogen bonds may be relevant.

Let us consider one new fact, see how it reminds us of much known and suggestive chemistry, and then apply this to the origin of the Pfeiffer effect. The new fact is due to Dr. K.W. Johns. When methyl-1,10-phenanthrolinium iodide(I) is added to a solution of an alkaloid salt, a very large increase of rotation occurs.

(I)

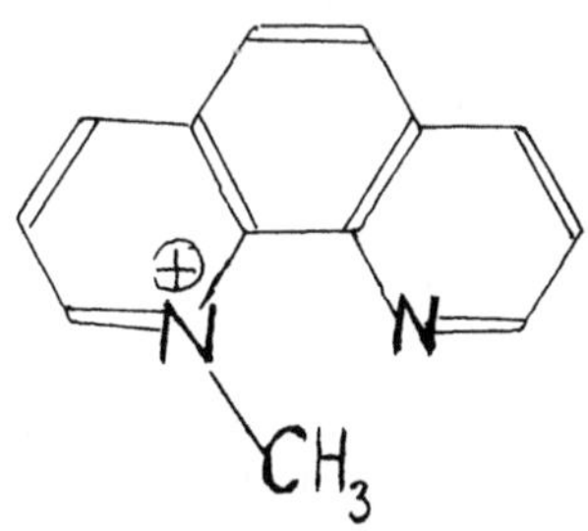

That is a new kind of Pfeiffer effect, where the added "racemate" is in fact not initially racemic, though it clearly must become so by some chemical change in solution. 1-methyl-1,10-phenanthrolinium cation possesses an em rrassing amount of reflection symmetry: it is planar. Yet, in the presence of alkaloid environment compounds it generates a Pfeiffer rotation second to none.

The origin of the chirality in I must lie in the equilibrium with the species shown in II - IV. The formulae denote the position of covalent hydration or carbinol formation. I shall restrict my attention to II not because it is the only possible species present, but because its complexed analogue has more obvious relationships to the metal when we return to considerations of $[M(phen)_3]^{n+}$.

II has plenty of new chiral elements: an asymmetric carbon atom (2), an asymmetric nitrogen atom (1) and a skewed diene system.

Such covalent hydrates among N-heterocycles are not new, although the observation of a 'Pfeiffer effect' arising from one most certainly is. The original suggestion of a covalent hydrate

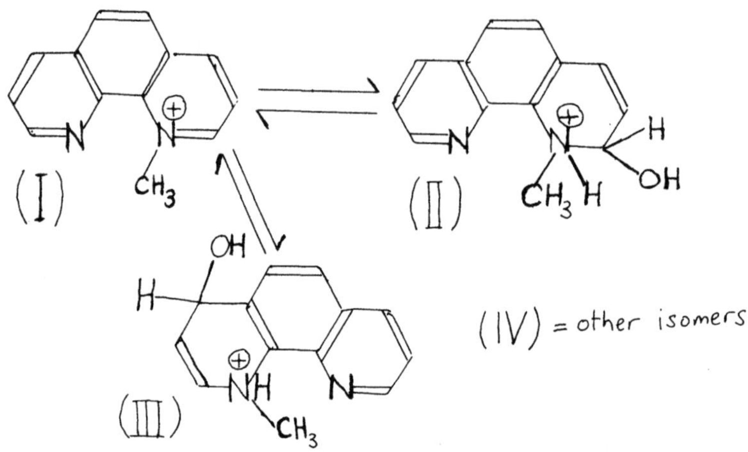

was made to explain an apparent change of optical activity. Lowry said that, because the molar extinction coefficient of nicotine was about 10 times greater in water than in cyclohexane the species present were V and VI respectively.

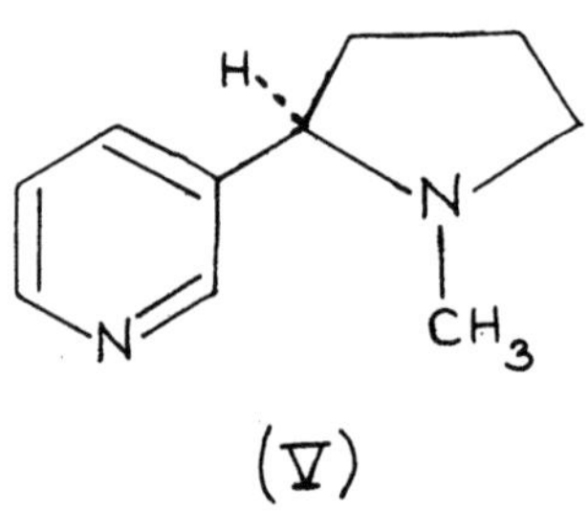

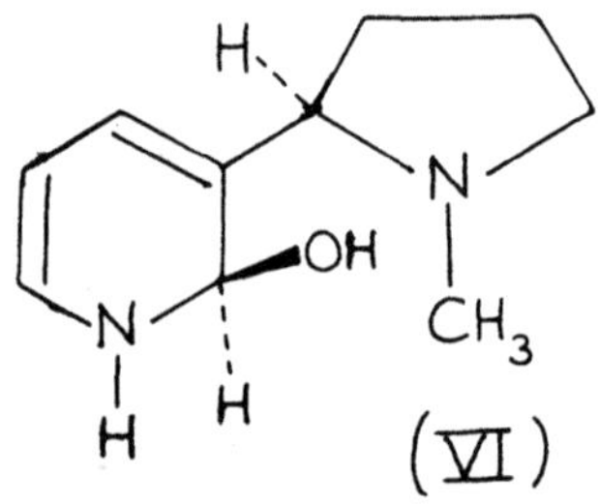

This was based on an error in measurement. (Dr. D.H. Vaughan showed recently that the difference is small). However, it is a most stimulating suggestion, reminiscent of the anomalous properties often shown by systems containing water and N-heterocycles like pyridine.

The most striking analogy to the new 'Pfeiffer' effect with the flat cation occurs in an enzymic system, adenosine aminohydrolase. This catalyzes stereoselectively both the hydration of pteridine(VII) (clearly a flat non-chiral molecule) and the reverse dehydration of VIII.

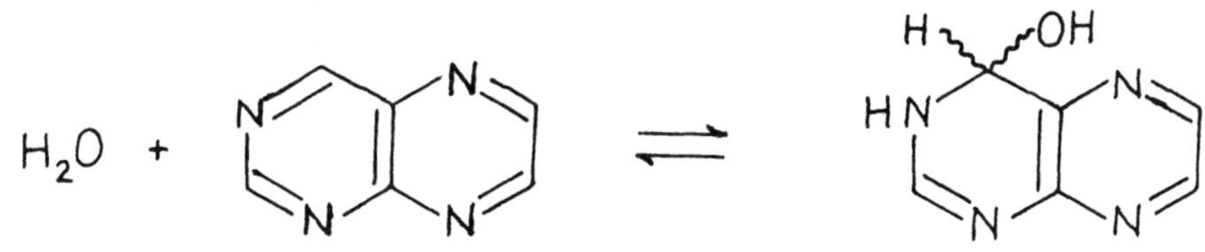

(VII) (VIII) (+) or (-)

The result [13] could be summarized as:-

(-)VIII

$K_{equ} = 6.9$ $k_1^{(-)}$ fast

$k_{-1}^{(-)}$ fast

VII

$k_1^{(+)}$ slow

$K_{equ} = 6.9$

(+)VIII $k_{-1}^{(+)}$ slow

Clearly, the equilibrium thermodynamic activities of (-)VIII and (+)VIII are equal, and, since the enzyme is but a catalyst, which does not alter the position of equilibrium, the final system (starting from VII or either enantiomer of VIII) is optically inactive. However, the kinetic stereoselectivity of the asymmetric catalyst (whereby $k_1^{(-)} > k_1^{(+)}$ leads to an initially faster formation of the (-)hydrate, (-)VIII.

The analogous scheme, not now kinetically but thermodynamically based for the enhanced optical rotation produced by I would indicate

K^+ (+)II

I

K^- (-)II

and here, because the environmental compound is not a catalyst but a reagent, $K^+ \neq K^-$, and the situation at equilibrium is optically active.

The new model for the cation:cation is as follows. The reaction of water with the N-heterocyclic ligands generates an equilibrium amount of covalent hydrate. These are enantiomers (where only one flat N-heterocycle like I is involved) or diastereoisomers (where two or more N-heterocycles are dihedrally arranged, as in $ML_3{}^{n+}$). This mixture of hydrates then (slowly or rapidly, depending on the particular system) changes to conform to the most stable mixture possible in the particular environment. That depends on the interaction (through hydrogen-bonding) of the asymmetric solvent with each of the diastereomeric covalent hydrates with its individual solvation sphere. Schematically, it is shown in the first scheme below for a flat molecule (like I) and in the second scheme for a tris-complex. For the sake of brevity, I have left out the acid-base equilibria from covalent hydrate to pseudobase, although these will be critically important in understanding the effects of pH, and, in particular mechanistic features. For example, with $[NiP_3]^{2+}$ and (-) malic acid, as the pH is increased, the magnitude of the Pfeiffer rotation decreases, but its rate of appearance increases [9] (presumably through participation of pseudo-basic species).

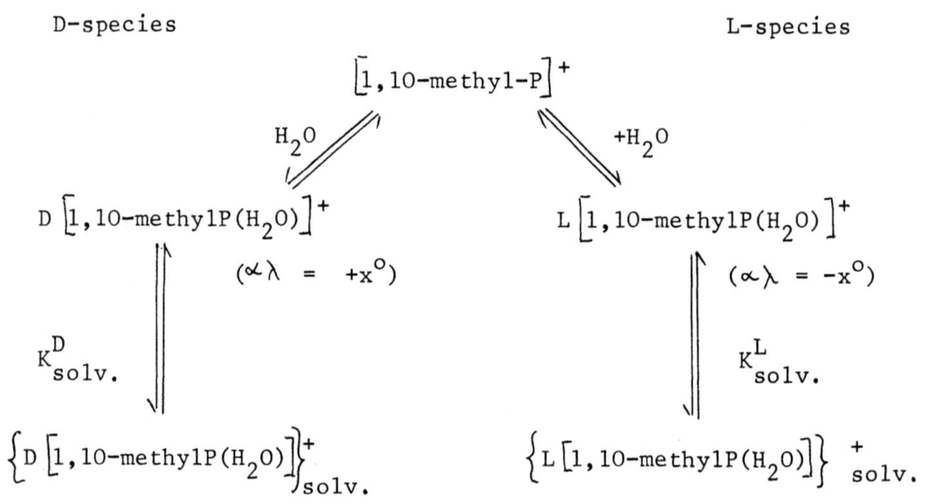

Providing that the solvation of the D hydrate in the asymmetric solvent generated by the presence of E differs from that of L, then the total concentration of $[D] \neq [L]$.

Returning to the analogy with adenosine aminohydrolase, there, the kinetic selection between (+) and (-) hydrates of pteridine is brought about by the asymmetric field presumably fairly localized around the enzyme active site. In the methyl-phenanthrolinium cation, thermodynamic selection is brought about by the asymmetric field localized around each alkaloidium cation, but capable of extending its chirality throughout the solution by hydrogen bonding.

I should comment further upon this question of solvation. When we lose optical purity, we expect to obtain, ceteris paribus, only the (small) entropy of mixing. However, Wynberg pointed out the significance of the cooling on mixing (+) and (-)coniine (at room temperature). The exothermicity means that other things are not equal. Crudely speaking, the solvation of (+) molecules by (+) is not nearly so good as that of (+) by (-). Extending this from liquids to solutions of optically active E in inactive liquid W (like water), we may expect that the solvation of (+) solute P by W (now asymmetric because it contains E) is different (even very different) from the solvation of (-)P by W.

Clearly, the solvation energies of (+) and (-) solutes by optically inactive W are equal, because there must be equal concentrations of right and left chiral solvation holes or sites in W. That is, when a racemic solute is dissolved in a solvent, each molecule is solvated by a sheath of geometry complementary to itself. Each such sheath fits around the asymmetric molecule as a glove fits a hand. (The solute may distort the normal distribution of solvation sheaths in the solvent, but for a racemic solute, the numbers of right-handed sheaths and left-handed are equal). When only one enantiomeric form of solute is present (the environment compound of the Pfeiffer effect) then only one chirality of solvation sheath is required, and the whole solvent becomes asymmetric, since the matching of the handed solvation spheres with the bulk solvent will dictate preferential chiralities throughout the medium.

This picture of the preferential solvation is like that proposed [14] by Bosnich and Watts, who said, "... the structure adopted by the first solvation sphere of one enantiomer will connect better with the chiral bulk solvent structure than will that of the other enantiomer. .... the solvating water molecules take on an inherently chiral structure, thus allowing the propagation of the dissymmetric interaction possibly over several intervening solvent molecules." Schipper has also discussed [7] the interaction of chiral solvent sheaths surrounding Pfeiffer and environmental cations and concludes by saying: 'It is difficult to see why the effect only occurs for unsaturated ligand systems if the solvent effect is responsible for the discrimination...'

This difficulty disappears once we recall the special reactivity toward water of these particular unsaturated ligands.

The unique feature of the new model is the presence of the asymmetric tetrahedral carbon atoms formed by covalent hydration and bearing a covalently attached hydroxyl group C-OH, as the basis for preferential chiral solvation (which, of course, removes any need to postulate hydrogen bonding between hydroxyl hydrogens of solvent and the aromatic rings of the ligands, the basis of a recent suggestion [9]).

As a final comment on chirality transmitted through water association (i.e. through groups of water molecules, linked, by hydrogen bonding, and possibly through "symmetrical" anions, into chiral chains, rings, and the like, on an averaged view) let us remember that a fair number of simple "octahedral" complex hydrates crystallize in enantiomorphous space groups. These include $\alpha[Ni(H_2O)_6]SO_4$ and $[Co(H_2O)_6]SO_4.H_2O$ and the water contained in these crystals forms exactly the kind of chiral hydrogen bonding pattern which is at issue.

Now consider the case of $[ML_3]^{n+}$. Again, I have omitted, for brevity, the second and successive covalent hydrations of the ligand and any acid-base equilibria generated (taken together, the number of possible species present - including isomers - in a simple aqueous solution of ML_3^{n+}, is surprisingly high).

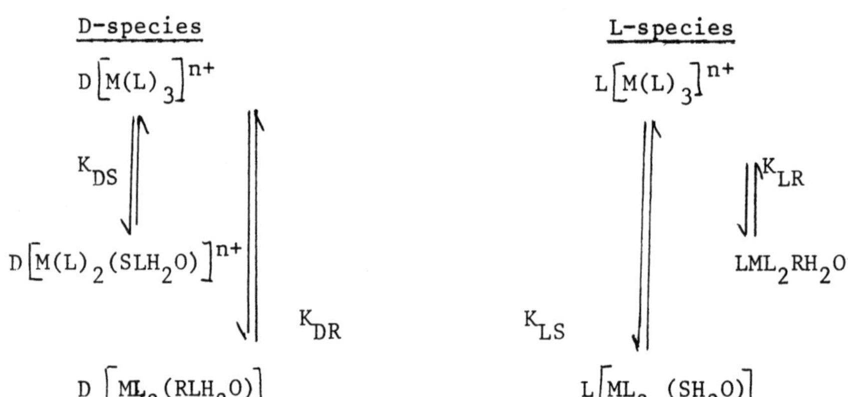

I use the labels D and L to signify chirality at the metal, and S and R to signify chirality at the new carbon centre (this implying the associated skewed diene and other elements of dissymmetry). In symmetrical solvents (water, alcohols and the like) the equilibrium constants K_{DS} and K_{LR} (denoting the formation of enantiomeric species) will be equal. Different, but again equal to each other, will be K_{DR} and K_{LS}. In the presence of optically active compounds, of the four diastereoisomers DS, DR, LS, LR, the asymmetric solvent environment will (by preferential solvation, chiefly through hydrogen-bonding) stabilize one (say DS) leading to a shift of the populations.

As more DS forms, so the equilibrium among the 4 diastereoisomers will shift, by water attack at the ligands, and this will be reflected in the inequality of total D-forms and total L-forms. Note further that, in the thermodynamic equilibrium constants of the type K_{DS}, K_{DR}, K_{LS}, K_{LR}, the activity of water appears, so that changes in water activities will lead (on the present view) to changes in hydrate populations, which will cause changes in Pfeiffer rotations, as observed.

What evidence is there for the reality of the equilibria involving covalent hydrate and pseudo-base of the complexed N-heterocycles? The Table lists a few recent results from Cardiff and elsewhere on the equilibrium constants for pseudo-base formation, calculated as

$$\text{Cation}^{n+} + OH^- \rightleftharpoons \{\text{Cation.OH}\}^{(n-1)+}$$

$$K = \frac{[\{\text{Cation.OH}\}^{(n-1)+}]}{[\text{Cation}^{n+}][OH^-]}$$

In another typical study, the existence of a covalent hydrate is established [15] in aqueous solutions of $[Ni(phen)_3]^{2+}$. While in strong acid (or anhydrous solvents) the complex is largely unsolvated, in water, over a large range of pH values, the covalent hydrate is a dominant species.

The novel mechanism for the Pfeiffer-active interaction involves the hydroxo-substituent at the carbon adjacent to the ligating nitrogen interacting strongly with solvent. This would account for the lack of Pfeiffer effect in alcoholic media. It seems likely (from our few studies of alkoxide attack upon co-ordinated N-heterocycles) that the formation of the "pseudo-alkoxide" species ($\diagdown\bar{N}-C\diagup(OR)H$) is favoured relative to the pseudo-base. e.g. [16] for $[Ru(5NO_2P)_3]^{2+}$ + X^- $\rightleftharpoons$ $[Ru(5NO_2P)_2(5NO_2XP)]^+$ in self-solvent at 37°C $K_{OH^-} = 31$, $K_{MeO^-} = 40$ K_{OEt^-} 250. This should compensate for the greater conjugate basicity of alkoxides (as against hydroxide) and the equilibrium constants for 'covalent alcoholation' should therefore be roughly comparable with those for covalent hydration of the same complex ion. So, the newly chiral centres formed by alcoholation are almost certainly present. However, there is no such possibility of strong hydrogen bonding of the resultant mixture of diastereoisomers, and so the asymmetric solvation sheath of the 'environment' compound cannot transmit its stereoselection nearly so well in alcohols as in water. The Pfeiffer effect is indeed weaker in alcohols than in water.

This new model makes some predictions.

1) The claim now made is that the Pfeiffer effect in cation:

cation systems (i.e. the diastereotopic or chirodiastaltic part of the whole interaction) arises not from the simple $[ML_3]^{n+}$ unhydrated, but only from these new species, the covalent hydrates. These will have slightly different spectra and circular dichroism, so the Pfeiffer circular dichroism should be that of the covalent hydrates.

Unfortunately, because of the high optical density and chiroptical strength of the alkaloids around 280 nm, we have not been able to study the "exciton" region directly, but there are indications that the spectroscopic origin of the Pfeiffer effect is not centred in the familiar exciton bands of $[ML_3]^{2+}$.

i) Drude plots of $\frac{\alpha}{\lambda^2 - \lambda_0^2}$ give values of λ_0 at around 310 nm for the colourless zinc and cadmium systems containing alkaloids, not coincident [22] with the 'exciton' absorption. Recalling the new absorption band at 312nm for the covalent hydrate of 2,2'-bipyridyl itself in water (that hydrate making up about 3% of the dissolved N-heterocycle [18]), that value of λ_0 seems reasonable for absorption due to a covalent hydrate of colourless $[M(phen)_3]^{2+}$.

ii) Japanese workers find that, for $[Ni(phen)_3]^{2+}$ and cinchonidine, in the presence of sulphate ions, the Pfeiffer c.d. in the d-d region of the nickel ion is not that of (+) $NiP_3^{\ 2+}$ under the same conditions, but is more akin to that of $NiP_3^{\ 2+}$ in mixed solvents. Similar effects are found for the same complex ion in the presence of strychninium in water [23]. In particular, for the longest wavelength band, the $E(^3T_{2g})$ component is larger than the $A_1(^3T_{2g})$ in the Pfeiffer c.d., whereas in the cd of an ordinary aqueous solution of resolved $[Ni(phen)_3]^{2+}$ the E component is not seen unless acetone or highly charged anions are added. Clearly, a simple explanation is that we are dealing with two nickel tris-chelated species in solution, one containing a covalently hydrated ligand. Recalling that we earlier showed [15] that the c.d. of $[NiP_3]^{2+}$ varied with water activity, and that an association constant (K = 8) for NiP_3^{2+} was measured [24] in 1,2-dichloroethane, these facts again support the suggestion that the Pfeiffer activity arises not from $[Ni(phen)_3]^{2+}$ but from $[Ni(phen)_2(phen.OH_2)]^{2+}$.

2) The establishment of Pfeiffer equilibrium requires the exchange of water in the covalent hydrate. We have already shown [15] that the racemization of $[NiP_3]^{2+}$ occurs via a covalent hydrate, and the generation of the Pfeiffer effect for this species is, of course, known [11] to occur at the same rate as does racemisation.

Further, since the new equilibrium between the original Pfeiffer compound and its covalent hydrate in water will show one value, and in the different situation in the presence of the environment solution (where water activity has altered) will show another, there may well be kinetic effects. These were noted, even by Pfeiffer himself. Thus, on adding tris-1,10-phenanthrolinezinc(II) sulphate to aqueous cinchoninium chloride (initial specific rotation at the sodium-D lines +5.29°) the value changed "immediately" to -1.89° and "finally" (after a 'slow' change) to -2.46°. Similar effects were noted by Pfeiffer and Nakatsuka [25] for the cation:anion system $[NiP_3]^{2+}$:(+)bromocamphorsulphonate and by Brasted and Mayer [26] using (+)glutamate (the latter authors refer to several other kinetic effects). Clearly, such observations suggest a process with a finite activation energy. This can hardly be substitution at zinc ion, but fits well with covalent hydration.

Finally, I should return to the interactions of environment compounds and Pfeiffer active ions of the unlike charge. These can be divided into two types which were designated 2a and 2b above. Let me say at once that for category 2b, such systems as alkaloid + $[Cr(ox)_3]^{3-}$, ion-pairing alone will suffice to give a reasonable explanation. Indeed, such effects have been known for many years, and there seems little justification for including them under the heading 'Pfeiffer effect'.

In systems of type 2a, where the cation is $[ML_3]^{n+}$, containing an N-heterocyclic ligand, and the anion is optically active, the enhancement of optical activity arises (in the new theory) from a complicated set of hydration and ion-pairing equilibria

$$\begin{array}{ccccc} [ML_3]^{2+} & + & X^- & \rightleftharpoons & \{[ML_3]X\}^+ \\ \updownarrow & & & & \\ [ML_2(LH_2O)]^{2+} & + & X^- & \rightleftharpoons & \{[ML_2(LH_2O)]X\}^+ \end{array}$$

Each of the four species containing metal ions will have its own concentration and specific optical activity, and the observed rotations will depend on conditions in a sensitive way. Studies of ion-pairing between cationic complex ions of N-heterocyclic ligands and anions (whether optically active or not) will need to be re-interpreted in terms of all the species present, including covalent hydrates, as will the effects on Pfeiffer rotations of adding substances like sodium sulphate.

What are the lessons of this new approach to the Pfeiffer effect. The first is that we should always be suspicious, particularly in water, of assuming that the nature of solutes is the same as that of the solids weighed out. The term 'formal' is, to my mind, more scientific than 'molar', and, as has so often been the case in solution studies, it is critically important to take into account <u>all</u> the species present, since even the most minor constituents may cause specific effects.

The second is a more general point, appropriate to a teaching gathering. Much of the chemistry underlying this new explanation of the Pfeiffer effect has been known for many years. However, the relevant work had appeared in widely separated journals and only a general chemical education will relate these separate areas.

TABLE. Equilibrium constants for pseudo-base formation by reaction of hydroxide ion with cations.

Species	Charge	K	Reference
$(N\text{-methyl-P})^+$	1	≤ 0.1	17
$(BipyH)^+$	1	~ 0.03	18,27
"Diquat"	2	9.3×10^3	19
$[Pt(bipy)_2]^{2+}$	2	10^5	20
$[Pd(5,5'\text{-di-}CH_3\text{-B})_2]^{2+}$	2	6.3×10^5	20
$[Fe(5NO_2P)_3]^{2+}$	2	260	21

REFERENCES

1. The International Dictionary of Physics and Chemistry, 2nd edition, 833: van Nostrand, Princeton, 1961.
2. The name was attached to the effect in 1954 by R.C. Brasted and F.P. Dwyer, at a meeting in Indiana: see reference [2] in my present reference [3]. Unfortunately, the name has been taken to include differing classes of interacting compounds by different groups of students. Arguably, it might be restricted to those systems studied by Pfeiffer himself (all containing 2,2'-bipyridyl or 1,10-phenanthroline).
3. R.C. Brasted, V.J. Landis, E.J. Kuhajek, P.E.R. Nordquist, and L. Mayer, "Coordination Chemistry", ed. by S. Kirschner, Plenum Press, N.Y., p42 (1969).
4. P. Pfeiffer and K. Quehl, Chem. Ber., 64, 2667 (1931).
5. E.E. Turner and M.M. Harris, Quart. Revs., 1, 299 (1947).
6. P.E. Schipper, Inorganica Chim. acta, 12, 199 (1975).
7. P.E. Schipper, J. Amer. Chem. Soc., 100, 1079 (1978) and references therein.
8. E.B.R. Prideaux and F.T. Winfield, J. Chem. Soc., 1587 (1930).
9. S. Kirschner, J. Indian Chem. Soc., 51, 28 (1974).
10. S. Kirschner, Record Chem. Progress, 32, 29 (1971).
11. S. Kirschner and K. Magnell, "Coordination Chemistry" (ed. S. Kirschner) Plenum Press, New York, p64 (1969).
12. H. Yoneda, K. Miyoshi, and S. Suzuki, Chem. Letters, 349 (1974): K. Miyoshi, K. Sakata, and H. Yoneda, Chemistry Letters, 1087 (1974): K. Miyoshi, Y. Kuroda, and H. Yoneda,

J. Phys. Chem., 80, 270 (1976): K. Miyoshi, K. Sakata, and H. Yoneda, J. Phys. Chem., 80, 649 (1976).
13. B.E. Evans and R.V. Wolfenden, Biochemistry, 12, 392 (1973).
14. B. Bosnich and D.W. Watts, Inorg. Chem., 14, 47 (1975).
15. R.D. Gillard and P.A. Williams, Transition Metal Chem., 2, 14, (1977).
16. R.D. Gillard, L.A.P. Kane-Maguire, and P.A. Williams, J. Chem. Soc. (Dalton) 1039 (1977).
17. J.W. Bunting and W.G. Meathrel, Canadian J. Chem.,52, 975 (1974)
18. M.S. Henry and M.Z. Hoffman, J. Amer. Chem. Soc., 99, 5201 (1977).
19. D.J. Norris, J.W. Bunting, and W.G. Meathrel, Canadian J. Chem., 55, 2601 (1977).
20. R.D. Gillard and J.R. Lyons, Chem. Comm., 873 (1973).
21. W.S. Walters, unpublished work.
22. E.J. Kuhajek, Ph.D. thesis, Univ. of Minnesota (1964).
23. K. Miyoshi, Y. Kuroda, H. Okazaki, and H. Yoneda, Bull. Chem. Soc. Japan, 50, 1476 (1977).
24. T. Fujiwara and Y. Yamamoto, XVII International Conference on Coord. Chem., Hamburg, 166 (1976).
25. P. Pfeiffer and K. Nakatsuka, Chem. Ber., 66, 410 (1933).
26. L.A. Mayer and R.C. Brasted, J. Coord. Chem., 3, 85 (1973).
27. This value refers to the ratio K = $[Bipy.H_2O]/[Bipy]$.

INDEX

Stephen F. Mason (ed.), Optical Activity and Chiral Discrimination. 369–372.

MIX
Papier aus verantwortungsvollen Quellen
Paper from responsible sources
FSC® C105338

If you have any concerns about our products,
you can contact us on
ProductSafety@springernature.com

In case Publisher is established outside the EU,
the EU authorized representative is:
Springer Nature Customer Service Center GmbH
Europaplatz 3, 69115 Heidelberg, Germany

Printed by Libri Plureos GmbH
in Hamburg, Germany